Green Energy and Technology

Climate change, environmental impact and the limited natural resources urge scientific research and novel technical solutions. The monograph series Green Energy and Technology serves as a publishing platform for scientific and technological approaches to "green"—i.e. environmentally friendly and sustainable—technologies. While a focus lies on energy and power supply, it also covers "green" solutions in industrial engineering and engineering design. Green Energy and Technology addresses researchers, advanced students, technical consultants as well as decision makers in industries and politics. Hence, the level of presentation spans from instructional to highly technical.

Indexed in Scopus.

Indexed in Ei Compendex.

Mohd Asyraf Kassim · Mardiana Idayu Ahmad ·
R. R. Dirgarini Julia Nurlianti Subagyono
Editors

Sustainable Biorefinery

Biomass Solutions for Energy, Environment and Food

Springer

Editors
Mohd Asyraf Kassim
School of Industrial Technology
Universiti Sains Malaysia
Gelugor, Pulau Pinang, Malaysia

Mardiana Idayu Ahmad
School of Industrial Technology
Universiti Sains Malaysia
Gelugor, Pulau Pinang, Malaysia

R. R. Dirgarini Julia Nurlianti Subagyono
Physical Chemistry Laboratory, Faculty
of Mathematics and Natural Sciences
Mulawarman University
Samarinda, East Kalimantan, Indonesia

ISSN 1865-3529 ISSN 1865-3537 (electronic)
Green Energy and Technology
ISBN 978-981-95-8180-1 ISBN 978-981-95-8181-8 (eBook)
https://doi.org/10.1007/978-981-95-8181-8

This Springer imprint is published by the registered company Springer Nature Singapore Pte Ltd.
The registered company address is: 152 Beach Road, #21-01/04 Gateway East, Singapore 189721, Singapore

Preface

The edited volume *Sustainable Biorefinery Approaches: Biomass Solutions for Environment, Food and Energy* presents state-of-the-art developments, emerging challenges, and transformative opportunities in the valorization of biomass for sustainable applications across the environmental, food, and energy domains. This book offers a holistic understanding of how diverse biomass resources, when integrated into innovative biorefinery systems, can be converted into high-value bioproducts that support global aspirations for a circular economy and sustainable development. By bridging fundamental principles with advanced technological insights, this volume expands current knowledge, stimulates new research directions, and strengthens the technological foundations required for greener, more resilient, and resource-efficient systems. It is intended as a comprehensive reference for students, researchers, industry professionals, and policymakers.

The chapters in this book, authored by leading experts in biomass utilization and green technologies, collectively showcase the diversity and potential of sustainable biorefinery pathways. Chapter "Bio-aviation Fuel: Exploring Feedstock Options and Production Pathways" provides a comprehensive overview of bio-aviation fuel development, emphasizing feedstock classification, production technologies, and the hydroprocessed esters and fatty acids (HEFA) pathway as a commercially mature solution for sustainable aviation fuel within a circular economy framework. Chapter "Black Soldier Fly Larvae: A Potential Biomass for Sustainable Biofuel, Feed, and Medicine" critically reviews black soldier fly larvae as a multifunctional sustainable biomass, highlighting their biological characteristics, ecological role in waste valorization, and applications in biofuel production, animal and aquaculture feed, and medical and pharmaceutical products. Chapter "Current Issues in Halal Certification of Food Products from Biomass" examines the complexities of halal certification in biomass-based food products, addressing challenges related to raw material sourcing, fermentation processes, enzymatic conversions, and global supply chain traceability, while proposing strategies for a science-based halal assurance system.

Chapter "Biodiesel Production and Techniques from Various Biomass Feedstocks" explores a wide range of biomass feedstocks for biodiesel production, discussing lipid extraction techniques and conversion methods to improve fuel yield, quality, and economic feasibility. Chapter "Role of Biomass as Carbon Neutral" reviews the role of biomass in achieving carbon neutrality, focusing on carbon capture, storage, utilization strategies, and integrated closed-loop carbon management systems. Environmental sustainability is further explored through Chapter "Addressing Pollution Through Green Processes for Environmental Mitigation and Remediation," which presents green process innovations involving biodegradable polymer reinforcement and waste-derived solvent extraction technologies for pollution mitigation within a circular economy context. Chapter "Fruits and Vegetable Waste: Harnessing Potential Toward Sustainable Bioenergy Production" investigates the conversion of fruit and vegetable waste into diverse biofuels through thermochemical, biochemical, and physical technologies, highlighting waste-to-energy pathways that reduce greenhouse gas emissions and fossil fuel dependency. Chapter "Utilizing Eggshell Waste to Develop Composite Materials for Sustainable Water Purification" discusses the development of eggshell-based composite materials for sustainable water purification, focusing on adsorption and photocatalytic degradation of dye-contaminated wastewater.

The principles of circular economy are reinforced in Chapter "Transforming Waste into Wealth: The Role of Biorefineries in Advancing a Circular Economy," which provides an in-depth analysis of biorefineries as integrated systems for transforming organic waste into high-value bioproducts, supported by environmental and economic assessments. Chapter "Biomass to Bioenergy: Thermochemical Conversion of Lignocellulosic and Microalgal Feedstock" reviews thermochemical conversion pathways of lignocellulosic and microalgal biomass, emphasizing the influence of feedstock characteristics, catalysts, and upgrading techniques on biofuel quality and performance. Chapter "Bioenergy Production from Biomass via Biochemical Technology" focuses on biochemical conversion technologies for liquid and gaseous biofuel production, highlighting the roles of microorganisms, enzymes, and process optimization in improving yield and sustainability. The final set of chapters highlights specialized and emerging innovations. Chapter "Improvement of Liquid Biofuel from Biomass Using Genetic Modified Microbes" reviews recent advances in genetic and metabolic engineering approaches for producing drop-in biofuels, discussing microbial pathway manipulation, technical challenges, and future prospects for sustainable fuel integration. Chapter "Microalgal Secondary Metabolites as Emerging Candidates for the Prevention and Management of Metabolic Diseases: A Review" examines microalgae as sustainable sources of bioactive secondary metabolites for food, nutraceutical, and pharmaceutical applications, highlighting their antioxidant, antimicrobial, anti-diabetic, anti-inflammatory, anti-obesity, and anti-cancer activities in relation to metabolic disease prevention and current translational challenges. Together, these chapters reflect the breadth and depth of biorefinery innovations that contribute to environmental protection, food security, renewable energy development, and the transition to low-carbon societies.

We extend our sincere appreciation to all contributing authors for their expertise, commitment, and high-quality scholarly work. Special thanks are due to the reviewers for their constructive comments, which have strengthened the rigor and clarity of this volume. We are also grateful to the Springer editorial team for their continuous support and professional guidance throughout the preparation and production of this book. It has been a privilege to collaborate on this project, and we hope that this volume will serve as a valuable reference for academics, scientists, engineers, policymakers, and practitioners seeking sustainable and innovative solutions grounded in biomass and biorefinery science.

Penang, Malaysia Mohd Asyraf Kassim
Penang, Malaysia Mardiana Idayu Ahmad
Samarinda, Indonesia R. R. Dirgarini Julia Nurlianti Subagyono
December 2025

Contents

Contributors

Mohamad Hafizi Abu Bakar Bioprocess Technology Division, School of Industrial Technology, Universiti Sains Malaysia, Gelugor, Penang, Malaysia

Mardiana Idayu Ahmad Environmental Technology Division, School of Industrial Technology, Universiti Sains Malaysia, Penang, Malaysia

Retno Aryani Faculty of Mathematics and Natural Sciences, Department of Biology, Mulawarman University, Samarinda, Indonesia

Qaisara Izzaty Mohd Azni Bioprocess Technology Division, School of Industrial Technology, Universiti Sains Malaysia, Penang, Malaysia

Fozy Binhweel Environmental Technology Division, School of Industrial Technology, Universiti Sains Malaysia, Penang, Malaysia;
Faculty of Environmental Science and Marine Biology, Hadhramout University, Hahdramout, Yemen

Alan L. Chaffee School of Chemistry, Monash University, Melbourne, Australia

Darnah Theoretical and Material Physics, Physics Department, FMIPA Mulawarman University, Samarinda, Indonesia;
Study Program of Statistics, Department of Mathematics, FMIPA Mulawarman University, Samarinda, Indonesia

Nurul Najihah Sham Faizal School of Industrial Technology, Universiti Sains Malaysia, Penang, Malaysia

Siti Fatimah Theoretical and Material Physics, Physics Department, FMIPA Mulawarman University, Samarinda, Indonesia

Dadan Hamdani Theoretical and Material Physics, Physics Department, FMIPA Mulawarman University, Samarinda, Indonesia

Sahara Hamas Intifadhah Theoretical and Material Physics, Physics Department, FMIPA Mulawarman University, Samarinda, Indonesia

Norli Ismail Environmental Technology Division, School of Industrial Technology, Universiti Sains Malaysia, Penang, Malaysia

Raihatul Jannah Faculty of Mathematics and Natural Sciences, Department of Biology, Mulawarman University, Samarinda, Indonesia

Thiruventhan Karunakaran Centre for Drug Research, Universiti Sains Malaysia, Gelugor, Penang, Malaysia

Mohd Asyraf Kassim Bioprocess Technology Division, School of Industrial Technology, Universiti Sains Malaysia, Gelugor, Penang, Malaysia

Aura Nur Kholifah Faculty of Mathematics and Natural Sciences, Department of Biology, Mulawarman University, Samarinda, Indonesia

Ratna Kusuma Theoretical and Material Physics, Physics Department, FMIPA Mulawarman University, Samarinda, Indonesia;
Faculty of Mathematics and Natural Sciences, Department of Biology, FMIPA Mulawarman University, Samarinda, Indonesia

Hetty Manurung Faculty of Mathematics and Natural Sciences, Department of Biology, Mulawarman University, Samarinda, Indonesia

Nawwar Mardianto Faculty of Mathematics and Natural Sciences, Department of Biology, Mulawarman University, Samarinda, Indonesia

Eva Marliana Theoretical and Material Physics, Physics Department, FMIPA Mulawarman University, Samarinda, Indonesia;
Department of Chemistry, FMIPA Mulawarman University, Samarinda, Indonesia

Marc Marshall School of Chemistry, Monash University, Melbourne, Australia

Kean Meng Tan Bioprocess Technology Division, School of Industrial Technology, Universiti Sains Malaysia, Penang, Malaysia

Muflihah Chemistry Education Study Program, Faculty of Teacher Training and Education, Mulawarman University, Samarinda, Indonesia

Rahmawati Munir Theoretical and Material Physics, Physics Department, FMIPA Mulawarman University, Samarinda, Indonesia

Nor Shafiqah Nor Shahril Bioprocess Technology Division, School of Industrial Technology, Universiti Sains Malaysia, Gelugor, Penang, Malaysia

Rudy Agung Nugroho Faculty of Mathematics and Natural Sciences, Department of Biology, Mulawarman University, Samarinda, Indonesia

Sulistyo Prabowo Food & Agricultural Products Technology Study Program, Faculty of Agriculture, Mulawarman University, Samarinda, Indonesia;
Center for Halal Science and Policy Research, Mulawarman University, Samarinda, Indonesia

Hassan Pyar Faculty of Environmental Science and Marine Biology, Hadhramout University, Hahdramout, Yemen

Nur Aliah Balqis Roslan Bioprocess Technology Division, School of Industrial Technology, Universiti Sains Malaysia, Penang, Malaysia

Rudianto Faculty of Mathematics and Natural Sciences, Department of Biology, Mulawarman University, Samarinda, Indonesia

Siti Nur Halifah Mohd Sani Bioprocess Technology Division, School of Industrial Technology, Universiti Sains Malaysia, Penang, Malaysia

Noor Aziah Serri Bioprocess Technology Division, School of Industrial Technology, Universiti Sains Malaysia, Penang, Malaysia

Mohammad Aliff Shakir Environmental Technology Division, School of Industrial Technology, Universiti Sains Malaysia, Penang, Malaysia

Fatin Shafiqa Shazili Bioprocess Technology Division, School of Industrial Technology, Universiti Sains Malaysia, Penang, Malaysia

R. R. Dirgarini Julia Nurlianti Subagyono Chemistry Department, Faculty of Mathematics and Natural Sciences, Mulawarman University, Samarinda, Indonesia

Amir Talebi Environmental Technology Division, School of Industrial Technology, Universiti Sains Malaysia, Penang, Malaysia

Wan Zafira Ezza Wan Zakaria Bioprocess Technology Division, School of Industrial Technology, Universiti Sains Malaysia, Penang, Malaysia

Bio-aviation Fuel Exploring Feedstock Options and Production Pathways

Mardiana Idayu Ahmad, Mohd Asyraf Kassim, and Mohammad Aliff Shakir

Abstract Amid rising environmental concerns, the aviation sector is under mounting pressure to decarbonize and transition toward more sustainable fuel alternatives. The limited availability of fossil-based jet fuels, coupled with global emissions targets, underscores the urgent need for a renewable aviation fuel solution. This chapter provides a comprehensive overview of bio-aviation fuel development, highlighting its classification, feedstock options, and production technologies from a circular economy perspective. It examines first- to fourth-generation biofuel sources including edible oils, non-edible oils, algae, and genetically modified organisms, while emphasizing their availability, economic viability, and environmental impact. The core emphasis is placed on the Hydroprocessed Esters and Fatty Acids (HEFA) pathway, which is currently the most commercially mature method for producing synthetic paraffinic kerosene (SPK). The chapter further outlines the essential process steps in HEFA fuel production, including deoxygenation, isomerization, and cracking. By integrating waste utilization and low-carbon processes, HEFA-based bio-aviation fuel offers a promising route to achieving both environmental sustainability and aviation fuel security.

Keywords Bio-aviation fuel · Biomass feedstock · Biodiesel · First-generation · Circular economy

1 Introduction

The increasing global demand for energy, coupled with growing environmental concerns, has stimulated in renewable and sustainable energy sources. Among these, bioenergy particularly biofuels has gained considerable attention as a viable alternative to fossil fuels [1]. The depletion of conventional fossil resources and the adverse

M. I. Ahmad (✉) · M. A. Kassim · M. A. Shakir
School of Industrial Technology, Universiti Sains Malaysia, 11800 Penang, Malaysia
e-mail: mardianaidayu@usm.my

M. A. Kassim et al. (eds.), *Sustainable Biorefinery*, Green Energy and Technology,
https://doi.org/10.1007/978-981-95-8181-8_1

environmental impacts associated with their combustion have compelled policymakers and researchers to accelerate the development of cleaner and more sustainable fuel options. Within this context, the aviation sector stands out as a critical area in need of transformation. Driven by population growth, economic development, and global connectivity, the aviation industry continues to expand at an unprecedented pace. This rapid growth comes at an environmental cost [2]. Aviation is one of the most energy-intensive transportation modes and a significant contributor to greenhouse gas (GHG) emissions. Traditional jet fuel derived from petroleum must meet strict technical specifications to ensure optimal engine performance and safety. Unlike road transportation fuels, jet fuels are subject to more rigorous standards for energy density, combustion efficiency, and cold flow properties [3]. As a result, the aviation sector heavily depends on high-quality petro-based fuels, which currently dominate the market.

According to the U.S. Energy Information Administration (EIA), the global consumption of aviation fuel surpassed 5.2 million barrels per day in 2010, and this figure has continued to rise [4]. The International Air Transport Association (IATA) has committed to reducing carbon emissions by 50% by 2050 compared to 2005 levels, demonstrating a clear need for transformative solutions. One such solution lies in the development and adoption of bio-aviation fuels, also referred to as sustainable aviation fuels (SAF). These renewable fuels have the potential to substantially reduce carbon emissions and enhance fuel security while maintaining compatibility with existing aviation infrastructure [5]. Jet fuel production represents a small fraction of each petroleum barrel, approximately 4 gallons per 42-gallon barrel, and yet it constitutes a significant portion of the operational costs of airlines. The fluctuating price of crude oil adds complexity to long-term aviation planning. Transitioning to bio-aviation fuels can mitigate reliance on a single energy source, promote price stability, and reduce environmental impact. Importantly, these fuels can be derived from a wide range of feedstocks including food and non-food biomass, waste oils, algae, and even carbon-rich gases [6]. Among these, non-edible and waste-derived feedstocks are especially promising, as they circumvent the food-versus-fuel dilemma and offer a more sustainable production pathway.

To ensure environmental and social sustainability, ideal feedstocks for bio-aviation fuels should be non-toxic, fast-growing, cultivable on marginal land, and require minimal agricultural input such as fertilizers or irrigation [7]. Using such underutilized plants can offer socio-economic benefits to rural communities while contributing to carbon neutrality through closed-loop systems. Advances in biofuel technology have enabled the conversion of various biomass sources into liquid fuels such as biodiesel, bioethanol, and more recently, bio-aviation fuel. Among the diverse production routes, hydroprocessed esters and fatty acids (HEFA) have shown the most promise for commercial-scale production. This method converts natural oils and fats into hydrocarbons that closely resemble conventional jet fuel. The resulting product is known as hydroprocessed renewable jet (HRJ) fuel. Other production pathways include alcohol-to-jet (ATJ), Fischer–Tropsch synthesis, and advanced thermal and biological processes. However, not all biofuels are suitable for aviation. Fatty acid methyl esters (FAME) are widely used in biodiesel is not compatible with jet engines due to their oxygen content, poor cold flow properties, and low energy density [8]. Even limited blending of FAME with jet fuel, such as 10–20%, can pose risks to engine performance and safety.

The development of bio-aviation fuel is not merely a technological challenge but a strategic opportunity to align the aviation industry with long-term global sustainability objectives. Central to this transition is the careful selection and utilization of appropriate feedstocks. Rather than relying on food-based raw materials, which can exacerbate the food-versus-fuel conflict, the focus has shifted toward non-edible, waste-derived, and genetically engineered feedstocks that offer greater environmental and social advantages [9, 10]. These include agricultural residues, used cooking oils, microalgae, municipal solid waste, and carbon-rich industrial gases. Such feedstocks are not only more abundant and cost-effective, but also reduce dependency on arable land, minimize input requirements like fertilizers and water, and enhance the overall carbon balance through waste valorization. Genetically modified organisms and carbon-utilizing microbes further expand the frontier by enabling high-yield production systems with tailored fuel properties. By harnessing this wide array of underutilized and renewable resources, the aviation sector can significantly lower its lifecycle emissions, reduce exposure to crude oil price volatility, and strengthen fuel supply chains. Moreover, feedstock diversification encourages localized production systems, which can spur rural economic development and integrate well with circular economy frameworks. Thus, sustainable feedstock development lies at the heart of creating a resilient, low-carbon aviation fuel ecosystem.

This chapter aims to provide a comprehensive analysis of bio-aviation fuel development by exploring the evolution, classification, and utilization of various feedstock types from first-generation edible crops to more advanced and sustainable fourth-generation sources such as genetically modified organisms and carbon-utilizing systems. Through detailed examination of the physicochemical characteristics, sustainability aspects, and technological readiness of these feedstocks, the chapter outlines their respective roles and potential in shaping the future of aviation fuel. In addition, it reviews a range of production pathways, highlighting their advantages, challenges, and applicability to large-scale implementation. By integrating discussions on circular economy principles, this chapter also evaluates how bio-aviation fuel development contributes to resource efficiency, waste valorization, and closed-loop systems. Ultimately, the chapter provides a holistic perspective on the technical, economic, and environmental dimensions of sustainable aviation fuel, offering insight for researchers and policymakers, seeking to transition toward a low-carbon aviation future.

2 Sustainable Feedstocks

The production of bio-aviation fuel relies on a wide range of biomass resources that can be systematically categorized based on their origin and characteristics. Selecting appropriate and sustainable feedstock is a critical step in determining the technical, economic, and environmental viability of bio-aviation fuel development. In general, renewable biomass sources suitable for aviation fuel synthesis can be derived from both plant-based and animal-based materials [11]. These include edible oils, non-edible oils, waste-derived oils, and algal biomass, each offering distinct advantages and challenges in terms of availability, cost, and conversion efficiency. Numerous studies have highlighted the potential of various biomass types for this purpose. Edible oils such as soybean and palm oil, non-edible oils like jatropha, used cooking oils, and microalgae have all been explored as viable feedstocks for producing bio-aviation fuel [12]. Globally, it has been estimated that more than 350 different oil-producing crops both edible and non-edible could be harnessed for this application. This broad diversity of feedstock options enhances the appeal of bio-aviation fuel as a renewable energy solution, particularly in regions with abundant agricultural or waste resources.

Feedstock availability plays a central role in determining production potential, especially since it directly influences both the yield and scalability of the fuel synthesis process. In assessing the feasibility of large-scale deployment, factors such as regional biomass availability, seasonal variations, land use, and logistical considerations must be accounted for. A sustainable feedstock supply chain is essential for ensuring a consistent and cost-effective production pipeline. To better understand the feedstock landscape, biomass resources for bio-aviation fuel are commonly classified into four main generations as shown in Fig. 1 such as first-generation (food-based), second-generation (non-food lignocellulosic biomass and waste), and third-generation (algae and other high-yielding organisms). A fourth-generation category has also emerged in recent years, encompassing genetically modified feedstocks and carbon-recycling systems. Each generation reflects a distinct technological approach and level of sustainability, which will be elaborated in the following subsections.

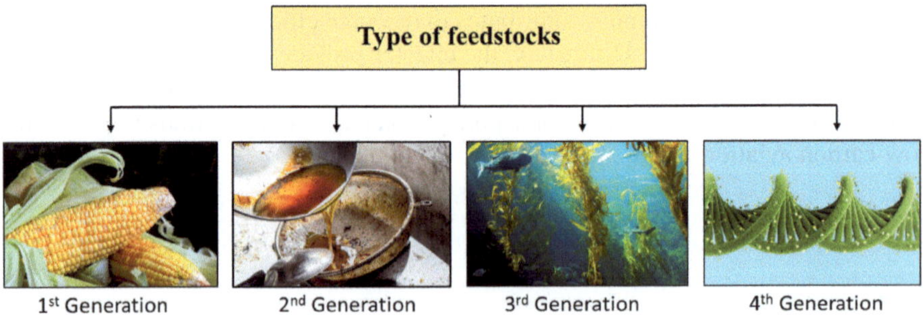

Fig. 1 Generational classification of Bio-aviation fuel feedstocks

2.1 First Generation

First-generation feedstocks refer to biomass resources derived primarily from food-based sources such as sugar, starch, and oil-rich crops. These materials were the earliest to be explored for biofuel production due to their widespread availability, established agricultural systems, and relatively straightforward conversion pathways [13]. Common examples include sugarcane, maize, wheat, cassava, and vegetable oils from crops such as palm, soybean, and sunflower. Their biochemical composition is rich in fermentable sugars or extractable oils, which makes them highly suitable for conventional fuel production processes such as fermentation and transesterification. In the context of aviation, some vegetable oils like soybean and palm oil have been evaluated for their potential to produce hydroprocessed esters and fatty acids (HEFA), which can be refined into synthetic paraffinic kerosene (SPK) compatible with jet engines. At the onset of the biofuel era, first-generation feedstocks gained significant attention due to their high energy yield, processing efficiency, and compatibility with existing fuel infrastructure. Their adoption was further supported by well-developed supply chains and market accessibility. As a result, the majority of early biofuels such as biodiesel and bioethanol were produced using food-grade agricultural products.

Despite initial interest in using first-generation feedstocks for aviation biofuel, their long-term use is limited due to socio-environmental concerns and compatibility issues with aviation-grade fuel requirements. One of the primary concerns is the "food versus fuel" dilemma, which highlights the competition between allocating agricultural land and resources for energy versus food supply. Increased biofuel demand can drive up food prices, limit food availability, and exert pressure on vulnerable communities, especially in developing regions [14]. Additionally, the environmental sustainability of expanding monoculture plantations for biofuel purposes has been questioned. Issues such as land degradation, biodiversity loss, deforestation, and high water and fertilizer usage have all been linked to the extensive cultivation of first-generation biofuel crops. Moreover, the production of biofuels from edible feedstocks is constrained by the availability of arable land and suitable climate conditions. In regions where agricultural land is scarce or under stress, prioritizing energy crops over food cultivation raises serious ethical and economic concerns. These challenges have led researchers and policymakers to reconsider the long-term sustainability of first-generation feedstocks.

Consequently, the drawbacks associated with food-based biofuel sources have encouraged a shift in research and industry focus toward non-food, waste-derived, and more environmentally benign alternatives. While first-generation feedstocks played a crucial role in initiating the global biofuel movement, they are less ideal for sustained bio-aviation fuel production, and their limitations have driven the exploration of second-, third-, and even fourth-generation biomass options that offer a more sustainable and socially responsible path forward for aviation fuels.

2.2 Second Generation

Second-generation feedstocks encompass a wide range of non-food biomass sources, including agricultural residues, forestry waste, lignocellulosic biomass, and non-edible oil-producing plants. These resources are not used for human consumption, making them a more sustainable and ethical alternative to first-generation (food-based) feedstocks. The transition to second-generation materials was driven by growing concerns over food security, land competition, and the environmental drawbacks of large-scale monoculture associated with edible biomass cultivation. For bio-aviation, second-generation feedstocks offer an important pathway for producing sustainable aviation fuel without compromising food supply or land availability. A prominent subcategory within this generation includes non-edible oil crops such as Jatropha curcas, jojoba, rubber seed, sea mango, and karanja [15]. These plants produce oils with similar chemical properties to edible oils which comprise saturated and unsaturated fatty acids but contain natural toxins or anti-nutritional compounds that render them unsuitable for consumption. Despite their inedibility, their oil profiles are compatible with the Hydroprocessed Esters and Fatty Acids (HEFA) pathway, making them valuable as precursors for bio-aviation fuel production.

Importantly, many non-edible oil crops can thrive on marginal or degraded lands unsuitable for food agriculture. Their cultivation does not compete directly with staple food production and can even contribute to land rehabilitation and rural development. This makes them especially attractive in regions where arable land is limited or where food-versus-fuel debates are more pronounced. Additionally, such crops often require minimal agricultural inputs, reducing the environmental footprint of their production systems [16]. Beyond oil-based feedstocks, second-generation resources also include lignocellulosic biomass as a major component of plant cell walls composed of cellulose, hemicellulose, and lignin. Sources such as corn stover, rice straw, wheat husk, bagasse, sawdust, and forest residues fall into this category [17]. These materials are abundant, renewable, and often considered waste by-products of agriculture and forestry, which makes them ideal for valorization in biofuel production [18]. Through thermochemical processes such as gasification and Fischer–Tropsch synthesis, lignocellulosic biomass can be converted into synthetic hydrocarbons, including jet fuel range molecules.

Second-generation feedstocks offer several advantages as they are biodegradable, have low sulfur content, produce fewer emissions during combustion, and can be sourced locally. As a result, it can reduce transportation costs and dependence on imported fuels. However, they are not without challenges. The complex structure of lignocellulosic materials often requires pre-treatment before conversion, which can add to processing costs and energy demand [19]. Despite these limitations, second-generation feedstocks play a critical role in the transition to sustainable aviation fuels. They bridge the gap between food security and fuel needs while supporting a circular bioeconomy, waste valorization, and low-emission fuel alternatives for the aviation industry.

2.3 Third Generation

Third-generation feedstocks primarily focus on the use of microalgae and other aquatic biomass, including certain macroalgae as renewable sources for biofuel production [13]. These organisms, which include both eukaryotic and prokaryotic strains, are cultivated for their high lipid, carbohydrate, and protein content, which can be converted into various forms of bioenergy. Unlike terrestrial crops, algae do not compete directly with food production and can grow rapidly under controlled conditions using non-arable land, saline water, or wastewater. Many microalgal species contain 20–50% lipid on a dry weight basis, with certain strains achieving even higher values under nutrient-stress conditions [20]. Additionally, microalgae exhibit rapid biomass accumulation, often reaching 1–2 cell doublings per day, which is significantly faster than terrestrial oil crops. These quantitative characteristics further emphasize their potential as high-yield feedstocks for aviation biofuel production. Microalgae possess the unique ability to utilize nutrients such as carbon, nitrogen, phosphorus, and sulfur from industrial waste streams, including effluents from power plants and wastewater treatment facilities [21]. This not only supports biomass growth but also contributes to waste mitigation and carbon capture, as microalgae can fix atmospheric CO_2 through photosynthesis. Certain strains, such as C*hlamydomonas reinhardtii* and *red seaweeds like Gracilaria sp.*, have demonstrated potential in bioethanol and biodiesel production especially under low-temperature conditions suitable for psychrophilic microalgae species [22, 23].

In aviation biofuel production, microalgae are particularly valuable due to their high lipid yield, which makes them a suitable candidate for the HEFA. Unlike transesterification which is typically used to produce biodiesel from vegetable oils or algal lipids, the HEFA process involves catalytic hydrotreatment, followed by isomerization and cracking to produce synthetic paraffinic kerosene (SPK) that meets jet fuel specifications [10]. In addition to lipids, microalgae contain carbohydrates and proteins that can be utilized for other bio-based applications. Simultaneous Saccharification and Fermentation (SSF) has been explored for extracting valuable bioproducts from red microalgal species such as *Porphyridium cruentum* [24]. Algae-based biofuels have attracted significant attention due to several inherent advantages. Algae exhibit high photosynthetic efficiency, fast growth rates, and the ability to thrive in a variety of environments including saline water, freshwater, and even industrial wastewater [25, 26]. Moreover, they have a high carbon dioxide fixation rate, making them excellent candidates for climate-responsive fuel production systems. Their polysaccharides can be converted into biofuels through enzymatic hydrolysis followed by fermentation. The relatively simple structure of algal cell walls marked by the absence of lignin and low hemicellulose content thereby facilitates more efficient processing compared to lignocellulosic biomass [27]. As a result, algal biomass typically yields higher outputs during saccharification and subsequent fermentation steps. Starch-rich algal strains are particularly well-suited for producing bioethanol or biobutanol under controlled fermentation conditions, including specific pH, temperature, and oxygen levels. Furthermore, anaerobic fermentation of starch-containing algae can yield not only bioethanol but also hydrogen and acetic acid.

In general, third-generation feedstocks especially microalgae offer a promising route for sustainable aviation fuel production. They combine environmental benefits such as CO_2 reduction and wastewater utilization with high productivity and biochemical versatility. Despite current commercialization challenges related to harvesting, dewatering, and conversion costs, their high lipid content and non-competitive land use profile position them as a viable future option for bio-aviation fuels aligned with circular economy goals.

2.4 Fourth Generation

Fourth-generation biofuels represent the most advanced frontier in biofuel development, incorporating genetically engineered organisms, solar energy-driven systems, and synthetic biology techniques [28]. The aim is to maximize fuel yield, resource efficiency, and environmental sustainability. This generation extends beyond traditional biomass sources by utilizing engineered microalgae, microbial cell factories, and photobiological systems designed to capture and convert carbon dioxide into liquid fuels with minimal environmental impact. One of the defining features of fourth-generation biofuel production is the use of photo-bio-solar and electrofuels as feedstocks [29]. These systems integrate advanced CO_2 capture technologies to enhance the adaptability and growth of microalgae in nutrient-limited environments such as wastewater, improving both productivity and sustainability. Genetically modified strains of microalgae and microbes have attracted increasing attention due to their ability to improve biomass yield, reduce processing costs, and optimize fuel quality.

Unlike earlier generations, fourth-generation feedstocks are not derived directly from conventional crops or wild-type organisms. Instead, they rely on solar energy-based systems, genetically modified algae, and engineered microorganisms to produce biofuels. Experimental processes have demonstrated the use of raw materials such as methanol, fresh oils, and sodium hydroxide, combined through a chemical cogeneration route at ambient conditions (25 °C and 100 kPa). In one such system, parabolic trough solar collectors enhanced with nanoparticles and specialized heat transfer fluids were used to maximize thermal efficiency. This approach enabled the simultaneous production of a significant amount of biodiesel and net power, highlighting the scalability and potential of solar-integrated biofuel systems. In parallel, engineered microbial strains such as *Clostridium, Pseudomonas putida,* and *Bacillus subtilis* are being explored for their biofuel-producing capabilities [30].

Genetic engineering techniques allow scientists to fine-tune microbial pathways, increasing lipid accumulation and enhancing substrate utilization, especially when converting complex biomass like agricultural residues [31]. These enhanced lipid profiles are particularly advantageous for producing aviation-grade fuels through the HEFA which is the most commercially mature technology for bio-aviation fuel synthesis. Moreover, the application of immobilized lipase enzymes has shown promising results in biodiesel synthesis by offering advantages such as enzyme stability, reusability, and lower operational costs [32]. For sustainable aviation, such innovations contribute to more efficient processing and higher-quality fuel outputs. The innovation in this generation is the formation of closed-loop production systems, whereby by-products such as glycerol are repurposed as feedstock, resulting in zero-waste, circular biofuel production. The fourth-generation approach represents a significant leap toward sustainable and high-efficiency fuel systems that directly support aviation decarbonization goals. It not only addresses the limitations of earlier generations such as land use conflicts and resource competition, but also aligns closely with the principles of green chemistry, renewable energy integration, and industrial ecology. The main benefits of fourth-generation biofuels include enhanced feedstock flexibility, reduced emissions, superior process efficiency, and lower long-term costs. However, challenges remain in scaling up these systems, reducing capital investment, and ensuring regulatory compliance. Nonetheless, this generation holds strong potential to redefine the landscape of bio-aviation fuel and other advanced biofuels, offering a technologically progressive, carbon–neutral, and environmentally resilient solution to the future of energy.

3 Bio-aviation Production Techniques

Bio-aviation fuel also known as synthetic paraffinic kerosene (SPK) is a renewable jet fuel produced from biomass and blended with conventional petroleum-based jet fuels to reduce greenhouse gas emissions and enhance fuel sustainability [33]. Various production platforms have been developed to synthesize SPK with each based on different types of biomass feedstocks and conversion technologies as shown in Fig. 2. One of the most commercially established methods is the Hydroprocessed Esters and Fatty Acids (HEFA). It is an oil-to-jet platform that involves the hydrodeoxygenation of oils and fats to produce SPK [34]. HEFA fuels are typically derived from plant oils, animal fats, or used cooking oils, offering a direct drop-in replacement for petro-jet fuel. Alternative oil-to-jet processes include hydrothermal liquefaction, where plant or algal oils are thermochemically processed under high pressure, and fast pyrolysis of cellulosic biomass followed by upgrading to jet fuel [35, 36].

Another key pathway is the gas-to-jet platform, which starts with biomass gasification to produce syngas which is a mixture of hydrogen and carbon monoxide. This syngas is then converted into hydrocarbons using the Fischer–Tropsch (FT) synthesis process [37]. The resulting product known as Fischer–Tropsch SPK (FT-SPK) is further hydroprocessed to meet jet fuel specifications. In some cases, alkylated and bio-based aromatics are incorporated into FT-SPK to create FT-SPK/A, which improves the aromatic content needed for compatibility with current aviation infrastructure [38]. The Alcohol-to-Jet (ATJ) platform involves hydrolyzing biomass to obtain fermentable sugars, which are then fermented into alcohols (such as ethanol

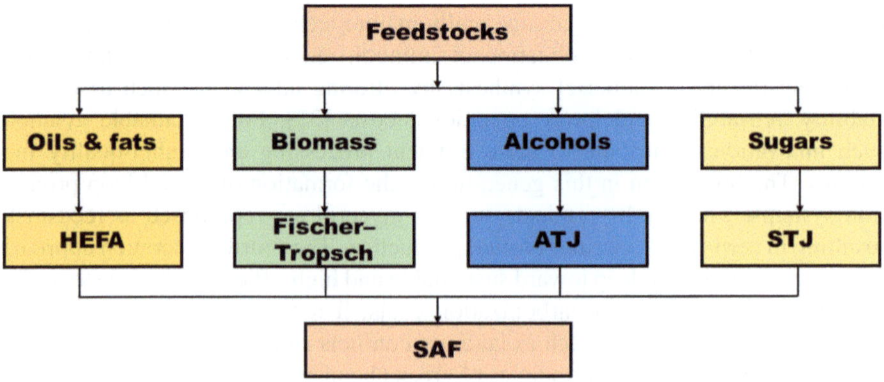

Fig. 2 Simplified process flow diagram illustrating major sustainable aviation fuel (SAF) production pathways

or butanol) [39]. These alcohols undergo a series of catalytic steps, including dehydration, oligomerization, hydrogenation, and fractionation to produce ATJ-SPK. This pathway is particularly versatile as it allows the use of a wide variety of lignocellulosic biomass.

Another advanced pathway is the Sugar-to-Jet (STJ) platform, which uses biomass to generate fermentable sugars through hydrolysis. These sugars are then fermented into farnesene, which is subsequently hydroprocessed and distilled to form SIP-SPK [40]. Alternatively, sugars or sugar-derived intermediates can undergo catalytic or biochemical reforming, followed by aqueous-phase reforming and upgrading to jet fuel hydrocarbons. Each of these platforms has unique advantages and limitations, which are typically evaluated in terms of feedstock flexibility, fuel quality, processing complexity, and environmental impact. Beyond their technical feasibility, bio-aviation fuels present broader socio-economic and environmental benefits. When produced using sustainable feedstocks, these fuels can lead to substantial emissions reductions during all phases of production, including extraction, refining, and transport. Additionally, they offer energy security, price stability, and the potential to stimulate rural development, especially by enabling the productive use of marginal lands and creating new agricultural and processing jobs.

However, despite these benefits, the commercialization of bio-aviation fuels remains limited. Large-scale deployment is hindered by insufficient investment, infrastructure costs, and policy uncertainties. Also, without financial incentives or regulatory support, production costs remain significantly higher than those of conventional fossil-based jet fuels. As such, targeted subsidies and enabling legislation are critical to accelerate the adoption of bio-aviation fuels and enhance their economic competitiveness. A comparative assessment of key sustainable aviation fuel (SAF) production pathways reveals notable differences in conversion efficiency, techno-economic feasibility, and environmental performance. The Fischer–Tropsch (FT) process generally offers high-quality synthetic paraffinic kerosene (FT-SPK) with excellent combustion characteristics and ultra-low sulfur content. However, it requires complex gasification infrastructure, high capital investment, and large processing scales to achieve cost competitiveness.

ATJ pathway provides significant feedstock flexibility, especially for ethanol and butanol from lignocellulosic biomass. ATJ-SPK production typically shows 40–60% conversion efficiency with jet fuel yields of 30–50 wt% [41, 42]. However, costs remain high (USD 1.50 to 2.00 per liter) due to the multiple catalytic steps required [43]. The Sugar-to-Jet (STJ/SIP) pathway, which produces farnesane-based hydrocarbons, offers good product selectivity but low-carbon efficiency typically 10–15%, and comparatively high costs (USD 1.79 to 6.80 per liter) [44, 45]. Its reliance on fermentable sugars also raises concerns over land use and supply chain sustainability. Overall, FT and HEFA routes remain the most mature pathways, often achieving 50–80% GHG reduction when using waste-derived feedstocks. ATJ and STJ provide valuable diversification but still require improvements in yield, hydrogen use, and cost efficiency.

3.1 Hydrotreating Esters and Fatty Acids

Hydrotreating esters and fatty acids is a well-established pathway for producing bio-aviation fuel-range hydrocarbons from triglycerides and unsaturated fatty acids. These compounds are typically sourced from vegetable oils and other lipid-based feedstocks. They contain long carbon chains (C_8–C_{22}) which after initial hydrotreatment are generally unsuitable for direct use in jet fuel applications due to their molecular structure and freezing point limitations [46]. As a result, the HEFA process is usually divided into two main stages to optimize the conversion into aviation-grade fuel as shown in Fig. 3. The first stage involves the hydrogenation of triglycerides under elevated temperature and pressure in the presence of a catalyst [47]. This step saturates the carbon double bonds and removes oxygen through hydrodeoxygenation, decarboxylation, or decarbonylation reactions. In typical HEFA operations, the process is carried out at temperatures of approximately 300–380 °C and under hydrogen pressures of 30–80 bar to ensure efficient oxygen removal and complete saturation of the feedstock.

In the second stage, the saturated molecules undergo isomerization and cracking to break down the long carbon chains into hydrocarbons within the jet fuel range [48]. Although non-catalytic hydrotreatment is possible, the presence of a catalyst significantly improves the selectivity of the reaction toward desirable products. Catalysts that are heterogeneous to the feedstock are often preferred because they can be more easily separated, regenerated, and reused. The isomerization and cracking stage typically operates at temperatures of 250–380 °C and pressures of 15–40 bar, depending on the catalyst composition and desired degree of branching. Catalyst performance in the HEFA process is heavily influenced by acidity, which plays a critical role in deoxygenation efficiency. Targeted acidic sites promote selective reactions and higher yields, but excessive catalyst loading or prolonged operation can lead to coke deposition, which blocks active sites and causes deactivation.

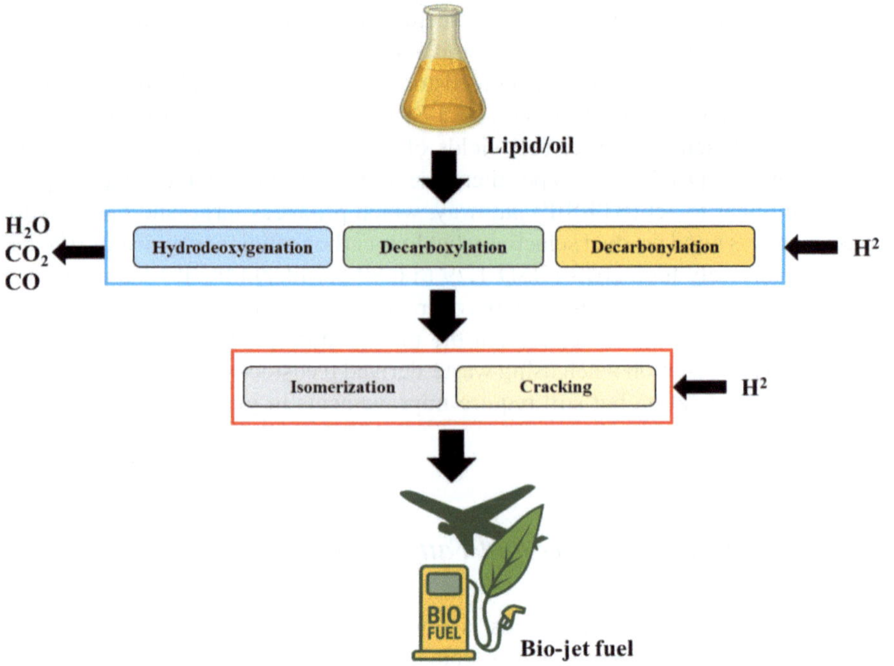

Fig. 3 Overview of the hydroprocessed esters and fatty acids pathway for bio-aviation fuel production

The use of basic catalysts often results in undesired side reactions such as over-cracking, which leads to the formation of light hydrocarbons outside the desired jet fuel range. As a solution, some studies have proposed combining acidic and basic catalysts to balance reactivity, enhance decarboxylation, and reduce coke formation [49]. The primary difference between diesel-range and jet fuel hydrocarbons lies in the carbon chain length, with diesel molecules typically being longer, resulting in a higher freezing point. This characteristic is critical in aviation, where fuels must perform under low-temperature, high-altitude conditions. For this reason, fatty acid methyl esters (FAME) commonly found in biodiesel are considered contaminants in bio-aviation fuels due to their poor cold flow properties. In addition to jet-range hydrocarbons, the HEFA process also yields co-products such as light gases and naphtha, which can be easily separated through distillation. This enhances the overall resource utilization and contributes to the economic viability of the process.

3.2 Deoxygenation Process

The deoxygenation stage is the initial and critical phase in the hydrotreating of esters and fatty acids (HEFA) for bio-aviation fuel production. This step aims to eliminate oxygen from triglycerides and fatty acid molecules, producing alkanes, which form the basis of synthetic paraffinic kerosene (SPK). These hydrocarbons are characterized by their low aromatic content which making them highly suitable for aviation applications. The three principal reaction pathways involved in deoxygenation are hydrodeoxygenation (HDO), decarboxylation, and decarbonylation. In hydrodeoxygenation, oxygen is removed from the molecules through a reaction with hydrogen gas in the presence of an appropriate catalyst. This process yields water as the byproduct while preserving the carbon chain length of the original molecule [50]. By contrast, decarboxylation involves the cleavage of the carbon–carbon (C–C) bond, resulting in the release of carbon dioxide (CO_2) without the need for hydrogen. Decarbonylation removes oxygen by breaking the carbonyl ($C = O$) bond using hydrogen, forming carbon monoxide (CO) and water. Both decarboxylation and decarbonylation reduce the carbon chain length of the resulting alkane by one carbon atom [51].

Among these methods, hydrodeoxygenation is considered the most carbon-efficient, as it avoids the loss of carbon atoms during conversion. Additionally, it improves key fuel properties such as energy content and stability. However, hydrodeoxygenation requires a significant amount of hydrogen, which poses economic and sustainability challenges, particularly since hydrogen is primarily produced from non-renewable natural gas sources.

The overall efficiency of the deoxygenation process is influenced by several factors, including the type of catalyst, the catalyst support material, and the reaction conditions. A wide variety of catalysts have been employed in this process, including pure metals, metal carbides, bimetallic and trimetallic combinations, metal phosphides, oxides, and nitrides [52]. Under severe operating conditions, platinum group metals (PGMs) such as Pd and Pt are especially valued for their resistance to oxidation and corrosion. These metals are commonly supported on materials like zeolites, activated carbon, or metal oxides to enhance activity and stability.

For decarboxylation and decarbonylation, noble or transition metals are typically preferred due to their ability to weaken carboxyl and carbonyl functional groups. Moderate-acidity transition metal oxides are more effective in promoting hydrodeoxygenation reactions by activating hydroxyl groups. Palladium (Pd) has demonstrated excellent selectivity in forming straight-chain paraffinic hydrocarbons [53]. However, it is costly and vulnerable to deactivation through active site poisoning, which occurs due to the accumulation of heavy organic intermediates, carbon deposits, or carbon monoxide. These limitations hinder its commercial viability despite its catalytic efficiency.

3.3 Isomerization and Cracking

Hydrogenated fuels produced through deoxygenation are primarily straight-chain paraffinic hydrocarbons, which offer high cetane numbers but exhibit poor cold flow properties. These properties has limiting their suitability for aviation applications. To overcome these limitations and enhance fuel performance, the isomerization and cracking stages are integrated into the hydrotreatment process. These additional conversion steps modify the hydrocarbon structure, significantly improving fuel characteristics, particularly in terms of freezing point and viscosity. Isomerization involves the rearrangement of straight-chain hydrocarbons into branched-chain structures, while cracking breaks down long-chain hydrocarbons into shorter-chain molecules that fall within the desired range for jet fuels [54]. These modifications improve the cold flow properties and combustion behavior of the fuel, ensuring its compliance with aviation fuel standards. However, care must be taken to avoid over-cracking, which typically occurs at temperatures exceeding 600 °C may lead to the formation of low-molecular-weight hydrocarbons (C_1–C_4). These products hold lower economic value compared to diesel- or jet-range hydrocarbons and reduce overall fuel yield. In typical HEFA upgrading, the isomerization and cracking stage produces a mixture of hydrocarbon fractions, with the majority falling within the jet fuel range (C_8–C_{16}), accompanied by smaller proportions of diesel-range hydrocarbons (C_{16}–C_{22}) and naphtha fractions (C_5–C_8). The exact distribution depends on catalyst acidity and reaction severity but is generally optimized to maximize jet fuel yield.

The efficiency of the isomerization process is largely influenced by the acidity of the catalyst and the selectivity toward desired hydrocarbon structures. Isomerization typically proceeds through dehydrogenation of paraffins on the metallic sites of the catalyst, followed by protonation on acid sites to form alkylcarbenium ions [55]. These ions undergo rearrangement into mono-, di-, or tri-branched intermediates, which are subsequently deprotonated and hydrogenated to form the final branched paraffinic products. A variety of catalysts have been explored for oil hydrocracking. Commonly used systems include sulfided Ni-Mo, Co-Mo, and Ni-W supported on silica–alumina, which provide both acidic and hydrogenation functionalities [56]. However, maintaining sulfur in the catalyst is necessary to retain activity, though it introduces drawbacks such as hydrogen sulfide (H_2S) emissions, corrosion, and sulfur contamination in the final fuel. Overall, isomerization and cracking are essential post-treatment steps in the HEFA process by transforming straight-chain hydrocarbons into aviation-compatible fuels with optimal flow and combustion characteristics. The selection of appropriate catalysts and reaction conditions is crucial for balancing fuel quality, product yield, and process economics.

4 Circular Economy Perspective of Bio-aviation Fuel

With increasing global urgency to address environmental challenges and reduce reliance on fossil fuels, the concept of the circular economy has gained traction especially in sectors with high carbon intensity such as aviation. The transition from a traditional linear economy, based on the ‘take-make-dispose’ model to a circular model emphasizes waste minimization, resource recovery, and renewable energy integration. In this context, the development of bio-aviation fuel from waste-based and renewable feedstocks aligns directly with circular economy principles by offering a sustainable pathway to decarbonize the aviation sector [57]. Bio-aviation fuel produced from waste lipids, including used cooking oil and other residual fats, exemplifies the waste-to-energy model within a circular framework. In urban and industrial settings, waste streams originating from residential, commercial, and food processing sectors are abundant and underutilized. Repurposing these lipid-rich wastes for aviation fuel not only addresses environmental waste burdens but also adds economic value to by-products that would otherwise be discarded [58]. This approach supports resource efficiency, contributes to emission reductions, and reduces dependence on conventional petroleum-derived jet fuel.

The aviation industry faces a dual challenge such as a rapidly growing demand for air travel and increasing pressure to reduce its carbon footprint. In response, waste-derived sustainable aviation fuels (SAFs) are emerging as a promising solution that aligns with circular economy strategies. These fuels can be integrated into existing aircraft systems and infrastructure without significant modification, supporting the concept of circular mobility where waste is repurposed into high-performance fuels, thereby closing the loop between consumption and resource regeneration. Within this closed-loop system, carbon captured from waste can be converted into usable energy for aviation. The residual outputs such as glycerol from transesterification processes can be redirected for further fuel synthesis or biochemical applications. Glycerol can be upgraded through several established valorization pathways, including catalytic conversion into value-added chemicals such as glycerol carbonate, propanediol, and glycerol ethers, or biochemical routes that produce organic acids and polyols. It can also be reformed into hydrogen or syngas, contributing to renewable energy generation and enhancing overall process circularity. Advanced catalytic and biological conversion technologies further enhance the efficiency of these systems, enabling low-carbon, low-waste production cycles. This synergy reinforces the value of integrated waste management, where circularity in fuel production leads to tangible environmental and economic benefits.

Several countries have already implemented supportive frameworks to incorporate circular economy principles in their sustainable aviation strategies. The European Union's Renewable Energy Directive includes provisions for SAFs derived from waste oils and non-food biomass, encouraging member states to scale up circular aviation initiatives [59]. In the Asia–Pacific region, nations are beginning to explore the inclusion of waste-based bio-aviation fuels within broader renewable energy strategies. By integrating circular economy principles into bio-aviation fuel development, the aviation sector can reduce its environmental footprint, diversify its energy sources, and foster a more resilient and sustainable air transportation system. As technologies evolve and policies align with climate targets, the circular bio-aviation fuel model stands as a viable and future-ready solution for a greener aviation industry.

5 Challenges in Bio-aviation Fuel Production

Bio-aviation fuel has gained global attention as a renewable and sustainable alternative to conventional jet fuel, offering significant potential to reduce greenhouse gas emissions and enhance energy security in the aviation sector. However, the development and deployment of sustainable aviation fuel (SAF) face multiple challenges across the entire production chain. Feedstock availability and quality are central issues affecting the scalability and consistency of SAF production. The accessibility of biomass feedstocks including vegetable oils, non-edible oilseeds, lignocellulosic residues, animal fats, and algal lipids varies significantly depending on geographical location, seasonal fluctuations, agricultural output, and market competition. Inconsistent feedstock supply can disrupt production planning and elevate operating costs. Moreover, the variability in feedstock composition, particularly in terms of fatty acid profiles, moisture levels, and contaminants, has complicated processing. These variations influence hydrogen consumption, catalyst lifespan, and conversion efficiency during HEFA or other production pathways.

Moreover, feedstocks used for bio-aviation fuel ranging from non-edible oils and lignocellulosic biomass to algal lipids are often contain impurities such as free fatty acids (FFAs), water, and particulates. These contaminants can pose substantial challenges to the conversion processes, particularly in hydrotreatment and catalytic upgrading. High levels of FFAs and water may impair catalyst activity and longevity during hydroprocessing, resulting in reduced reaction efficiency and increased hydrogen consumption. FFAs tend to neutralize catalysts, reducing their effectiveness and impeding the complete conversion of feedstock into hydrocarbon fuel [60]. Similarly, water can interfere with catalyst performance, lower reaction rates, and promote undesirable side reactions which complicate downstream separation.

Furthermore, feedstock costs remain the most significant contributor to the total expense of bio-aviation fuel production. While the use of low-cost feedstocks such as lignocellulosic residues, animal fats, and industrial waste oils presents economic advantages, these materials often come with elevated FFA content, moisture, and impurities. Feedstocks with high FFA levels may require pre-treatment steps, such as esterification or degumming, prior to hydroprocessing to prevent catalyst deactivation and ensure compatibility with aviation fuel standards. Last but not least, environmental hurdles in sustainable aviation fuel production also arise from the need to manage impurities and ensure consistent feedstock quality. The energy intensity of feedstock conversion processes also contributes to the environmental burden, particularly when high temperatures, pressure, and hydrogen are required. Additionally, transporting bulky biomass over long distances may increase the overall carbon footprint. Addressing these issues calls for integrated solutions such as advanced feedstock pre-treatment, decentralized biorefinery models, efficient supply chain logistics, and life cycle-based sustainability assessments to ensure that the production of bio-aviation fuel remains both economically viable and environmentally responsible. Overcoming these multifaceted challenges is essential to unlock the full potential of bio-aviation fuel as a commercially viable and climate-resilient solution for the future of sustainable air travel.

6 Conclusion

The growing imperative to decarbonize aviation has intensified global interest in bio-aviation fuels, prompting a shift from fossil-based jet fuels toward renewable alternatives. This chapter presented a detailed examination of various feedstock classifications, ranging from first- to fourth-generation sources, while highlighting their availability, cost-effectiveness, and environmental implications. Among the production technologies discussed, the Hydroprocessed Esters and Fatty Acids (HEFA) pathway emerged as the most commercially mature and technically feasible option for producing synthetic paraffinic kerosene (SPK). The HEFA process which includes deoxygenation, isomerization, and cracking has enables the efficient transformation of triglyceride-rich feedstocks into high-quality jet fuel with improved cold flow and combustion characteristics. The findings of this chapter underscore that non-edible oils and waste lipids offer a more sustainable and economically viable feedstock base compared to edible oils. Meanwhile algal biomass stands out for its high carbon fixation potential, albeit with scalability challenges. Hydrodeoxygenation provides high carbon efficiency with minimal atom loss, though its high hydrogen demand raises concerns regarding cost and sustainability. The success of HEFA processing is also closely tied to catalyst selection and reaction optimization, as these factors influence coke formation, hydrocarbon yield, and overall fuel properties. Ultimately, by aligning with circular economy principles through waste utilization and low-carbon conversion processes, HEFA-based bio-aviation fuels represent a critical pathway toward achieving both fuel security and long-term environmental sustainability in the aviation sector.

References

1. Cherwoo L et al (2023) Biofuels an alternative to traditional fossil fuels: a comprehensive review. Sustain Energy Technol Assess 60:103503
2. Balogun SA et al (2022) Renewable aviation fuel: review of bio-jet fuel for aviation industry. Eng Sci Lett 1(01):7–11
3. Benavides A et al (2021) Chemical composition and low-temperature fluidity properties of jet fuels. Processes 9(7):1184
4. Mazzone D, Smith A, Witcover J (2022) Pass-through of alternative fuel policy incentives: evidence from diesel and biodiesel markets, the us renewable fuel standard, and low carbon fuel standards in California and Oregon
5. Ogunkunle O, Ahmed NA (2021) Overview of biodiesel combustion in mitigating the adverse impacts of engine emissions on the sustainable human–environment scenario. Sustainability 13(10):5465

6. Ahmed M et al (2023) Bio-jet fuel: an overview of various feedstock and production routes. In: 7th International conference on environment 2021 (ICENV2021). AIP Publishing LLC
7. Senusi W et al (2024) Biodiesel production and characteristics from waste frying oils: sources, challenges, and circular economic perspective. Environ Sci Pollut Res 31(23):33239–33258
8. Shakir MA et al (2024) Effect of activation parameters on palm kernel shell-based activated carbon as catalyst support for biodiesel from discarded beef tallow. In: IOP conference series: earth and environmental science. IOP Publishing
9. Binhweel F et al (2024) Kinetics, thermodynamics, and optimization analyses of lipid extraction from discarded beef tallow for bioenergy production. Sep Sci Technol 59(16):1601–1620
10. Shalfoh E et al (2025) Biomass to biofuel: optimizing sustainable biodiesel production from fish waste and thermodynamic-kinetic analysis. Biofuels, Bioprod Biorefin 19(3):654–677
11. Senusi W et al (2024) Biodiesel production and characteristics from waste frying oils: sources, challenges, and circular economic perspective. Environ Sci Poll Res 1–20
12. Chong JW, Chemmangattuvalappil NG, Thangalazhy-Gopakumar S (2022) Aviation biofuels: conversion routes and challenges. Sustainable technologies for the oil palm industry: latest advances and case studies. Springer, pp 33–85
13. Senusi W et al (2024) Comparative assessment for biodiesel production from low-cost feedstocks of third oil generation. Renew Energy 236:121369
14. Suhara A et al (2024) Biodiesel sustainability: review of progress and challenges of biodiesel as sustainable biofuel. Clean Technol 6(3):886–906
15. Pydimalla M et al (2023) Sustainable biodiesel: a comprehensive review on feedstock, production methods, applications, challenges and opportunities. Mater Today: Proc 92:458–464
16. Shakir MA et al (2023) Effect of fibrillation degree of empty fruit bunch fiber on sandwich composite incorporate with spent mushroom substrate. J Ind Text 53:15280837231153660
17. Aggarwal NK, Kumar N, Mittal M (2022) Bioethanol production. Green Chem Sustain Technol 47–63
18. Shakir MA et al (2024) Development of natural waste fiber desiccant paper coated with biopolymeric material for energy efficient dehumidification. Clean Technol Environ Policy 1–13
19. Shakir MA, Yhaya MF, Ahmad MI (2017) Physical effects of pre-hydrolysis and alkali treatment of empty fruit bunch (EFB) fibers. Int J Chem Eng Appl 8(6):387–391
20. Babu SS et al (2022) Microalgae biomass and lipids as feedstock for biofuels: sustainable biotechnology strategies. Sustainability 14(22):15070
21. Gan YY et al (2020) Effects of dry and wet torrefaction pretreatment on microalgae pyrolysis analyzed by TG-FTIR and double-shot Py-GC/MS. Energy 210:118579
22. Azizi K, Moraveji MK, Najafabadi HA (2018) A review on bio-fuel production from microalgal biomass by using pyrolysis method. Renew Sustain Energy Rev 82:3046–3059
23. Zabed HM et al (2019) Recent advances in biological pretreatment of microalgae and lignocellulosic biomass for biofuel production. Renew Sustain Energy Rev 105:105–128
24. Kim HM, Oh CH, Bae H-J (2017) Comparison of red microalgae (Porphyridium cruentum) culture conditions for bioethanol production. Biores Technol 233:44–50
25. Binhweel F et al (2025) Biodiesel production from Ulva lactuca: insights into kinetic and thermodynamic evaluation of lipid transesterification. Biocatal Agric Biotechnol 70:103856
26. Shakir MA, Ahmad MI (2025) Biodiesel production from Ulva lactuca oil extracted by supercritical carbon dioxide using a biogenic activated carbon catalyst. Bioresour Technol Rep 32:102393
27. Md Nadzir S et al (2023) A review of microalgal cell wall composition and degradation to enhance the recovery of biomolecules for biofuel production. Biofuels 14(9):979–997
28. Osman AI et al (2021) Bioethanol and biodiesel: bibliometric mapping, policies and future needs. Renew Sustain Energy Rev 152:111677
29. Zambare V et al (2021) Recent advances in feedstock and lipase research and development towards commercialization of enzymatic biodiesel. Processes 9(10):1743

30. Satheesh S et al (2024) Bioprospecting of microorganisms for biofuel production: metabolic engineering, applications, and challenges. In: Catalytic applications of biochar for environmental remediation: valorization of lignocellulosic waste biomass into bioenergy, vol 3, pp 91–107
31. Chaturvedi V, Goswami RK, Verma P (2021) Genetic engineering for enhancement of biofuel production in microalgae. Biorefineries: a step towards renewable and clean energy. Springer, pp 539–559
32. Parandi E et al (2022) Biodiesel production from waste cooking oil using a novel biocatalyst of lipase enzyme immobilized magnetic nanocomposite. Fuel 313:123057
33. Farooqui SA, Sinha AK, Ray A (2022) Progress and trends in renewable jet fuels. Advanced biofuel technologies. Elsevier, pp 333–364
34. Monteiro RRC et al (2022) Production of jet biofuels by catalytic hydroprocessing of esters and fatty acids: a review. Catalysts 12(2):237
35. Gollakota ARK, Thandlam AK, Shu CM (2021) Biomass to bio jet fuels: a take off to the aviation industry. In: Liquid biofuels, pp 183–213
36. Bishai M (2024) Upgrading biomass-derived pyrolysis bio-oil to biojet fuel through catalytic cracking and hydrodeoxygenation. In: Kuila A (ed) Biojet fuel: current technology and future prospect. Springer Nature Singapore, Singapore, pp 117–145
37. Okolie JA et al (2023) Multi-criteria decision analysis for the evaluation and screening of sustainable aviation fuel production pathways. iScience 26(6)
38. Aburto J, Castillo-Landero A (2025) Development of advanced sustainable processes for aviation fuel production. In: Aslam M, Mishra S, Aburto Anell JA (eds) Sustainable aviation fuels: recent advances and future challenges. Springer Nature Switzerland, Cham, pp 181–195
39. Detsios N et al (2024) Techno-economic evaluation of jet fuel production via an alternative gasification-driven biomass-to-liquid pathway and benchmarking with the state-of-the-art fischer-tropsch and alcohol-to-jet concepts. Energies 17(7):1685
40. Redda ZT et al (2024) Biomass conversion to synthetic aviation fuels. In: Neelancherry R, Gao B, Wisniewski A (eds) Agricultural waste to value-added products: bioproducts and its applications. Springer Nature Singapore, Singapore, pp 183–205
41. Mansy AE et al (2025) Catalytic production of aviation jet biofuels from biomass: a review. Environ Chem Lett 23(2):419–461
42. de Jong S et al (2017) Life-cycle analysis of greenhouse gas emissions from renewable jet fuel production. Biotechnol Biofuels 10(1):64
43. Detsios N et al (2023) Recent advances on alternative aviation fuels/pathways: a critical review. Energies 16(4):1904
44. Kumar D, Long SP, Singh V (2018) Biorefinery for combined production of jet fuel and ethanol from lipid-producing sugarcane: a techno-economic evaluation. GCB Bioenergy 10(2):92–107
45. Lim M, Luckert MK, Qiu F (2023) Economic opportunities and challenges in biojet production: a literature review and analysis. Biomass Bioenerg 170:106727
46. Hong E et al (2025) Valorization of palm-based fatty acid distillate byproducts through coupling ketonization with hydrotreating: changing the long-chain distribution of paraffinic hydrocarbons. Ind Crops Prod 225:120574
47. Brandão RD et al (2021) The conversion of coconut oil into hydrocarbons within the chain length range of jet fuel. Biomass Convers Biorefinery 11(3):837–847
48. Chong JW, Chemmangattuvalappil NG, Thangalazhy-Gopakumar S (2023) Aviation biofuels: conversion routes and challenges. In: Foo DCY, Tun Abdul Aziz MK, Yusup S (eds) Sustainable technologies for the oil palm industry: latest advances and case studies. Springer Nature Singapore, Singapore, pp 33–85
49. Kawai E et al (2022) Maximizing margins and optimizing operational conditions for residue fluid catalytic cracking with an artificial intelligence hybrid reaction model. J Adv Manuf Process 4(3):e10118
50. Wei J et al (2023) Hydrodeoxygenation of oxygen-containing aromatic plastic wastes to liquid organic hydrogen carriers. Angew Chem Int Ed 62(46):e202310505

51. Chen B-S et al (2022) Advances in catalytic decarboxylation of bioderived fatty acids to diesel-range alkanes. Renew Sustain Energy Rev 158:112178
52. Al-Ali LI et al (2022) Transition metal phosphides (TMP) as a versatile class of catalysts for the hydrodeoxygenation reaction (HDO) of oil-derived compounds. Nanomaterials 12(9):1435
53. Mussa N-S, Toshtay K, Capron M (2024) Catalytic applications in the production of hydrotreated vegetable oil (HVO) as a renewable fuel: a review. Catalysts 14(7):452
54. Aljajan Y et al (2023) Hydroisomerization catalysts for high-quality diesel fuel production. Catalysts 13(10):1363
55. Han Y et al (2023) Shape selectivity of zeolite for hydroisomerization of long-chain alkanes. New J Chem 47(3):1401–1412
56. Farooqui SA et al (2022) Green diesel production by hydroprocessing technology. In: Aslam M, Shivaji Maktedar S, Sarma AK (eds) Green diesel: an alternative to biodiesel and petrodiesel. Springer Nature Singapore, Singapore, pp 109–148
57. Shakir MA et al (2023) From waste to wealth: converting rubber wood sawdust into green mycelium-based composite. Biomass Convers Biorefinery
58. Binhweel F, Ahmad MI, Shakir MA (2025) Harnessing waste lipids for biodiesel: optimization and thermodynamic investigation of beef tallow transesterification. Int J Green Energy 1–13
59. O'malley J, Pavlenko N, Searle S (2021) Estimating sustainable aviation fuel feedstock availability to meet growing European Union demand. International Council on Clean Transportation, Berlin, Germany
60. Giannopoulos D et al (2022) An Assessment of liquid biofuel value chains from heavy-metal contaminated feedstock. Fuels 3(3):509–532

Black Soldier Fly Larvae: A Potential Biomass for Sustainable Biofuel, Feed, and Medicine

Rudy Agung Nugroho, Ratna Kusuma, Retno Aryani, Hetty Manurung, Rudianto, Nawwar Mardianto, Aura Nur Kholifah, and Raihatul Jannah

Abstract Black Soldier Fly Larvae (BSFL; *Hermetia illucens*) have emerged as a highly promising sustainable biomass due to their rapid growth, exceptional adaptability to organic waste substrates, and rich nutritional and biochemical composition. This chapter critically reviews the biological characteristics, ecological functions, and emerging applications of BSFL within the framework of a circular bioeconomy. BSFL are distinguished by their high protein and lipid contents, enriched with essential amino acids, lauric acid, and bioactive compounds, positioning them as a viable alternative feedstock for sustainable biofuel production, animal and aquaculture feed, and medical and pharmaceutical applications. Recent advances demonstrate that BSFL-derived oil can be efficiently converted into high-quality biodiesel with favorable fuel properties, while BSFL meal has shown strong potential to replace conventional protein sources such as fishmeal and soybean meal without compromising animal growth or health. Beyond nutritional value, BSFL are also a source of antimicrobial peptides, chitin, and functional lipids with promising applications in wound healing, antimicrobial therapies, and nutraceutical development. Environmental assessments consistently indicate that BSFL farming offers a low carbon footprint, minimal land and water requirements, and effective waste valorization, contributing significantly to waste reduction and resource efficiency. However, challenges related to substrate safety, contaminant accumulation, regulatory compliance, scalability, and social acceptance remain. Overall, this chapter highlights BSFL as a multifunctional and sustainable biomass with significant potential to support global efforts in renewable energy production, sustainable agriculture, waste management, and biomedical innovation, provided that technological, regulatory, and ethical considerations are carefully addressed.

R. A. Nugroho (✉) · R. Kusuma · R. Aryani · H. Manurung · Rudianto · N. Mardianto · A. N. Kholifah · R. Jannah
Faculty of Mathematics and Natural Sciences, Department of Biology, Mulawarman University, Samarinda, Indonesia
e-mail: rudyagung.nugroho@fmipa.unmul.ac.id

M. A. Kassim et al. (eds.), *Sustainable Biorefinery*, Green Energy and Technology,
https://doi.org/10.1007/978-981-95-8181-8_2

1 Biology and Significance of BSFL

Black soldier fly (BSF) (*Hermetia illucens* L.) larvae are emerging as a remarkable organism for sustainable biomass conversion. Native to the Americas but now found globally, BSF larvae (BSFL) grow rapidly (20–22 days from egg to pre-pupae) and thrive on a wide range of organic wastes [1]. BSFL are non-pestiferous: adults do not bite or transmit disease [2]. On a dry-weight basis, BSFL are nutrient-rich, with crude protein content ranging roughly 32–53%, and fat 18–33% of dry weight [3]. In fact, analyses have shown BSFL average about 414.7 g/kg protein (range 216–655 g/kg) and 353.2 g/kg fat (range 294–515 g/kg) [4, 5]. These values exceed those of many conventional feed ingredients. They also contain essential amino acids, minerals, and bioactive lipids such as lauric acid [6]. This nutrient profile makes BSFL not only a high-value feed ingredient but also a candidate feedstock for biofuel and health products [7, 8].

The BSFL's ecological role further underpins its appeal. The larvae efficiently bioconvert organic wastes; for example, they can reduce the mass of chicken manure by over 50%, converting it into larval biomass rich in protein and fat [9]. Numerous studies note that BSFL can consume roughly their own weight in waste each day and, under optimal conditions, convert a large fraction of the substrate into insect biomass [10, 11]. This rapid growth and waste-assimilation ability make BSFL ideal for circular bioeconomy strategies. Importantly, BSFL production has a relatively low environmental footprint. One life-cycle study found that producing 100 kg of dried BSFL biomass generates only about 6.7 kg CO_2-equivalent emissions [12], a tiny fraction compared to livestock protein production. In short, BSFL combine fast growth, waste valorization, and high nutritional yield, positioning them as a potent resource for sustainable agriculture, energy, and health sectors [1].

Given these features, BSFL has attracted attention worldwide. Recent reviews highlight BSFL's advantages: rapid life-cycle, adaptability to diverse organic wastes, and minimal environmental demands [13]. Their larvae are already used commercially for composting organic waste [14] and producing insect meal [15]. For example, commercial facilities in Europe and Asia raise BSFL on agricultural byproducts to produce meal for aquafeed and livestock. Research shows that consumer attitudes and regulation are evolving to accept insects as feed, especially in aquaculture and poultry [16, 17]. Altogether, BSFL serve as a sustainable biomass that addresses waste reduction, feeds humans and animals, and even offers unique bioactive compounds for medicinal uses [18–20] (Fig. 1).

The biological adaptability and ecological efficiency of BSFL underscore their potential as a cornerstone of sustainable biomass conversion. Recent advancements in research have further elucidated the mechanisms behind their rapid growth and waste-assimilation capabilities. For instance, studies have identified specific enzymatic pathways in BSFL that facilitate the breakdown of complex organic polymers, such as cellulose and lignin, into simpler compounds [21, 22]. This enzymatic versatility enables BSFL to thrive on diverse substrates, including agricultural residues, food

Fig. 1 Application of sustainable biomass from insect larvae in animal feed, medicine, and biofuel

waste, and even industrial byproducts, making them a highly flexible resource for waste valorization.

Moreover, the nutritional profile of BSFL is not only rich in macronutrients but also contains micronutrients like zinc, iron, and magnesium, which are critical for animal and human health [23]. These trace elements are often deficient in conventional feed ingredients, highlighting BSFL as a functional food source. The presence of bioactive compounds, such as antimicrobial peptides and medium-chain fatty acids, further enhances their value beyond basic nutrition. For example, lauric acid, which constitutes a significant proportion of BSFL lipids, has been shown to modulate gut microbiota in livestock, reducing pathogenic bacteria while promoting beneficial strains [24].

The ecological benefits of BSFL extend beyond waste reduction. Their ability to process organic waste into nutrient-dense biomass aligns with the principles of the circular economy, minimizing resource extraction and landfill dependence. A recent life-cycle assessment demonstrated that BSFL farming could reduce water usage by up to 90% compared to traditional livestock farming, owing to their low hydration requirements and efficient nutrient cycling [25]. Additionally, the residual frass from BSFL production serves as a high-quality organic fertilizer, rich in nitrogen and phosphorus, which can replace synthetic fertilizers in agricultural systems [26].

Commercial scalability of BSFL production is another area of growing interest. Innovations in automated rearing systems have significantly improved yield consistency and reduced labor costs. For example, modular bioreactors equipped with IoT-based monitoring systems now allow real-time optimization of environmental conditions, such as temperature and humidity, to maximize larval growth rates [26]. These technological advancements are critical for meeting the rising demand for sustainable protein sources in aquaculture and livestock industries.

Despite these advantages, challenges remain in standardizing substrate quality and ensuring biosafety. Contaminants like heavy metals and pesticides in waste streams can accumulate in BSFL biomass, posing risks to downstream applications. Recent studies advocate for pre-treatment protocols, such as microbial detoxification, to mitigate these risks [27]. Regulatory frameworks are also evolving to address these concerns, with the European Food Safety Authority (EFSA) recently establishing guidelines for safe insect-based feed production [28].

The global adoption of BSFL technology is further supported by shifting consumer perceptions. Surveys indicate increasing acceptance of insect-derived products, particularly in regions facing protein shortages or environmental crises. Educational campaigns highlighting the sustainability and nutritional benefits of BSFL have played a pivotal role in this shift. For instance, partnerships between research institutions and industry stakeholders have facilitated pilot projects in urban waste management, demonstrating the practicality of BSFL systems [29]. The following sections explore BSFL applications in biofuel, feed, medicine, and the broader environmental and socio-economic context.

2 Emerging Applications of Black Soldier Fly

2.1 Black Soldier Fly Larvae for Sustainable Biofuel

One promising application of BSFL is as a feedstock for biofuel production. BSFL accumulate substantial lipid reserves, often 30–40% of their dry weight in fat, depending on diet [30, 31]. In particular, BSFL oil is uniquely high in medium-chain fatty acids, especially lauric acid (C12:0). Lauric acid can constitute over 70% of the fatty acids in larvae reared on certain substrates [32]. Lauric acid is known for antimicrobial [33] and even antiviral [34], anticancer [35] properties, but it also produces biodiesel with favorable properties (good cetane number and oxidative stability) [36]. Thus, BSFL oil can be chemically transformed into a high-quality biodiesel. In laboratory transesterification tests, BSFL oil gave excellent biodiesel yields: e.g., using KOH catalyst yielded $\pm$ 92.7% conversion to biodiesel, and NaOH yielded $\pm$ 91.0% [37]. These yields meet or approach commercial biodiesel standards. Importantly, BSFL can be fed on waste streams (food waste, manure, etc.), meaning the fat for biodiesel comes from waste valorization rather than cropland. For instance, one study used the residual fraction of restaurant food waste to rear larvae, then extracted lipids and produced biodiesel [38, 39]. The waste-fed larvae yielded biodiesel with 85–93% fatty acid methyl ester content using simple base-catalyzed transesterification [30].

Beyond traditional biodiesel, BSFL biomass can be processed via other thermochemical routes. The high energy content of dried larvae (owing to fats and proteins) suggests potential for pyrolysis or gasification. Studies on insect biomass pyrolysis indicate that insect frass or exoskeleton can be converted to biochar and biogas. In fact, the hard exoskeleton (chitin-rich puparia) of BSF has been shown to be suitable for biochar production with unique properties (potential soil amendment) [40]. If integrated with insect farming, a portion of BSFL waste (dead pupae, leftover substrate) could be thermochemically processed to yield syngas or char. Early research suggests the energy yield from BSFL biomass via pyrolysis is comparable to other biomass types, though more study is needed. Likewise, anaerobic digestion of BSFL residue (frass) can produce methane-rich biogas; one insect school report noted BSF frass achieving ± 140 m^3 CH_4 per ton organic matter [2, 41].

The biofuel pathway aligns well with the circular economy theme. By converting organic waste into larval fat, and then into fuel (Fig. 2), BSFL systems can close resource loops. Compared to conventional biodiesel feedstocks (soybean, palm), BSFL-derived biodiesel does not compete with food crops and requires far less land. A life-cycle assessment comparison found that BSFL biodiesel could have a much smaller carbon footprint than vegetable-oil biodiesel, especially when the larvae are fed on field residues [12]. Thus, BSFL biodiesel is an emerging sustainable fuel technology: current research demonstrates high conversion efficiencies and suggests great potential for scale-up, provided sufficient organic waste feedstock is available [42, 43]. Key challenges remain in scaling extraction and refining processes, but progress in enzyme and microbial transesterification is making BSFL oil-to-fuel increasingly viable.

The potential of Black Soldier Fly Larvae (BSFL) as a feedstock for biofuel production is increasingly recognized due to their high lipid content and efficient waste-to-energy conversion capabilities. Recent studies highlight that BSFL accumulate lipids constituting 30–40% of their dry weight, with lauric acid (C12:0)

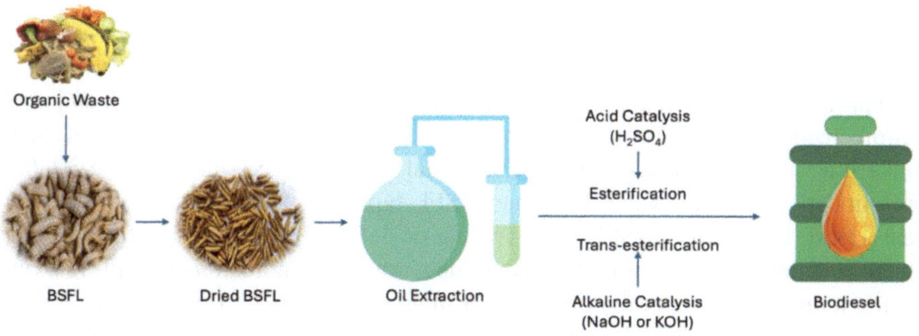

Fig. 2 Biodiesel derived from the black soldier fly larvae

dominating their fatty acid profile [32]. Lauric acid is particularly valuable for biodiesel production due to its favorable oxidative stability and cetane number, which enhance fuel quality [36]. Transesterification of BSFL oil using alkaline catalysts like KOH and NaOH has demonstrated conversion efficiencies exceeding 90%, meeting commercial biodiesel standards [37]. This positions BSFL-derived biodiesel as a viable alternative to conventional vegetable oils, which often compete with food crops for arable land. Figure 2 illustrates biodiesel derived from the black soldier fly larvae.

The environmental advantages of BSFL-based biofuels are underscored by life-cycle assessments. Compared to palm or soybean biodiesel, BSFL-derived fuels exhibit a significantly lower carbon footprint, particularly when larvae are reared on organic waste [12]. For example, a study by Kim et al. [29, 42, 43] demonstrated that BSFL biodiesel production from food waste reduced greenhouse gas emissions by 70% relative to fossil diesel. Furthermore, BSFL systems require minimal land and water, addressing critical sustainability challenges associated with conventional biofuel feedstocks.

Despite these benefits, scalability remains a challenge. Large-scale extraction and refining processes must be optimized to ensure cost-effectiveness. Innovations in enzymatic and microbial transesterification are promising avenues to improve efficiency [30]. Additionally, feedstock consistency is critical, as variations in waste composition can affect lipid yields. Pre-treatment protocols to remove contaminants like heavy metals are essential to ensure biosafety.

The integration of BSFL biofuel production with waste management systems offers a dual benefit: reducing landfill dependence while generating renewable energy. Pilot projects in Europe and Asia have demonstrated the feasibility of this approach, with commercial facilities producing BSFL biodiesel from restaurant waste. Regulatory frameworks are evolving to support such initiatives, with the European Union recently approving insect-based biofuels under stringent quality controls.

2.2 *Black Soldier Fly Larvae Meal for Animal and Aquaculture Feed*

As a protein source, BSFL meal is rapidly gaining acceptance in livestock and aquaculture feed. The high-protein content and favorable amino acid profile of BSFL make it a valuable replacement for traditional feeds. Indeed, insects have 30–65% protein depending on species [44], and BSFL typically provide ±40% protein (dry basis) [45]. In practice, BSFL meal (whole or defatted) has been tested in poultry, swine, and fish diets. A comprehensive review notes that inclusion of BSFL in monogastric animal diets generally does not harm growth or health. For example, feeding trials with chickens have shown that up to ~20% replacement of soybean meal with BSFL meal leads to no reduction in weight gain or feed efficiency. In another case, pigs fed diets containing 5–10% BSFL meal performed as well as control pigs, with normal feed conversion and meat quality. Overall, numerous trials conclude "no negative effects on growth, meat quality or immunity" from moderate BSFL inclusion. This is likely because BSFL provides essential nutrients and is highly digestible for poultry and swine.

In aquaculture, BSFL have been extensively studied as fishmeal replacers. Fishmeal is expensive and limited, so many researchers have tested partial replacement with BSFL. The larval meal's crude protein (±40%) and lipid (±30%) make it roughly comparable to fishmeal. Trials in tilapia, carp, trout, and marine species show promising results. For instance, one study in juvenile turbot (a carnivorous marine fish) replaced 14% of fishmeal with BSFL meal. The fish fed diets with defatted BSFL (14%) grew just as well as the all-fishmeal control, whereas full-fat BSFL at 14% slightly reduced growth [46]. That suggests defatted BSFL (high-protein) can substitute fishmeal successfully at modest levels. In Nile tilapia (a major farmed fish), complete replacement of up to 60% of fishmeal with BSFL meal was tested. Remarkably, tilapia fed 40–60% BSFL diets grew as well or better than control fishmeal-fed fish, and had improved feed conversion ratios. The economic conversion ratio (ECR) analysis showed that the 50–60% BSFL diets gave the best profitability. In short, tilapia farmers could replace half or more of expensive fishmeal with insect meal and reduce costs, without loss of performance [47]. Similar benefits were seen in other fish: perch and bass also grew well on BSFL-based feeds [48, 49].

For poultry and pigs, BSFL use is often limited by regulatory constraints, but acceptance is growing. In 2021, the EU allowed insect protein (including BSFL) in poultry and pig feeds under strict conditions. Studies now show that up to 10–15% BSFL in chicken diets can be used without negative effects, improving feed intake and immunity [50]. One review found that supplementation of $< 20\%$ BSFL meal in broiler diets had no adverse impact on performance or blood parameters [51, 52]. Indeed, some data suggest BSFL oil and protein may even enhance gut health due to antimicrobial fatty acids and chitin. However, palatability and cost remain considerations. In aquaculture, especially, fish often accept BSFL feed readily due to the larvae's naturally strong attractant scent.

Furthermore, the use of BSFL meal as a sustainable protein source in animal and aquaculture feed has gained significant traction due to its high nutritional value, environmental benefits, and economic viability. Recent research underscores the potential of BSFL meal to replace conventional protein sources like fishmeal and soybean meal, addressing global challenges of feed scarcity, environmental degradation, and rising costs (Fig. 3).

The BSFL meal is distinguished by its high-protein content (40–50% on a dry-weight basis) and balanced amino acid profile, which includes essential amino acids such as lysine, methionine, and leucine [4, 5]. These nutrients are critical for the growth and development of monogastric animals, including poultry, swine, and fish. Additionally, BSFL meal contains bioactive lipids, such as lauric acid (C12:0), which exhibit antimicrobial properties, enhancing gut health and immunity in livestock. The presence of chitin, a polysaccharide derived from the larval exoskeleton, further contributes to immune modulation and pathogen resistance.

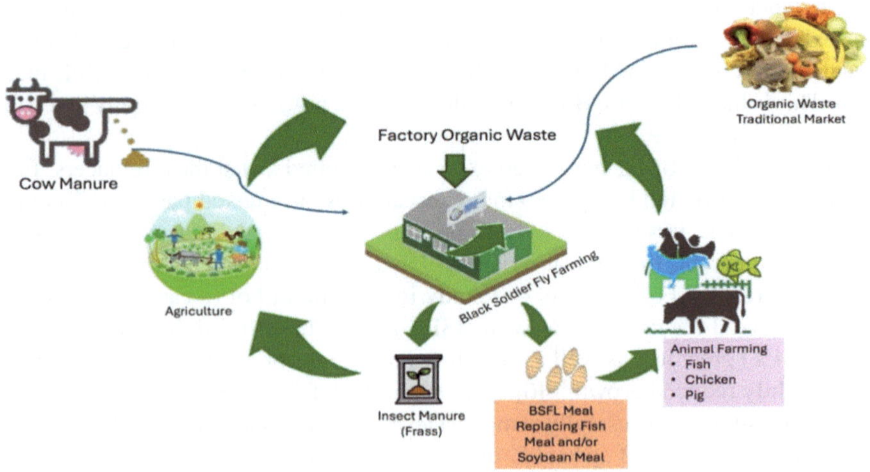

Fig. 3 The potential of BSFL meal to replace conventional protein sources such as fishmeal and soybean meal

Feeding trials across poultry and swine have demonstrated the efficacy of BSFL meal as a partial or complete replacement for traditional protein sources. For instance, broiler chickens fed diets with up to 20% BSFL meal exhibited comparable growth rates, feed conversion ratios, and meat quality to those fed conventional soybean-based diets [50]. Similarly, pigs supplemented with 5–10% BSFL meal showed no adverse effects on growth performance or carcass traits, while exhibiting improved gut health due to the prebiotic effects of chitin [8]. These findings highlight the adaptability of BSFL meal in monogastric animal nutrition, though regulatory frameworks in some regions still limit its inclusion levels [51].

In aquaculture, BSFL meal has emerged as a promising alternative to fishmeal, which is increasingly scarce and expensive. Studies on carnivorous and omnivorous fish species, such as turbot, tilapia, and Atlantic salmon, reveal that defatted BSFL meal can replace up to 60% of fishmeal without compromising growth or feed efficiency [46, 47]. Notably, Nile tilapia fed diets with 50–60% BSFL meal not only matched the performance of fishmeal-fed controls but also demonstrated improved feed conversion ratios and economic viability [47]. The high palatability of BSFL meal, attributed to its natural attractants, further enhances its suitability for aquafeeds [49].

The production of BSFL meal aligns with circular economy principles by valorizing organic waste streams, such as food waste and agricultural byproducts, into high-quality protein. Life-cycle assessments indicate that BSFL farming generates minimal greenhouse gas emissions and requires significantly less land and water compared to conventional feed production [12]. Economically, while BSFL meal is currently more expensive than fishmeal or soybean meal, its cost is projected to decline with advancements in rearing technologies and economies of scale [16]. Integrated farming systems, where BSFL are reared on-site using local waste, further reduce transportation costs and enhance sustainability [10].

Despite its potential, the widespread adoption of BSFL meal faces challenges, including regulatory hurdles, consumer acceptance, and the need for standardized processing methods. Contaminants such as heavy metals and pesticides in waste substrates pose risks of bioaccumulation, necessitating stringent quality control measures. Future research should focus on optimizing larval diets to enhance nutrient profiles, developing cost-effective defatting techniques, and expanding market outreach to foster consumer confidence [1].

Overall, BSFL are an excellent alternative protein and fat source for animals. Compared to fishmeal and soy, insect meal is more sustainable and has a lower land and water footprint. Moreover, BSFL production can valorize local organic waste into feed on-site, reducing dependence on imported grains. Ongoing research focuses on optimizing diets (e.g., defatting to increase protein) and balancing amino acids. But the trend is clear: BSFL meal is a credible, sustainable ingredient in poultry, swine, and especially aquafeeds, backed by numerous feeding trials showing normal or improved animal growth.

2.3 Black Soldier Fly Larvae for Medical and Pharmaceutical Applications

Beyond nutrition, BSFL are a source of bioactive compounds with medicinal interest. In particular, the innate immune system of BSF larvae produces many antimicrobial peptides (AMPs) that protect the larvae in filthy environments. Insects are renowned for their diverse AMPs [53], and BSF is exceptional: genomic analyses reveal the BSF genome encodes about 50 distinct AMPs—the largest repertoire known for a single insect. These peptides include defensins, cecropins, attacins, sarcotoxins, and others. Laboratory extracts of BSFL have demonstrated broad-spectrum antimicrobial activity against human and animal pathogens [54]. For example, purified AMP fractions from BSFL hemolymph can inhibit Gram-negative and Gram-positive bacteria at micromolar concentrations. Notably, BSF AMPs show low propensity for inducing microbial resistance—an advantage of peptide antimicrobials. Some AMPs (e.g., a cecropin-like peptide from BSF) have activity against *Pseudomonas* and *E. coli*. These features suggest BSFL could be tapped for novel antibiotics or preservatives. Researchers are already testing synthetic BSF-derived peptides as veterinary antibiotics. One review highlights BSF AMPs as promising antibiotic alternatives in livestock: they offer "extensive antimicrobial properties and a lower tendency to induce resistance" compared to conventional antibiotics.

Besides AMPs, BSFL contain high levels of lauric acid (C12:0) in their oil. Lauric acid has known medicinal properties, including antibacterial, antifungal, antiviral, and even anticancer effects. For instance, lauric acid can disrupt lipid membranes of enveloped viruses and certain bacteria. The remarkably high lauric acid content in BSFL oil (reported up to 76% under some diets) suggests potential for pharmaceutically active fats or nutraceutical extracts [32]. Hypothetically, BSFL oil could be processed into antimicrobial topical treatments or incorporated into health supplements. Indeed, lauric acid is already used in some topical formulations for acne and eczema. BSFL-derived lauric acid could become a specialty ingredient in cosmetics or functional foods, pending safety testing.

Another bioactive component of BSFL is chitin and its derivative chitosan, from the exoskeleton. Chitin from crustaceans is known for wound-healing and hemostatic properties; insect chitin is similar. Chitin/chitosan biopolymers are "excellent biomatrices" for wound dressings, as they are biocompatible, antimicrobial, and accelerate healing [55]. For example, chitosan fibers and membranes have been developed into bandages and tissue scaffolds. BSFL chitin (after processing) could be used similarly. While specific studies on BSF chitin are limited, the general biopolymer literature shows that chitin materials significantly promote dermal regeneration and collagen formation. Thus, BSFL waste from insect farms (pupal casings and frass exoskeleton) could be valorized into medical-grade chitosan for wound-care products.

The BSFL possesses a "pharmacy" of bioactive substances. Their AMPs could yield new antimicrobial drugs for livestock and perhaps humans. Their lauric-rich fat has antimicrobial and anti-inflammatory potential. Their chitin enables advanced biomaterials in wound healing. These medicinal prospects are still under investigation, but they underscore that BSFL biomass is not just "protein and fat", it is a complex mix with novel pharmaceutical potential. As antibiotic resistance rises, BSF-derived antimicrobials offer an exciting avenue for research. Further work is needed to isolate specific peptides, test clinical safety, and scale extraction, but the early data are promising that "insect-derived medicine" could become a reality.

The medicinal and pharmaceutical applications of Black Soldier Fly Larvae (BSFL) (Fig. 4) represent a burgeoning area of research, driven by their unique bioactive compounds, including antimicrobial peptides (AMPs), lauric acid, and chitin. These components exhibit significant potential for addressing global challenges in antimicrobial resistance, wound healing, and nutraceutical development.

In addition, the BSFL produce a diverse array of AMPs as part of their innate immune system, which enables them to thrive in pathogen-rich environments. Genomic analyses reveal that the BSF genome encodes approximately 50 distinct AMPs, the largest repertoire known for any single insect [53]. These peptides include defensins, cecropins, attacins, and sarcotoxins, which exhibit broad-spectrum activity against Gram-positive and Gram-negative bacteria, fungi, and even some viruses [54]. For instance, purified cecropin-like peptides from BSFL hemolymph have demonstrated potent inhibitory effects against Pseudomonas aeruginosa and Escherichia coli at micromolar concentrations [53]. Unlike conventional antibiotics, BSFL-derived AMPs show a low propensity for inducing microbial resistance, making them promising candidates for novel antimicrobial therapies [54].

Recent studies highlight the potential of BSFL AMPs as alternatives to antibiotics in livestock farming, where antimicrobial resistance is a critical concern. For

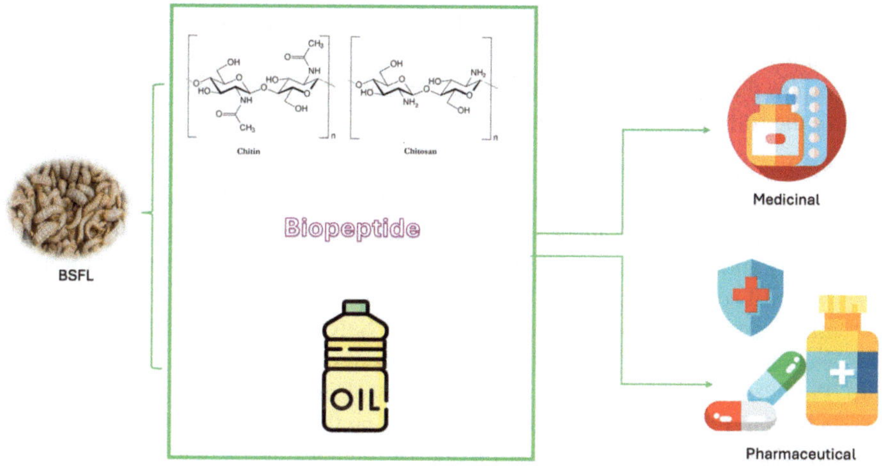

Fig. 4 Black soldier fly larvae for medicinal and pharmaceutical purposes

example, synthetic analogs of BSF defensins have been tested as veterinary antibiotics, showing efficacy in reducing bacterial loads in poultry without adverse effects on gut microbiota [19]. Further research is needed to optimize the production, stability, and delivery of these peptides for clinical and agricultural use.

The BSFL oil is exceptionally rich in lauric acid (C12:0), constituting up to 76% of total fatty acids under certain dietary conditions [32]. Lauric acid is renowned for its antimicrobial, antiviral, and anti-inflammatory properties. It disrupts the lipid membranes of enveloped viruses and bacteria, making it effective against pathogens such as Staphylococcus aureus and herpes simplex virus [34, 35].

In pharmaceutical applications, BSFL-derived lauric acid holds promise for topical formulations. For instance, lauric acid is already used in dermatological treatments for acne and eczema due to its sebum-regulating and antimicrobial effects [6]. The BSFL oil could be processed into specialized nutraceuticals, or functional foods aimed at enhancing immune function and gut health. Preliminary studies suggest that dietary supplementation with BSFL oil improves antioxidant capacity and immune responses in aquatic species, such as swimming crab (*Portunus trituberculatus*) [33].

Meanwhile, the chitin-rich exoskeleton of BSFL is another valuable resource for biomedical applications. Chitin and its derivative, chitosan, are biocompatible polymers with demonstrated wound-healing, hemostatic, and antimicrobial properties [55]. Chitosan-based dressings accelerate tissue regeneration by promoting collagen synthesis and reducing bacterial colonization in wounds [20].

While most research on chitin biomaterials has focused on crustacean sources, BSFL chitin offers a sustainable alternative. For example, ethanolic extracts of BSFL exoskeletons have shown significant wound-healing efficacy in murine models, comparable to commercial chitosan products [20]. The valorization of BSFL waste (e.g., pupal casings) into medical-grade chitosan could reduce reliance on shellfish-derived chitin, which is limited by allergenicity and supply chain constraints.

Despite these promising applications, several challenges must be addressed to realize the full potential of BSFL in medicine:

i. Scalability and Extraction: Efficient methods for large-scale AMP isolation and lauric acid purification are needed. Enzymatic and microbial extraction techniques are under investigation to enhance yield and reduce costs.
ii. Safety and Regulation: Rigorous clinical trials are required to evaluate the safety of BSFL-derived compounds for human use. Regulatory frameworks, such as those established by the European Food Safety Authority (EFSA), must be expanded to include insect-based pharmaceuticals.
iii. Public Acceptance: Consumer perceptions of insect-derived medicines remain a barrier. Educational campaigns highlighting the ecological and health benefits of BSFL products could foster broader acceptance.

3 Environmental, Economic, and Ethical Considerations

Large-scale BSFL farming involves multiple sustainability dimensions. **Environmentally**, BSFL production is generally low-impact. As noted, life-cycle analyses find very low greenhouse gas emissions per unit of BSFL biomass, far lower than traditional livestock. Water and land use are also minimal; BSFL can be reared vertically in modular reactors, requiring little water beyond that in the feed substrate. A huge environmental benefit is waste reduction: turning organic residues into insect protein reduces landfill use and methane emissions. For example, feeding food waste to BSFL can cut organic waste mass by 50–80%, while sterilizing pathogens in the process. The residual "frass" (insect manure) is a nutrient-rich fertilizer. Thus, BSFL systems create a closed-loop: waste → larvae → feed/biofuel, with byproduct fertilizer, closing nutrient cycles and reducing pollution. However, careful management is needed. Feedstocks may contain contaminants (pesticides, heavy metals, mycotoxins) that can bioaccumulate in larvae. Regulations typically restrict the use of hospital waste, feces from diseased animals, or certain kitchen wastes. Operators must monitor feed and larvae for toxins. If done well, BSFL farms can dramatically improve waste management and reduce the environmental footprint of agriculture.

Economically, BSFL farming is still developing but shows great promise. The insect protein market is growing rapidly (market reports forecast a multi-billion dollar value by 2030). BSFL meal and oil are currently more expensive than conventional soy or fishmeal, but costs are falling. On the revenue side, farmers earn not just from larvae products but also from waste disposal fees (communities pay to have food waste eaten by insects). Case studies show that substituting BSFL for fishmeal in aquafeed can substantially cut costs. For instance, Nile tilapia feeding trials demonstrated that replacing fishmeal with 50–60% BSFL meal improved profit, because BSFL meal was cheaper and conversion efficient. Modeling studies also suggest that in poultry or swine, BSFL inclusion can reduce feed costs if larvae are produced cheaply (e.g., on local waste). Yet economies of scale matter: capital costs (incubators, dryers, milling equipment) are high for new insect farms. Market uncertainty and competition with established feeds delay investment. Nevertheless, public–private partnerships and start-ups are rapidly expanding production capacity worldwide. In the future, integrated insect farms (co-located with agri-businesses or waste treatment plants) will improve economic viability.

Ethically and socially, insect farming raises novel issues. Some public concern centers on animal welfare: do insects feel pain, and how should they be killed humanely? Ethical frameworks for insects are nascent. Some advocates argue that insect farming is more humane than vertebrate livestock, as insects likely have simpler nervous systems. Others urge research on humane killing methods (e.g., freezing to immobilize larvae). Social acceptance is another factor. Research surveys indicate that two-thirds of consumers accept insects as feed, especially for fish and poultry [56, 57]. Attitudes are more cautious about insects in human food, but acceptance is growing (in Europe, novel food approvals now include some insect products). Education and marketing will shape how society views BSFL products.

Regulatory issues also loom. In many countries, laws on feeding insects to food animals are evolving. The EU, for example, has allowed insect meal (including BSFL) in aquafeed since 2017 and in poultry/swine feed since 2021. However, safety standards require controlling substrates and preventing contamination. Ethical sourcing of feedstock (avoiding the use of protected habitats or unsustainable palm kernel meal) is another concern.

The large-scale farming of Black Soldier Fly Larvae (BSFL) presents a multifaceted opportunity to address global sustainability challenges, but it also necessitates careful consideration of environmental, economic, and ethical dimensions. These factors are critical for ensuring the long-term viability and societal acceptance of BSFL-based solutions. The BSFL farming is distinguished by its minimal environmental footprint compared to conventional livestock and feed production systems. Life-cycle assessments (LCAs) reveal that producing 100 kg of dried BSFL biomass generates approximately 6.7 kg of CO_2-equivalent emissions, a fraction of the emissions associated with traditional protein sources like soybean meal or fishmeal [12]. This low carbon footprint is attributed to the larvae's ability to thrive on organic waste streams, such as food waste and agricultural byproducts, which would otherwise contribute to landfill methane emissions [9].

A key environmental advantage of BSFL farming is its role in waste valorization. BSFL can reduce the mass of organic waste by 50–80% while converting it into nutrient-rich biomass [10]. For example, feeding food waste to BSFL not only diverts waste from landfills but also sterilizes pathogens, reducing the risk of contamination [11]. The residual frass (insect manure) from BSFL production serves as an organic fertilizer, rich in nitrogen and phosphorus, further closing nutrient loops and reducing reliance on synthetic fertilizers [14]. However, environmental risks must be managed. Feedstocks for BSFL farming may contain contaminants such as heavy metals, pesticides, or mycotoxins, which can bioaccumulate in larval biomass. Regulatory frameworks, such as those established by the European Food Safety Authority (EFSA), mandate stringent monitoring of feedstock quality to mitigate these risks.

At the point of economic viability and market potential, the economic potential of BSFL farming is underscored by its dual revenue streams: the sale of larval products (meal, oil) and waste disposal services. The global insect protein market is projected to exceed $1 billion by 2030, driven by demand for sustainable feed ingredients [16]. The BSFL meal, though currently more expensive than conventional protein sources, offers cost savings in aquaculture and poultry feed. For instance, replacing 50–60% of fishmeal with BSFL meal in Nile tilapia diets improved profitability due to lower feed costs and comparable growth performance [47]. Despite these advantages, scaling BSFL production faces challenges. High capital costs for infrastructure (e.g., incubators, dryers) and economies of scale are barriers for small-scale producers [1]. Integrated farming models, where BSFL are co-located with waste treatment facilities or agri-businesses, can enhance economic viability by reducing transportation and feedstock costs [10].

Further, in terms of ethical and social considerations, the ethical implications of insect farming are an emerging area of debate. While BSFL are invertebrates with simpler nervous systems than vertebrates, questions remain about humane rearing and killing practices. Current methods, such as freezing larvae, are considered low-stress, but further research is needed to establish best practices [56]. Social acceptance of BSFL products also varies by region and application. Surveys indicate that two-thirds of consumers accept insect-based feed for aquaculture and poultry, though hesitancy persists for direct human consumption [57]. In Europe, regulatory approvals for insect-based feed and food (e.g., BSFL meal in poultry diets since 2021) reflect growing acceptance [17]. Educational campaigns highlighting the sustainability and nutritional benefits of BSFL can further shift public perception.

Based on the fact about regulatory and policy frameworks, regulatory frameworks for BSFL farming are evolving to address safety and sustainability concerns. The EU's approval of insect meal in animal feed under strict substrate controls exemplifies this trend. However, inconsistencies in global regulations pose challenges for international trade. For example, while BSFL meal is widely accepted in aquafeed, its use in poultry and swine feed is still restricted in some regions [51]. Ethical sourcing of feedstock is another regulatory priority. Avoiding substrates linked to deforestation (e.g., unsustainable palm kernel meal) ensures that BSFL farming aligns with broader sustainability goals [1].

4 Conclusions

BSFL farming offers significant sustainability benefits: less waste and pollution, lower emissions, and new economic opportunities. But success requires addressing risks (contaminants, regulatory compliance) and societal concerns (animal welfare, consumer perceptions). If these are managed, BSFL can become a cornerstone of a sustainable bioeconomy, transforming waste into wealth, closing nutrient loops, and even contributing to health care.

Black Soldier Fly Larvae (*Hermetia illucens*) have emerged as a versatile and sustainable biomass resource with applications spanning biofuel production, animal feed, and medicinal uses. BSFL are distinguished by their rapid growth cycle and ability to thrive on diverse organic waste substrates, making them a key player in waste valorization and circular economy initiatives. Their nutritional profile is particularly noteworthy, with protein content ranging from 30 to 65% and fat content between 18 and 33%, enriched with essential amino acids and bioactive lipids such as lauric acid. These attributes not only make BSFL a viable alternative to conventional feed ingredients like fishmeal and soybean meal but also position them as a promising feedstock for biodiesel production. Studies highlight their efficacy in improving animal health and growth metrics while reducing feed costs, underscoring their economic and ecological advantages.

Despite their potential, challenges remain in scaling BSFL production, ensuring substrate quality, and navigating regulatory frameworks for their use in animal feed and pharmaceuticals. Contaminants such as heavy metals and pesticides in waste streams pose risks of bioaccumulation, necessitating stringent quality control measures. Additionally, BSFL-derived antimicrobial peptides and lauric acid offer innovative solutions for combating antimicrobial resistance and developing health supplements, though clinical validation and public acceptance are still evolving. The document emphasizes that consumer perceptions and regulatory approvals are critical for mainstream adoption. As research advances and policies adapt, BSFL are poised to play a transformative role in sustainable agriculture, waste management, and biomedical innovation, contributing to global efforts in ecological preservation and resource efficiency.

This analysis underscores the need for interdisciplinary collaboration to address existing barriers, ensuring the safe and scalable integration of BSFL into commercial and industrial applications. Future research should focus on optimizing production technologies, expanding regulatory harmonization, and enhancing public awareness to fully realize the potential of BSFL as a cornerstone of sustainable development.

References

1. Siddiqui SA, Süfer Ö, Çalışkan KG, Lutuf H, Rahayu T, Castro-Muñoz R, Fernando I (2025) Enhancing the bioconversion rate and end products of black soldier fly (BSF) treatment—a comprehensive review. Environ Dev Sustain 27:9673–9741
2. Lu S, Taethaisong N, Meethip W, Surakhunthod J, Sinpru B, Sroichak T, Archa P, Thongpea S, Paengkoum S, Purba RAP, Paengkoum P (2022) Nutritional composition of black soldier fly larvae (*Hermetia illucens* L.) and its potential uses as alternative protein sources in animal diets: a review. Insects 13
3. Nugroho RA, Aryani R, Hardi EH, Manurung H, Rudianto R, Wirawan NA, Syalsabillah N, Jati WN (2023) Nutritive value, material reduction, biomass conversion rate, and survival of black solider fly larvae reared on palm kernel meal supplemented with fish pellets and fructose. Int J Trop Insect Sci 43:1243–1254

4. Nugroho RA, Aryani R, Hardi EH, Manurung H, Rudianto R (2024a) Nutritional benefits of black soldier fly larvae oil in animal feed. Bioeksperimen: Jurnal Penelitian Biologi 10:30–39
5. Nugroho RA, Aryani R, Hardi EH, Manurung H, Rudianto R, Jati WN (2024) Fermented palm kernel waste with different sugars as substrate for black soldier fly larvae. GJESM 10:503–516
6. Nugrahati, C.N.A., Nugroho R.A. (2024) Asam laurat minyak maggot (*Hermetia illucens* L.) sebagai bahan dasar kosmetik. In: Proceeding biology education conference: biology, science, environmental, and learning, pp 14–19
7. Mohan K, Sathishkumar P, Rajan DK, Rajarajeswaran J, Ganesan AR (2023) Black soldier fly (*Hermetia illucens)* larvae as potential feedstock for the biodiesel production: recent advances and challenges. Sci Total Environ 859:160235
8. Phaengphairee P, Boontiam W, Wealleans A, Hong J, Kim YY (2023) Dietary supplementation with full-fat *Hermetia illucens* larvae and multi-probiotics, as a substitute for antibiotics, improves the growth performance, gut health, and antioxidative capacity of weaned pigs. BMC Vet Res 19:7
9. Siddiqui SA, Ristow B, Rahayu T, Putra NS, Yuwono NW, Mategeko B, Smetana S, Saki M, Nawaz A, Nagdalian A (2022) Black soldier fly larvae (BSFL) and their affinity for organic waste processing. Waste Manage 140:1–13
10. Gautam B, Tiwari S, Pokhrel MR, Tomberlin JK, Khanal P (2025) Expanding black soldier fly (BSF; *Hermetia illucens*; Diptera: Stratiomyidae) in the developing world: use of BSF larvae as a biological tool to recycle various organic biowastes for alternative protein production in Nepal. Biotech Rep e00879
11. Salam M, Shahzadi A, Zheng H, Alam F, Nabi G, Dezhi S, Ullah W, Ammara S, Ali N, Bilal M (2022) Effect of different environmental conditions on the growth and development of black soldier fly larvae and its utilization in solid waste management and pollution mitigation. Environ Technol Innov 28:102649
12. Nugroho RA, Rofiq MN, Santoso AD, Yanuar AI, Hanifa R, Nadirah N (2023b) Bioconversion of biowaste by black soldier fly larvae (*Hermetia illucens* L.) for dried larvae production: a life cycle assessment and environmental impact analysis. F1000Res 12:814
13. Bataglia L, Conversano A, Di Bona D, Sogni D, Voccia D, Mazzoni E, Lamastra L (2025) Recent developments, challenges, and environmental benefits of using *Hermetia illucens* for bioenergy production within a circular economy approach. Energies 18:2826
14. Chen L, Wang G, Song H, Yang Q, Fu J, Liu J, Sun H, Wang Y, Tian Q, Sun Y (2025) Adding fruit fermentation liquid improves the efficiency of the black soldier fly in converting chicken manure and reshapes the structure of its intestinal microbial community. Insects 16:472
15. Cheng H, Thorsteinsdottir A, Dalsgaard TK, Danielsen M, Zatti K, Feyissa AH (2025) Evaluation of the physical and chemical quality of atlantic salmon feed with inclusion of full fat black soldier fly or mealworm meal: extrusion trials and modelling. Anim Feed Sci Technol 320:116201
16. Chidozie Ogwu M (2025) Market trends and business opportunities in edible insects edible insects: nutritional benefits, culinary innovations and sustainability. Springer, pp 189–214
17. Roccatello R, Endrizzi I, Aprea E, Dabbou S (2024) Insect-based feed in aquaculture: a consumer attitudes study. Aquaculture 582:740512
18. Iqbal W, Elahi U, Hai-jun Z, Ahmad S, Usman M, Yaqoob MU (2025) Clean and green bioconversion–a comprehensive review on black soldier fly (*Hermetia illucens*) larvae for converting organic wastes to quality products. Ann Anim Sci 25:57–81
19. Salahuddin M, Abdel-Wareth AA, Hiramatsu K, Tomberlin JK, Luza D, Lohakare J (2024) Flight toward sustainability in poultry nutrition with black soldier fly larvae. Animals 14:510
20. Sandakila DA, Hadi GD, Gultom R, Samalele S, Nugroho RA (2024) Ethanolic extract of black soldier fly (*Hermetia illucens* L.) larvae for wound healing in mice (*Mus musculus* L.). Aceh J Anim Sci 9
21. Diener S, Zurbrügg C, Gutiérrez FR, Nguyen DH, Morel A, Koottatep T, Tockner K (2011) Black soldier fly larvae for organic waste treatment-prospects and constraints. Proc WasteSafe 2:13–15

22. Shao M, Zhao X, Rehman KU, Cai M, Zheng L, Huang F, Zhang J (2024) Synergistic bioconversion of organic waste by black soldier fly (*Hermetia illucens*) larvae and thermophilic cellulose-degrading bacteria. Front Microbiol 14:1288227
23. Shelomi M (2020) Nutrient composition of black soldier fly (*Hermetia illucens*). In: African edible insects as alternative source of food, oil, protein and bioactive components, pp 195–212
24. Dabbou S, Lauwaerts A, Ferrocino I, Biasato I, Sirri F, Zampiga M, Bergagna S, Pagliasso G, Gariglio M, Colombino E (2021) Modified black soldier fly larva fat in broiler diet: effects on performance, carcass traits, blood parameters, histomorphological features and gut microbiota. Animals 11:1837
25. Halloran A, Roos N, Eilenberg J, Cerutti A, Bruun S (2016) Life cycle assessment of edible insects for food protein: a review. Agron Sustain Dev 36:1–13
26. Abd Manan F, Yeoh YK, Chai TT, Wong FC (2024) Unlocking the potential of black soldier fly frass as a sustainable organic fertilizer: a review of recent studies. J Environ Manage 367:121997
27. Truzzi C, Annibaldi A, Girolametti F, Giovannini L, Riolo P, Ruschioni S, Olivotto I, Illuminati S (2020) A chemically safe way to produce insect biomass for possible application in feed and food production. Int J Environ Res Public Health 17:2121
28. Meijer N, Safitri R, Tao W, Hoek-Van den Hil E (2025) European union legislation and regulatory framework for edible insect production–safety issues. Animal 101468
29. Kim C-H, Ryu J, Lee J, Ko K, Lee J-Y, Park KY, Chung H (2021) Use of black soldier fly larvae for food waste treatment and energy production in Asian countries: a review. Processes 9:161
30. Jung S, Jung J-M, Tsang YF, Bhatnagar A, Chen W-H, Lin K-YA, Kwon EE (2022) Biodiesel production from black soldier fly larvae derived from food waste by non-catalytic transesterification. Energy 238:121700
31. St-Hilaire S, Sheppard C, Tomberlin JK, Irving S, Newton L, McGuire MA, Mosley EE, Hardy RW, Sealey W (2007) Fly prepupae as a feedstuff for rainbow trout, *Oncorhynchus mykiss*. J World Aquaculture Soc 38:59–67
32. Suryati T, Julaeha E, Farabi K, Ambarsari H, Hidayat AT (2023) Lauric acid from the black soldier fly (*Hermetia illucens*) and its potential applications. Sustainability 15:10383
33. Zhan W, Peng H, Xie S, Deng Y, Zhu T, Cui Y, Cao H, Tang Z, Jin M, Zhou Q (2024) Dietary lauric acid promoted antioxidant and immune capacity by improving intestinal structure and microbial population of swimming crab (*Portunus trituberculatus*). Fish Shellfish Immunol 151:109739
34. Lieberman S, Enig MG, Preuss HG (2006) A review of monolaurin and lauric acid: natural virucidal and bactericidal agents. Altern Complement Ther 12:310–314
35. Sheela DL, Narayanankutty A, Nazeem PA, Raghavamenon AC, Muthangaparambil SR (2019) Lauric acid induce cell death in colon cancer cells mediated by the epidermal growth factor receptor downregulation: an in silico and in vitro study. Hum Exp Toxicol 38:753–761
36. Burmana AD, Tambun R, Haryanto B, Sarah M, Alexander V (2023) Recycling heterogeneous catalyst waste in biodiesel production using methanol and hydrochloric acid: a case study on the washing effect with lauric acid as raw material. Case Stud Chem Environ Eng 8:100510
37. Sarwono A, Aryani E, Rahman A, Zahra NL, Harmayanti A (2024) Exploring black soldier fly larvae as a sustainable biodiesel source with base catalysts: an analytical and characterization study. In: Proceedings of the 2nd international conference on nature-based solution in climate change, RESILIENCE 2023, 24 Nov 2023, Jakarta, Indonesia 1
38. Saha S, Ghosh S, Pal H (2024) Resource recovery from food waste through conversion to value-added products sustainable clean energy production using waste biomass: sustainable energy production and utilization. Springer, pp 43–73
39. Zheng L, Li Q, Zhang J, Yu Z (2012) Double the biodiesel yield: rearing black soldier fly larvae, *Hermetia illucens*, on solid residual fraction of restaurant waste after grease extraction for biodiesel production. Renew Energy 41:75–79
40. Bulak P, Proc-Pietrycha K, Kaczor M, Złotko K, Polakowski C, Wiącek D, Waniak-Nowicka H, Zięba E, Waśko A, Oleszczuk P, Bieganowski A (2023) A novel type of biochar from chitinous *Hermetia illucens* waste with a built-in stimulating effect on plants and soil arthropods. Sci Rep 13:8306

41. Prachumchai R, Cherdthong A (2023) Black soldier fly larva oil in diets with roughage to concentrate ratios on fermentation characteristics, degradability, and methane generation. Animals 13:2416
42. Kim JY, Park G, Choi Y, Park W-K, Koo B, Park K, Tsang YF, Kwon EE (2024) Strategic approach for converting fat-rich food waste into high-quality biodiesel using black soldier fly larvae for sustainable bioenergy. Sci Total Environ 951:175651
43. Kim JY, Park W-K, Park G, Choi Y, Kwon EE (2024) Feed-shifting strategy for increasing biodiesel production from black soldier fly larvae. Biores Technol 414:131633
44. de Castro RJS, Ohara A, dos Santos AJG, Domingues MAF (2018) Nutritional, functional and biological properties of insect proteins: processes for obtaining, consumption and future challenges. Trends Food Sci Technol 76:82–89
45. Yildirim-Aksoy M, Eljack R, Beck BH (2020) Nutritional value of frass from black soldier fly larvae, *Hermetia illucens*, in a channel catfish, *Ictalurus punctatus*, Diet. Aquac Nutr 26:812–819
46. Zhao J, Pan J, Zhang Z, Chen Z, Mai K, Zhang Y (2023) Fishmeal protein replacement by defatted and full-fat black soldier fly larvae meal in juvenile turbot diet: effects on the growth performance and intestinal microbiota. Aquac Nutr 2023:8128141
47. Vodounnou JV, Iko R, Okou G, Kpogue D, Ahouansou Montcho S, Micha J-C (2025) Complete substitution of fish meal with black soldier flies *Hermetia illucens* (L. 1758) larvae meal at varying incorporation rates for feeding *Oreochromis niloticus* Raised in Captivity. Aquac Sci Manag 2:1
48. Hu J, Wang G, Huang W, Zhao H, Mo W, Huang Y (2019) Effects of fish meal replacement by black soldier fly *(Hermetia illucens)* larvae meal on growth performance, body composition, serum biochemical indexes and antioxidant ability of juvenile *Litopenaeus vannamei*. Chin J Anim Nutr 31:5292–5300
49. Weththasinghe P, Hansen JØ, Nøkland D, Lagos L, Rawski M, Øverland M (2021) Full-fat black soldier fly larvae (*Hermetia illucens*) meal and paste in extruded diets for Atlantic salmon (*Salmo salar*): effect on physical pellet quality, nutrient digestibility, Nutrient Utilization and Growth Performances. Aquaculture 530:735785
50. de Souza Vilela J, Andronicos NM, Kolakshyapati M, Hilliar M, Sibanda TZ, Andrew NR, Swick RA, Wilkinson S, Ruhnke I (2021) Black soldier fly larvae in broiler diets improve broiler performance and modulate the immune system. Anim Nutr 7:695–706
51. El-Kaiaty AM, Dawa DT, El-Sayed TR (2022) The impact of black soldier fly (*Hermetia illucens*) as feed supplementation on productive and physiological performance of broiler chickens. World's VetY J 133–140
52. Mat K, Kari ZA, Rusli ND, Rahman MM, Harun HC, Al-Amsyar SM, Nor MFM, Dawood MA, Hassan AM (2022) Effects of the inclusion of black soldier fly larvae (*Hermetia illucens*) meal on growth performance and blood plasma constituents in broiler chicken (*Gallus gallus* domesticus) production. Saudi J Biol Sci 29:809–815
53. Pimchan T, Hamzeh A, Siringan P, Thumanu K, Hanboonsong Y, Yongsawatdigul J (2024) Antibacterial peptides from black soldier fly (*Hermetia illucens*) larvae: mode of action and characterization. Sci Rep 14:26469
54. Xia J, Ge C, Yao H (2021) Antimicrobial peptides from black soldier fly (*Hermetia illucens*) as potential antimicrobial factors representing an alternative to antibiotics in livestock farming. Animals (Basel) 11
55. Ribeiro JC, Gonçalves ATS, Moura AP, Varela P, Cunha LM (2022) Insects as food and feed in Portugal and Norway–cross-cultural comparison of determinants of acceptance. Food Qual Prefer 102:104650
56. Singh R, Shitiz K, Singh A (2017) Chitin and chitosan: biopolymers for wound management. Int Wound J 14:1276–1289
57. Sogari G, Menozzi D, Mora C, Gariglio M, Gasco L, Schiavone A (2022) How information affects consumers' purchase intention and willingness to pay for poultry farmed with insect-based meal and live insects. J Insects Food Feed 8:197–206

Current Issues in Halal Certification of Food Products from Biomass

Sulistyo Prabowo and Muflihah

Abstract This chapter explores the complexities in halal certification of biomass-based food products. With rapid innovations in biotechnology and food processing, the emergence of biomass as a food ingredient introduces novel challenges for halal compliance. These challenges include the source of raw materials, microbial fermentation processes, enzymatic conversions, and the traceability of ingredients in the global supply chain. Comparative analysis of global halal standards from Indonesia, Malaysia, GCC countries, and Western certifiers reveals gaps and opportunities for harmonization. Case studies of microbial enzymes, dietary fibers, and fermentation-derived flavor enhancers are presented to demonstrate real-world issues. The chapter concludes with strategic recommendations to build an adaptive, science-based halal assurance system.

Keywords Halal certification · Biomass food · Microbial fermentation · Traceability · Halal regulation · Food biotechnology

1 Introduction

The biomass-based food industry is now showing a significant growth trend, along with the increasing needs of the global community for functional, sustainable food and the results of technological innovation. Behind these opportunities, there is a major challenge, namely ensuring that these products are in line with halal principles, given the diversity of certification systems at the global level that are not yet uniform.

S. Prabowo (✉)
Food & Agricultural Products Technology Study Program, Faculty of Agriculture, Mulawarman University, Samarinda, Indonesia
e-mail: sprabowo@faperta.unmul.ac.id

Center for Halal Science and Policy Research, Mulawarman University, Samarinda, Indonesia

Muflihah
Chemistry Education Study Program, Faculty of Teacher Training and Education, Mulawarman University, Samarinda, Indonesia

M. A. Kassim et al. (eds.), *Sustainable Biorefinery*, Green Energy and Technology,
https://doi.org/10.1007/978-981-95-8181-8_3

Biomass is organic material derived from living things, whether plants, animals, or microorganisms, which can be used as a source of energy or raw material for various products. Biomass consists of material made of carbon absorbed by plants through the process of photosynthesis. In addition, biomass can also be used as food, industrial raw materials, and fertilizers.

In the context of food, biomass plays a very important role because it is the main source of human nutritional needs. Biomass food products are biological products that have nutritional value and can be consumed directly or after undergoing processing. Biomass food products also include microbial fermentation results, enzymatic engineering, and extraction results from agricultural and agro-industrial waste.

Amidst the increasing global population and challenges to food security, efficient utilization of food biomass is a strategic solution to ensure sustainable food availability [1, 2].

Due to the unusual source of ingredients and the complex production process, various halal issues arise, especially related to the origin of ingredients, microbial growth media, use of additives, and genetic engineering status. Therefore, an in-depth study of halal critical points, cross-country regulatory frameworks, and strategic approaches to strengthening the certification system is very important. The framework in Fig. 1 illustrates interconnections between technological, regulatory, and ethical factors throughout the biomass-based food production chain that will be discussed in this chapter.

2 Characteristics of Food Biomass Products

Food biomass products have a number of unique characteristics that distinguish them from other types of biomass (such as energy biomass or industrial biomass). These characteristics relate to their origin, composition, use, and processing methods.

1. Sourced from Living Things (Organic): Biomass food products come from living things such as plants (e.g., grains, tubers, fruits) or animals (meat, milk, eggs). This material consists of organic compounds such as carbohydrates, proteins, fats, vitamins, and minerals that are essential for the human body [2].
2. Contains Nutritional Value: The main characteristic of food biomass is its nutritional content. This product is designed or harvested to meet human nutritional needs. The nutritional content varies greatly depending on the type—for example, cereals are rich in carbohydrates, soybeans are rich in protein, and fruits are rich in vitamins and antioxidants [3].
3. Can be Consumed Directly or After Processing: Food biomass can be eaten in its fresh form (e.g., fruits or vegetables) or after going through a processing process (such as fermentation, drying, milling). This process can increase nutritional value, extend shelf life, or improve taste and texture [4].

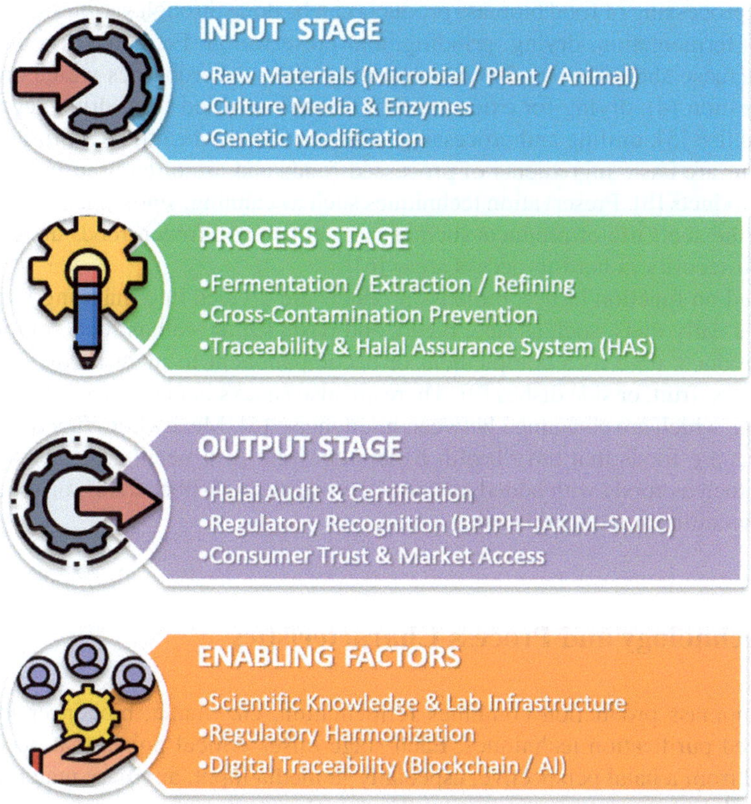

Fig. 1 Conceptual framework of halal certification challenges for biomass-based foods

4. Easily Decomposed and Biodegradable: Food biomass products have a limited shelf life because they are easily decomposed by microorganisms. This property makes them biodegradable, but also poses a risk of food loss if not handled properly [5].
5. Affected by Season and Climate: Food biomass production is highly dependent on environmental factors such as weather, rainfall, temperature, and soil conditions. Therefore, its productivity and quality can be very volatile, depending on the season or climate change [6].
6. Potential to Become Waste or Reprocessed: Food biomass products produce organic waste (such as skin, pulp, or uneaten parts) that can still be used as feed, compost, or even reprocessed into alternative food products through certain technology [1].
7. Role in Economy and Food Security: In addition to being a source of nutrition, food biomass is also an important economic commodity that supports the agricultural sector and the food industry. On a national scale, this product contributes directly to food security and public income [7].

The processing of food biomass products can be done through various techniques, such as fermentation, drying, grinding, and preservation. Fermented products not only increase shelf life but also increase the content of probiotics which are good for digestion [4], drying for extend shelf life, such as dried fruit, dried vegetables, or dried fish [5], milling and processing of flour such as wheat flour, corn flour, and rice flour are basic ingredients of processed foods and are often used in bread and pasta products [8]. Preservation techniques such as canning, smoking, or freezing to extend the shelf life of products such as canned fish and frozen foods are essential for food security in hard-to-access areas [6].

Based on function or use, biomass products are divided into main ingredients in people's daily diets, such as rice, bread, and pasta [3] as well as food complementary. Products used as a complement or accompaniment to main foods, including vegetables, fruit, or side dishes [9]. There are also snacks in the form of chips, cakes, and nuts, which are consumed between main meals [5]. Meanwhile, functional food products are foods that have health benefits beyond just meeting basic nutritional needs, such as foods with added probiotics (yogurt), high-fiber foods, or foods with added vitamins and minerals [4].

3 Technology and Process Characteristics

Food biomass production combines fermentation, enzymatic, bioconversion, and advanced purification techniques. Each stage raises critical points that need to be studied from a halal perspective, especially in media input, auxiliary materials, and the possibility of cross-contamination. The food biomass production process involves various technologies and methods that aim to convert organic materials into food products that can be consumed by humans. Technology in food biomass production focuses on improving the quality, efficiency, and safety of the resulting product.

3.1 Fermentation and Enzymatic Processes in Food Biomass Products

Fermentation is a crucial biological transformation in food processing, utilizing microorganisms like bacteria, yeast, and fungi to biochemically convert food components. This process significantly improves the flavor, aroma, nutritional value (often adding probiotics, vitamins, and minerals), digestibility, and shelf life of food biomass products through the production of compounds such as acids or alcohol. Common steps involve raw material preparation, inoculation, the fermentation stage, and termination.

Enzymatic processes, on the other hand, employ enzymes (proteins that act as biological catalysts) to chemically modify food materials. Examples include the hydrolysis of starch, proteolysis of proteins, and the breakdown of lactose. Enzymes improve taste, texture, and optimize raw material use, contributing greatly to the quality and value of the final food product.

The success of fermentation relies heavily on the specific microorganism used and the culture media that support its growth. Table 1 summarizes the roles of the main microbial groups.

Culture media play a vital role in providing the nutrients required for microorganisms to grow and produce the desired fermentation products. Different types of media are specifically designed to support the growth of certain groups of microorganisms. De Man, Rogosa, and Sharpe (MRS) and agar media are commonly used for bacteria, while Yeast extract Peptone Dextrose (YPD) and Wort Liquid Medium (WLM) media support the growth of yeast. For fungi, Sabouraud Dextrose Agar (SDA) and fungi agar media are often used. In addition to standard media, culture media can also be customized with natural ingredients such as those used in kimchi or vinegar fermentation. The quality of the culture media directly affects the final characteristics of the fermentation product, including taste, texture, and nutritional content, so a good understanding of the source of microorganisms and culture media is key to the success of the fermentation process of food biomass products.

Meanwhile, enzymatic processes utilize enzymes, which are proteins that speed up chemical reactions, to modify food materials. Some types of enzymatic processes include the hydrolysis of starch into simple sugars, the proteolysis of proteins into amino acids, and the fermentation of lactose into glucose and galactose. Enzymes have various applications in the manufacture of food products, such as the use of amylase in making bread, rennet in the production of cheese, and protease and amylase in the manufacture of fermented drinks.

Table 1 The roles of the main microbial groups

Microorganism group	Examples	Fermentation type & key product	Food applications
Bacteria	*Lactobacillus spp., Bifidobacterium spp.* (LAB)	Lactic acid fermentation (Lactic Acid)	Yogurt, Kimchi, Sauerkraut
	Acetobacter spp.	Acetic fermentation (Acetic Acid)	Vinegar
	Pediococcus	Lactic acid fermentation	Sauerkraut, Kimchi
Yeast	*Saccharomyces cerevisiae*	Alcoholic fermentation (Ethanol, CO_2)	Bread, Beer, Wine
	Saccharomyces bayanus	Alcoholic fermentation	Wine, Kefir
Fungi	*Rhizopus oligosporus*	Mold fermentation (Breakdown of components)	Tempeh (converting soybeans)
	Aspergillus oryzae	Mold fermentation (Breakdown of proteins/starches)	Soy Sauce, Miso

The use of enzymes in food biomass processing offers a number of advantages. Enzymes can improve the taste and texture of products by changing the structure of proteins and carbohydrates. Enzymatic processes can also optimize the use of raw materials by converting less valuable components into more useful products. Furthermore, enzymes can increase nutritional content by breaking down difficult-to-digest compounds, making nutrients more easily absorbed by the body. Overall, both fermentation and enzymatic processes contribute greatly to improving the quality and value of food biomass products.

3.2 *Cross-Contamination and Traceability*

Cross-contamination and traceability are two crucial aspects in food biomass production to ensure the quality, safety, and sustainability of products that reach consumers. Cross-contamination, which is the transfer of pathogenic microorganisms, chemicals, or allergens between materials or surfaces, can occur at various stages of production. The causes vary, ranging from microorganisms such as bacteria and fungi, hazardous chemicals from agriculture, to allergens and non-sterile processing processes. In the context of the halal assurance system, contamination can come from non-halal ingredients.

The impact of cross-contamination is significant, especially in relation to the health risks of consumers who may experience food poisoning, infection, or allergic reactions. In addition, contamination can reduce the quality of food products in terms of taste, odor, and texture, which leads to economic losses for producers due to products that must be discarded or reprocessed.

Cross-contamination prevention efforts involve a variety of strategic actions. The use of appropriate processing technologies, such as pasteurization and sterilization, plays a key role in minimizing risk. Separation of raw and ready-to-eat products, as well as strict hygiene practices at all stages of production, are also key. Furthermore, training the workforce in food hygiene principles and cross-contamination prevention techniques is an important investment in maintaining product quality and safety.

With careful attention to preventing cross-contamination and implementing effective traceability systems, food biomass producers can ensure that their products are safe to consume, of high quality, and meet established standards. Both of these aspects together build consumer confidence and support the sustainability of the food industry as a whole.

3.3 *Traceability in Food Biomass Products*

Traceability, or the ability to track the movement of raw materials to the final product in the supply chain, plays a crucial role in the food industry. This system provides complete transparency and provides assurance to consumers that the products they consume are safe, of high quality, and produced in accordance with applicable regulations. Thus, traceability is not just a matter of recording data, but an important foundation for building trust and compliance in the food sector.

The implementation of traceability brings a variety of significant benefits. From a food safety perspective, the system allows for rapid identification of the source of the problem and the withdrawal of problematic products from the market, protecting consumer health. In addition, traceability helps manufacturers comply with increasingly stringent international food safety regulations. For consumers, the transparency offered by traceability increases confidence in the products they consume, especially regarding their origin and production process. Finally, in cases of contamination or damage, traceability minimizes waste by allowing for more targeted product recalls.

The implementation of a traceability system involves several important stages. First, the collection of accurate and relevant data at every stage of production, from raw material sourcing to distribution. Second, effective risk management, where the traceability system helps identify and address potential issues such as cross-contamination. Third, transparent communication with consumers, allowing them to access detailed product information.

Cross-contamination and traceability are two inseparable elements in the food biomass industry. Managing cross-contamination through strict hygiene practices and product separation is a proactive step to maintain safety. Meanwhile, traceability is a reactive mechanism that allows for quick handling if problems occur. The combination of the two ensures food products are safe, of high quality, meet consumer expectations, and build sustainable trust.

4 Sharia Interpretation of Food Biomass Products

Biomass food products used in everyday life, especially those made from natural materials such as plants and animals, certainly have Islamic legal implications that need to be considered. Although Islam provides clear guidelines regarding the principles of halal and haram in food consumption, sharia interpretations of biomass food products can vary based on several factors, including the approach to Islamic law, the views of scholars, and the social and cultural context.

Biomass food products derived from natural materials, such as vegetables, fruits, meat, and other processed products, are generally permissible in Islam as long as they meet the criteria of *halal* and *tayyib* (both physically and spiritually). However, there are some areas where sharia interpretations may vary, especially when involving materials or processes not explicitly described in classical texts.

Plant-based Food Products: In general, foods derived from plants such as vegetables, fruits, and grains are considered halal. However, interpretations may vary for products such as vegetable alcohol (such as beer from malt) or products that use ingredients contaminated with haram substances, such as pesticides containing forbidden materials [10].

Animal Food Products: In the case of animal products such as meat, fish, and eggs, the law of slaughtering becomes very important. Meat that is slaughtered without following the standards of the Shari'a (for example, without mentioning the name of Allah during slaughtering or using a non-halal method) is considered haram. Interpretations of these methods of slaughtering vary, with some scholars considering more practical modern methods to be permissible, while others are more strict in this matter [11].

Processed Products: The processing of food products also affects the halal status of the product. Products containing additives (such as preservatives or artificial sweeteners) can cause differences of opinion, depending on whether the ingredients come from halal or haram sources [12].

5 Use of Enzymes and Microorganisms in Biomass Food Products

Some biomass food products, such as cheese, tempeh, and yogurt, use a fermentation process involving microorganisms or enzymes. This process often triggers debate in the world of Islamic jurisprudence because some of the microorganisms or enzymes used can come from sources whose halal status is questionable.

Microorganisms and Enzymes Derived from Haram Materials: The use of microorganisms or enzymes derived from haram materials (such as pork or alcohol) in the process of making food products can cause differences of opinion. Some scholars believe that the use of haram raw materials in fermentation, even if there are no direct traces of them in the final product, can still make the product haram [13]. However, there is a moderate view which argues that if the ingredient is not directly present or does not function in the final product, then the product can be considered halal.

Halal Microorganism-Based Fermentation Process: In contrast, products that use halal microorganisms, such as those found in fermented products such as tempeh, cheese, or yogurt, are considered shariah-compliant if the microorganisms are not contaminated with haram materials. A more moderate interpretation allows the use of enzymes or microorganisms produced through modern biotechnology, provided that no haram materials are involved in the production processes [4].

6 Traceability and Halal Certification of Biomass Food Products

In the modern era, the concept of traceability (product trace) becomes important in ensuring that biomass food products meet the expected halal standards. The traceability system and halal certification make it easy to ensure that the product is truly free from haram elements. However, here too, there are varying interpretations depending on the certification body and its implementation in a particular country or region.

7 Halal Critical Points in Biomass Derivative Products

In, the food biomass industry, the halalness of a product is an important aspect, especially for products that will be consumed by Muslims. Halalness in food biomass products involves various aspects that need to be understood, both in terms of raw materials, production processes, and additional materials used. Critical halal points that need to be considered in the production of biomass derivatives:

1. Sources of Animal and Vegetable Raw Materials:

Biomass-derived food products can come from animal or plant sources. For animal products, it is important to ensure that the source of meat, milk, eggs, or other animal products comes from animals that are halal and slaughtered according to Islamic law. Slaughter must be carried out by competent slaughterers and in accordance with sharia provisions, for example, slaughtering by Muslims while mentioning the name of Allah and being kind to the slaughtered animal. For plant-based food ingredients, food crops are generally considered halal, but some plant products may be contaminated with non-halal materials (such as alcohol or chemicals from processing). Therefore, strict supervision is needed to ensure that plant products are free from contamination that could damage their halal status [3].

2. Processing of Preservatives and Additives:

In the processing of biomass derivative products, the use of additives such as preservatives, dyes, or flavors must meet halal criteria. Some synthetic chemicals or materials from non-halal sources, such as alcohol, can be used in the processing process. Therefore, monitoring the additives and chemicals used is very important to ensure a halal product [2].

3. Fermentation Process:

In products involving fermentation processes, it is important to ensure that the innovation in fermentation and enzymatic technology must ensure that the microorganisms or enzymes used in the fermentation process or halal food processing are derived from sources that comply with Islamic law. For example, the use of microbes obtained from non-halal materials should be avoided in food products intended for the Muslim market [4]

4. Cross-Contamination:

During processing, the risk of cross-contamination between halal and non-halal ingredients is one of the critical points. For example, using the same tools or facilities to process halal and non-halal products can result in contamination of halal products, which can damage their halal status.

5. Storage and Packaging Halal Packaging:

Food product packaging also needs to be considered so that it is not contaminated with non-halal materials. Improper packaging or the use of non-halal packaging materials can damage the halal status of the product. Therefore, it is important to use packaging materials that meet halal standards and are safe for Muslim consumers. The product storage system must also minimize the risk of cross-contamination with non-halal products. Separate storage for halal and non-halal products is a recommended step to maintain halal products [6].

8 Halal Certification Process for Biomass Food Products

Biomass food products are the result of converting biological materials, such as microorganisms, agricultural waste, and fermentation products, into food ingredients or food additives. In a global context, the use of biomass in the food industry continues to increase due to sustainability and production efficiency. However, for Muslim consumers, it is important to ensure that biomass products meet halal standards. The halal certification process is an important tool in ensuring compliance with Islamic law throughout the food biomass production chain.

Halal certification begins with the identification of raw materials. In the context of food biomass, raw materials can come from microorganisms such as *Saccharomyces cerevisiae*, *Spirulina*, or agro-industrial waste such as sugarcane bagasse. These materials must be traced to ensure that they do not come from haram substances, such as enzymes from non-halal animals or alcohol from uncontrolled fermentation. Halal certification bodies require supporting documents such as Material Safety Data Sheets (MSDS) and Certificates of Analysis (CoA) from raw material producers [14].

The biomass production process is also strictly checked. The fermentation process, microbial harvesting, and drying must be free from cross-contamination with haram or najis materials. Machines, production tools, and storage areas must be cleaned according to *thaharah* (Islamic cleanliness rules). In addition, the use of fermentation media, such as peptone, gelatin, or enzymes, must be verified for halalness. In some cases, enzymes come from pigs or animals that are not slaughtered according to sharia, so their use is not allowed [15].

Halal certification at the global level requires mutual recognition between world halal institutions. For example, if a biomass product is certified by The Department of Islamic Development Malaysia (JAKIM, Malaysia), then other countries, such as Indonesia through The Halal Product Assurance Organizing Agency (BPJPH Indonesia), can recognize the certification if it is already on the list of recognized foreign halal certification bodies. This is important because food biomass is often an export commodity in the multinational industry [16].

Global halal certification registration follows the procedures of each country, but the principles are similar. Business actors register their products with an official halal certification body, submit technical documents, and agree to undergo a halal audit.

The audit is carried out by certified auditors who verify all aspects of raw materials, production processes, facilities, and halal quality management. Products that pass will receive a halal certificate and can use the official halal logo on their packaging [17].

Biomass products that are often submitted for halal certification include fermented amino acids, polysaccharides such as xanthan gum, and single-cell proteins (SCPs). These products are commonly used in the food and beverage industry as thickeners, stabilizers, or sources of nutrients. Because of their nature derived from microorganisms or fermentation processes, their halal aspects are highly dependent on the supporting materials and processes. The accuracy of halal audits is crucial in this matter [18].

In some cases, biomass products are rejected for certification because they use substrates from non-halal alcohol or enzymes. For example, xanthan gum fermented with alcohol as a precipitation agent will be considered non-halal if the alcohol comes from a *khamr* source (beer and alcoholic beverage industry) and cannot be completely removed. Therefore, industry players have begun to innovate using halal substrates and growth media that comply with international halal standards [19].

In Indonesia, Law No. 33 of 2014 concerning Halal Product Assurance, and Government Regulation No. 39 of 2021 are the legal basis for halal certification. In a global context, the SMIIC (Standards and Metrology Institute for Islamic Countries, OIC—The Organisation of Islamic Cooperation) standards and guidelines from the World Halal Food Council (WHFC) are important references in formulating cross-country halal certification policies. The harmonization of these standards makes it easier for the food biomass industry to penetrate the world's Muslim market [20].

Challenge in Global food biomass halal certification is the limited number of halal testing laboratories and certified auditors who understand food biotechnology and microbiology. Therefore, increasing the capacity of halal human resources and integrating digital technology into the traceability system is important to ensure the accuracy and efficiency of the halal certification process [21].

Overall, the global halal certification process for food biomass products is a combination of scientific, legal, and sharia aspects. This requires collaboration between industry, academics, regulators, and halal certification institutions across countries. With a transparent, credible, and internationally standardized certification process, Muslim consumer confidence in biomass products can increase, while encouraging the growth of the global halal industry.

With the increasing awareness of sustainability and the trend of consuming biomass-based products, halal certification becomes a strategic instrument in ensuring the access of these products to the global Muslim market estimated at 2 billion people. The food biomass industry that integrates halal principles in its innovation will have a significant competitive advantage in the world trade map.

9 Differences in Halal Certification Systems

Countries with large Muslim populations such as Indonesia, Malaysia, and Saudi Arabia have strict halal certification systems, but the assessment methods and requirements can differ. For example, Indonesia applies halal standards issued by the MUI (Indonesian Ulema Council), while other countries may have their own certification bodies with slightly different procedures [8].

The use of technology such as blockchain or Radio Frequency Identification (RFID) in the biomass food product traceability system also raises various interpretations among scholars. Some scholars argue that this technology can be relied on to ensure the halalness of products, while others may doubt the reliability of this system, especially if there is no adequate supervision of each stage in the supply chain [22].

10 Sustainability and Ethics Issues in Biomass Food Products

Contemporary challenges such as sustainability and ethics of food production also influence the sharia interpretation of biomass food products. In a world that is increasingly concerned with environmental and sustainability issues, the question of the halalness of products produced through environmentally damaging or unethical methods, such as the use of hazardous chemicals or farming methods that damage ecosystems, becomes relevant.

Some Islamic scholars and scholars believe that sharia principles do not only apply to halal and haram aspects, but also to the principles of social justice and environmental protection. Therefore, biomass food products that are produced in an environmentally unfriendly manner or involve excessive exploitation of natural resources can be considered inconsistent with Islamic teachings regarding sustainability [6].

The varied interpretations of sharia on biomass food products reflect the flexibility in the application of Islamic principles to technological developments and social dynamics. Although there are clear basic guidelines regarding what is halal and haram in Islam, factors such as production methods, use of enzymes and microorganisms, traceability, and sustainability and ethical issues influence the perspectives of Islamic scholars and scholars in determining the halal status of biomass food products. As science and technology advance, there will continue to be room for *ijtihad* (Islamic legal term that refers to the process of independent reasoning or intellectual effort used by a qualified scholar (*mujtahid*) to interpret and derive legal rulings from the sources of Islamic law (*Sharia*), especially in matters where no clear text exists in the Qur'an or Hadith) and adjustment to new challenges that arise.

Halal Certification Flow for Food Biomass Products

The certification process is a holistic system that follows the product's journey from raw materials to the final distribution, with critical checks at each stage.

1. **Input: Raw Material Traceability (Upstream)**

   This initial stage focuses on the origin and sourcing of all components.

- **Supporting Documentation (Evidence)**

  - **Biomass Sources:** Tracking the origin of all biomass inputs (e.g., microorganisms like *Saccharomyces cerevisiae*, or agro-industrial waste like sugarcane bagasse).
  - **Animal/Plant Sources:** Ensuring animal products are from halal animals slaughtered according to Islamic law, and plant products are free from non-halal contamination (e.g., alcohol, forbidden chemicals).

- **Raw Material Identification and Sourcing**

  - Collection of essential documents such as Material Safety Data Sheets (MSDS) and Certificates of Analysis (CoA) from raw material producers to verify the source's halal status.
  - Verification that raw materials are not from forbidden sources (e.g., enzymes from non-halal animals or alcohol from uncontrolled fermentation).

2. **Process: Critical Control Points and Halal Assurance System (SJH)**

   The process stage involves stringent checks on all manufacturing steps to prevent contamination.

- **Processing of Additives and Auxiliary Materials**

  - Ensuring all additives (preservatives, colors, flavors) and auxiliary materials (e.g., solvents like ethanol) meet halal criteria.
  - Fermentation Media Verification: Strict verification of culture media (e.g., peptone, gelatin) used for microorganisms/enzymes to ensure they are not derived from haram sources (e.g., pork, non-halal animal products).

- **Production Process & Cross-Contamination Control**

  - Verification that all steps (fermentation, microbial harvesting, drying) are free from contamination with haram or *najis* (unclean) materials.
  - Implementing strict hygiene practices and separation of tools/facilities to prevent cross-contamination between halal and non-halal ingredients.
  - Machines and production areas must be cleaned according to *thaharah* (Islamic cleanliness rules).

- **Storage and Packaging**

– Separate storage for halal and non-halal products to maintain halal status.
– Ensuring packaging materials are halal and do not cause contamination.

3. **Audit: Verification by Certified Auditors**

This is the independent assessment stage to verify compliance with the Halal Assurance System (SJH).

- **Halal Audit and Verification**

– Certified auditors verify all documented aspects: raw materials, production processes, facilities, and the overall Halal Quality Management System.
– The audit must be thorough, particularly for complex products like microbial enzymes, to trace the source of microorganisms and the culture media used.

- **Risk Management and Residue Testing**

– The traceability system helps identify and address potential issues like cross-contamination.
– Testing for residues, such as alcohol content (e.g., checking if ethanol is below the 0.5% limit for food/beverage products and not from a *khamr* source).

4. **Approval: Certification and Global Recognition**

The final stage where successful compliance is recognized and made transparent.

- **Issuance of Halal Certificate**

– Products that pass the audit receive a **Halal Certificate** and can use the official halal logo on their packaging.

- **Traceability and Transparency**

– The traceability system ensures complete transparency and provides assurance to consumers that the product is safe and high-quality, building consumer trust.

- **Global Harmonization and Mutual Recognition**

– Certification from recognized bodies (e.g., BPJPH Indonesia, JAKIM Malaysia) is important for mutual recognition, which facilitates the export of food biomass products in the multinational industry.

11 Perspective of Global Halal Certification Regulation on Food Biomass Products

Halal certification is a procedure to ensure that products produced are in accordance with the principles of Islamic law. Biomass food products involving plant and animal materials must meet the requirements set out in Islamic law, and this is often translated into an internationally recognized halal certification process. Although the basic principles of halal can be widely applied, the implementation of halal regulations and certification for biomass food products varies from country to country. This depends on the legal, cultural, and infrastructure factors that exist in each country or region.

Halal certification of biomass food products is an important aspect in ensuring that the food products produced are in accordance with Islamic principles. Although there are variations in the implementation of halal regulations and standards in different countries, global certification bodies such as MUI, JAKIM, and OIC play a vital role in regulating and facilitating the halal market. Challenges in the certification process include differences in standards between countries, food safety, and the use of advanced technology to increase transparency and traceability of products. With the increasing global demand for halal products, halal certification is becoming more than just religious compliance, but also an assurance of quality and sustainability.

The diversity of interpretations of halal, legal systems, and geopolitical dynamics hinders the unification of global halal standards. The efforts of OIC-SMIIC and WHFC are important for international accreditation and harmonization. Biomass food products derived from biological sources such as plants, animals, and microorganisms are increasingly being used in the global food industry. Along with the increasing demand for sustainable, halal, and safe food products, the need for uniform standards (harmonization) becomes important. However, reality shows that the harmonization of global standards faces various complex challenges, both from technical, legal, and cultural aspects.

12 Case Study of Biomass Products and Halal Certification

As technological innovation in the food industry increases, many new products are being developed from biomass. These products often pose challenges in the halal certification process due to the complexity of the source of the material, the production process, and the use of biotechnology. Here are some real-life case studies that show how halal certification deals with biomass-based products.

12.1 Microbial Enzymes in Bread and Cheese

Enzymes such as rennin (for cheese) and amylase (for bread) are often produced from genetically engineered microorganisms or fermentation media containing animal elements. If the fermentation media uses substrates from pigs or animal media that are not slaughtered in a halal manner, then the enzyme product can be considered haram or doubtful. Some countries, such as Malaysia through JAKIM, require audits of the sources of microorganisms and culture media for the purposes of halal certification [23]. A bakery in Indonesia that used imported enzymes had to replace its raw materials after an LPPOM MUI audit found that the enzymes were cultured on media from pig pancreas [24].

The popularity of probiotic supplements containing lactic acid bacteria such as *Lactobacillus* and *Bifidobacterium* continues to increase globally. These products are widely marketed for their significant benefits for digestive health, from maintaining the balance of gut microbiota to boosting the immune system. However, behind this health potential, there is a halal aspect that needs to be considered, especially for Muslim consumers. The main concern lies in the fermentation media used in the cultivation process of these microorganisms.

The production process of probiotics involves growing bacteria on a large scale using nutrient-rich culture media. Some commercial culture media are known to contain peptone, which is a mixture of peptides and amino acids obtained from the hydrolysis of proteins. The protein sources for these peptones vary, including animal tissues. This is a critical point for halalness, because peptone derived from pork or blood (which is forbidden in Islam) can contaminate the final product.

In the United Arab Emirates, a European probiotic product was denied certification for failing to provide complete documentation of the source of the culture material. The Gulf Standardization Organization (GSO) emphasizes the importance of traceability from the microorganism level to the end product [25].

According to the Indonesian Ulema Council (MUI), the use of haram materials in the production process of food and medicine can make the final product haram, even though the materials are not detected in the final product. MUI Fatwa Number 4 of 2003 concerning Halal Standardization of Food and Medicine Products explains that "haram and/or impure materials and products contaminated with them are haram to consume and/or use" [16].

These concerns are not without foundation. Several studies and industry reports have shown that pork peptone is still used in the production of microbiological culture media due to its availability and relatively low cost. An article in *the Journal of Islamic Marketing* highlights the importance of transparency in the halal product supply chain, including the culture media used in probiotic production. The article emphasizes that Muslim consumers have the right to know the source of the ingredients used in the products they consume [26].

Halal alternative culture media are actually available. Peptone derived from plant sources (such as soybeans or peas) or halal animals (such as cows or chickens slaughtered in an Islamic manner) can be used as a substitute for pork peptone. Several probiotic producers who recognize the importance of halal for Muslim consumers have switched to using these alternative culture media and obtained halal certification from authorized institutions.

However, the challenge is the lack of transparency and clear information about the culture media used by manufacturers globally. Muslim consumers often find it difficult to verify the halalness of probiotic products based only on the final product label. Therefore, halal certification from a credible institution is very important. Halal certification institutions will conduct a thorough audit of the entire production process, including the source of raw materials for culture media, to ensure there is no contamination of haram materials.

As a smart Muslim consumer, it is important to look for probiotic supplement products that have clear and trusted halal certification. In addition, consumers can also seek further information about the production process and raw materials used by the manufacturer. With increasing awareness of the importance of halal products, it is hoped that more probiotic manufacturers will switch to using halal and transparent culture media in their production processes. This will provide certainty and peace of mind for Muslim consumers in meeting their digestive health needs.

12.2 Ethanol Controversy in Final Refining

The use of alcohol solvents such as ethanol and methanol plays a significant role in the extraction and purification process of food biomass, especially to obtain high-value bioactive compounds such as polyphenols, flavonoids, plant proteins, and soluble dietary fiber. Ethanol is commonly used in the extraction of antioxidants from various natural sources, while methanol is often utilized in laboratory-scale research. This process is crucial in producing functional food ingredients with high health benefits and economic value.

However, the use of alcohol, especially ethanol derived from fermentation, poses challenges regarding the halalness of products for Muslim consumers. In Islamic law, fermented ethanol is categorized as *khamr* (intoxicating beverages), so its use in food products or raw materials to be certified halal requires special consideration. Although ethanol residues can be removed through physical processes such as evaporation or distillation, the halal issue involves more than just the presence of residues, including the source and intended use.

The National Sharia Council of the Indonesian Ulema Council (DSN-MUI) and the Halal Department of JAKIM (Malaysia) have issued guidelines regarding the use of alcohol in the food and pharmaceutical industries. DSN-MUI Fatwa No. 11/2009 and the LPPOM MUI guidelines (2020) permit the use of alcohol if it comes from non-*khamr* (e.g., synthetic alcohol), is used for technical purposes, and does not leave residue in the final product. If the alcohol comes from fermentation, it must be ensured that the purpose is not for consumption as a beverage, and its use is not excessive.

The halal certification process by institutions such as LPPOM MUI now demands full transparency regarding the origin and use of alcohol solvents in the production process. Every ingredient, including solvents, must be documented in detail in the Halal Assurance System (SJH). Verification includes audits of raw materials, production processes, and residue testing if necessary. The industry is also increasingly turning to solvents that are considered more halal such as hot water, purified food-grade vegetable ethanol, or supercritical CO_2 to avoid sharia doubts.

The increasing awareness of Muslim consumers and increasingly stringent global halal regulations encourage the food and biopharmaceutical industries to pay attention to halal aspects throughout the entire production process, not just to the final product. The development of halal-friendly extraction technology is an important focus in current food innovation, as emphasized in the LPPOM MUI Halal Assurance System Guidelines (2020) and DSN-MUI Fatwa No. 11/DSN-MUI/III/2009 concerning alcohol.

In the context of final product purification, the use of ethanol often raises questions regarding its halalness from an Islamic perspective. Although effective as a solvent and cleaning agent, the presence of ethanol residue, even in small amounts, can be a concern for Muslims. Islamic law deems *khamr* as haram and najis, and the majority of scholars agree that ethanol from fermented grapes or dates falls into this category. Differences of opinion arise when ethanol is used not as the main ingredient of intoxicating beverages, but as a solvent in the production process.

Some scholars adhere to the principle of caution and forbid the use of ethanol if there are still residues in the final product, referring to the *fiqh* principle "anything that intoxicates in large quantities, even a small amount is forbidden". However, a more permissive view allows the use of ethanol if its original properties are lost or if the residue is very small and does not intoxicate. MUI through Fatwa Number 10 of 2018 provides a limit on the ethanol content in the final product, which is less than 0.5% and is not intended for consumption as a drink, so that the product is considered halal.

However, the 0.5% limit applies to food and beverage products, and considerations for pharmaceutical or cosmetic products may differ. Some scholars still emphasize the importance of avoiding ethanol whenever possible and seeking halal alternatives. In addition, the final purification process using ethanol must ensure that it does not cause cross-contamination with other haram ingredients, and transparency of information to Muslim consumers regarding the use of ethanol is crucial in maintaining trust and the halalness of the product.

In the food biomass industry, vegetable solvents are now a promising alternative to replace alcohol in the extraction process of active ingredients, especially in the context of the halal industry. The use of alcohol such as ethanol in the extraction process is often in the spotlight due to potential non-compliance with halal principles, especially if obtained from the fermentation of non-halal materials. As a solution, vegetable solvents such as coconut oil and soybean oil are used to extract lipophilic compounds such as vitamins and bioactive components from biomass. In addition, vegetable glycerol, which is non-toxic and non-intoxicating, has been shown to be able to replace alcohol in various chemical and enzymatic reactions, making it a superior halal solvent [27].

In addition to conventional vegetable solvents, the latest innovations such as plant-based Deep Eutectic Solvents (DES) are also starting to be widely used. DES is made from a mixture of natural compounds such as choline chloride and organic acids (for example, citric acid), which produce biodegradable, safe, and halal-compliant solvents. Paiva et al. (2014) stated that the use of plant-based DES is not only environmentally friendly, but also provides religious advantages because it does not come from haram or impure materials. This makes DES a very relevant solvent in the halal food industry, especially in processing active ingredients from natural sources that are sensitive to alcohol-based chemical processes.

Overall, the use of vegetable solvents in the food industry not only supports the principles of Islamic law, but is also in line with the concept of green chemistry. These solvents are safe, do not leave harmful residues, and do not have an intoxicating effect, which makes them very suitable for halal and thayyib standards. In addition to supporting environmental sustainability, these solvents also strengthen the position of halal-organic products in the global market. Therefore, the adoption of vegetable solvents and DES in the biomass extraction industry is a strategic step toward an ethical, safe, and environmentally friendly food production environment [27, 28].

12.3 Halalness of Hydrogenated Vegetable Oil

Hydrogenated Vegetable Oil (HVO) is a crucial ingredient in the food industry, especially in margarine, ice cream, and various processed products. The hydrogenation process that turns liquid oil into solid involves the use of catalysts or chemical agents. The halal aspect of HVO is a serious concern, considering the potential use of catalysts sourced from animals.

The halalness of HVO does not only depend on the raw material of vegetable oil, but also on the entire supply chain and production process. If the catalyst used comes from non-halal animals or the slaughtering process is not in accordance with Islamic law, then the HVO produced can be contaminated with impure or haram elements. In fact, animal fatty acid-based catalysts have the potential to contaminate products according to Islamic law, making them unfit for consumption by Muslims.

A case study in Malaysia is a vivid illustration of the importance of halal audits in the HVO supply chain. A margarine manufacturer was forced to change its catalyst supplier after an audit by the Department of Islamic Development Malaysia (JAKIM) found that the catalyst used contained fatty acids from non-halal sources. This incident highlights the need for thoroughness and transparency in the selection of materials and production processes to ensure the halalness of the final product.

Therefore, HVO can only be considered halal if all stages of production, from raw material sourcing to catalyst use, are guaranteed to be free from haram and najis elements. Credible halal certification is important for Muslim consumers to ensure that the products they consume are in accordance with sharia principles.

12.4 Examining the Halalness of Cultured Meat

Cultured meat innovation, or meat grown from animal cells in a laboratory, offers a promising alternative biomass solution. However, amid the enthusiasm for its sustainability potential, its halal aspect is a crucial concern, especially for Muslim consumers. Two main challenges emerge in this regard.

First, the commonly used initial culture medium, *fetal bovine serum* (FBS), is derived from the blood of calves that are not necessarily slaughtered according to Islamic law. The unclear source and method of obtaining it raise doubts about the halalness of the final product. Alternative culture media that are halal and *animal-free* are the focus of intensive research to overcome this obstacle.

Second, the halal status of the original cells used in the culture process is also debated. Referring to the view of Alzeer and Abou Hadeed (2016), questions arise regarding the need for sharia slaughter of the animal that is the source of the initial cells. Some scholars argue that cells taken from living animals do not require slaughter, while others hold the principle that animal-derived products remain bound by the halal provisions of the animal of origin.

Nevertheless, innovation continues. The development of halal culture media and in-depth Islamic jurisprudence studies is expected to provide clarity on the legal status of cultured meat. If these challenges can be overcome, cultured meat has the potential to become a halal and sustainable meat alternative for Muslims in the future.

In Singapore, the fatwa is not yet final on the halal status of cultured meat due to the lack of evidence on the origin of the media and the method of culture ending. The MUI is also still delaying the halal recognition of this product [16].

12.5 Protein Isolates from Microalgae

Efforts by Japanese spirulina products to obtain halal certification from the Halal Product Guarantee Agency (BPJPH) have encountered obstacles. The results of the BPJPH laboratory examination found the presence of pork gelatin-based *carrier content* in the spirulina capsules [29]. This automatically invalidates the potential for halal certification because pork gelatin is included in the category of prohibited materials in Islam [30].

This finding certainly surprised many, considering that spirulina is known as a popular natural health supplement. The presence of non-halal ingredients in products marketed to Muslim consumers raises serious questions about the transparency and integrity of the supply chain. BPJPH emphasized that halal certification cannot be granted if there is evidence of contamination or the use of haram ingredients in the product [29].

This case is an important reminder for manufacturers, both domestic and foreign, to ensure that all product components, including additional materials such as capsule *carriers*, meet halal standards. Muslim consumers are also advised to be more careful in choosing products and always check for the existence of halal certification issued by authorized institutions. The failure of this certification is a valuable lesson on the importance of audits and strict supervision in the production process in order to maintain consumer Muslims' trust [31].

Microalgae such as Spirulina and Chlorella are gaining popularity as a sustainable and nutrient-rich alternative protein source. In the context of a large global Muslim population, the halal aspect of microalgae-derived products, including their protein isolates, becomes of utmost importance [32]. Halal Food and beverage products in Islam are based on the fulfillment of certain criteria derived from religious teachings. Therefore, the production process of microalgae protein isolate needs to be researched and its halalness verified to ensure that it can be consumed by Muslims without doubt [33].

One of the crucial aspects in determining the halalness of microalgae protein isolates is the source of the microalgae itself. In general, microalgae are considered halal because they come from nature and are not included in the category of animals or plants that are forbidden in Islam. However, it is important to ensure that the microalgae species used are safe and do not contain harmful toxins that can harm the health of Muslim consumers [9]. In addition, the method of microalgae cultivation also needs to be considered. The growth media used must not contain unclean or haram materials, such as animal extracts that are not slaughtered in an Islamic manner [34].

The process of protein extraction and isolation from microalgae is the next stage that requires attention to halal aspects. Chemicals used in this process, such as solvents and precipitating agents, must be ensured to be halal. The use of enzymes in the protein hydrolysis process must also come from halal sources [35]. In addition, the equipment used in the production process must be clean and free from contamination of haram materials. Halal certification from a trusted institution is important to provide assurance to Muslim consumers that the entire production process has met halal standards [36].

In addition to the materials and production process, the storage and packaging aspects of microalgae protein isolates also need to be considered. Packaging materials must not contain any haram elements or release hazardous substances that can contaminate the product. The storage process must be carried out in a clean place and avoid cross-contamination with non-halal products. Product labels must also include clear information regarding the halal status and composition of the product in a transparent manner [37].

With notice all over aspects from the source of microalgae, cultivation methods, extraction and isolation processes, to storage and packaging, protein isolates from microalgae have the potential to be a sustainable source of halal protein. Halal certification from authorized institutions will provide confidence to Muslim consumers and open up wider market opportunities for this product. Further research on the standardization of halal production processes for microalgae protein isolates is also needed to ensure their quality and halalness in a consistent way [38].

Case study shows that the challenges of halal certification for food biomass products are closely related to *traceability*, fermentation media, sources of active ingredients, and the commitment of producers to provide transparent documentation. Halal certification does not only assess the final product, but also the entire upstream-downstream process that accompanies it. Therefore, synergy between halal authorities, food scientists, and producers is needed to support the development of halal and *thayyib food biomass* [39].

12.6 Microfibers from Agricultural Waste

Microfibers from agricultural waste are microfibers produced from plant residues such as rice straw, corn cobs, coconut fiber, and banana stems. These materials are rich in cellulose, hemicellulose, and lignin, which are then processed through chemical, physical, or enzymatic processes into microfibers or nanofibers. The use of plant sources makes these microfibers potentially halal, as long as the processing process is in accordance with halal principles [40].

In the field of biomass, microfibers are widely used. In functional foods, microfibers act as a source of dietary fiber that can improve digestive health and the nutritional value of foods such as bread and cereals [41]. In addition, microfibers are used in the manufacture of bioplastics and environmentally friendly food packaging. Fibers from waste such as corn cobs have been shown to be able to replace plastic synthetic [42]. In the bakery industry, microfibers function as fillers to improve the texture, moisture, and stability of products.

However, the halal aspect of microfiber requires special attention. Although it comes from halal plants, chemical processes such as the use of sodium hydroxide or bleaching agents can raise doubts. In addition, the risk of cross-contamination with haram materials and the lack of clarity of the origin of the material (traceability) pose their own challenges. Therefore, comprehensive monitoring of the production process is essential [24].

To ensure the halal status of microfibers, the industry must implement a strict halal audit system. This includes tracing the supply chain of materials, monitoring additional materials used in the process, and certification from a trusted halal institution. This step is especially important if microfibers are used in food, considering that Muslim consumers are very concerned about the halal and hygiene aspects of products.

Overall, microfibers from agricultural waste have great potential to support a sustainable and environmentally friendly biomass industry. However, the success of this utilization depends on solid halal and food safety assurance. The development of international halal standards for plant-based microfibers is an urgent need to ensure their widespread acceptance, especially in countries with large Muslim populations.

12.7 Food Biomass Product Enzyme Sources

Enzymes are key components in various biotechnological processes, especially in the bioconversion of food biomass into value-added products. In the food industry, enzymes are used to hydrolyze starch into glucose, degrade proteins into bioactive peptides, and decompose lignocellulosic fibers into compounds that can be digested or used as nutritional raw materials. The efficiency and specificity of enzymes in accelerating chemical reactions make them very important in the development of modern biotechnology-based food products.

However, the main challenge that arises in the context of halal products is the source and production process of the enzymes. Enzymes can come from various sources, such as microorganisms (bacteria, fungi), animals, or genetically modified organisms (GMOs). Among the three, enzymes derived from animals, especially from pig pancreas (such as pig lipase or trypsin), are declared non-halal and are forbidden to be used in the processing of food products according to the fatwa of the Indonesian Ulema Council (MUI) and other halal institutions such as JAKIM (Malaysia) and MUIS (Singapore) [43].

Meanwhile, enzymes from microorganisms are often considered as a safer and halal alternative. However, the halal aspect of microbial enzymes is not only determined by the type of microorganism that produces them, but also by the culture or fermentation media used. For example, if the fermentation media contains peptone, gelatin, or pork extract as a source of nitrogen or nutrients, then the resulting enzyme can be of *syubhat* (doubtful) or even haram [44]. Therefore, the production process must be thoroughly audited to ensure that the media and additional materials used are free from haram and impure elements.

The problem becomes more complex when genetically engineered enzymes are used. For example, an enzyme-producing gene from an animal can be inserted into a microbe such as *Escherichia coli* or *Aspergillus niger* to mass-produce the enzyme. In this context, even though the microorganism itself is halal, the genetic source and fermentation medium remain important considerations. According to MS1500:2019 (Malaysia's halal standard), the use of GMOs is permitted as long as they do not come from haram sources and the production process is supervised to ensure that it is not contaminated by harmful and or unclean substances [23].

The halalness of enzymes in food products is also complicated by the fact that enzymes fall into the category of *processing aids*—processing aids that are usually undetectable in the final product. Although they do not leave residues, the process of using them can still affect the halal status of the product. Therefore, the presence of enzymes in the production stage must still be listed and traced in the halal certification process [45]. This is in line with the principle of *halalan thayyiban*, where not only the final product is considered, but also the entire production process chain.

To ensure the halalness of biomass-based food products involving enzymes, manufacturers must conduct a thorough audit of the enzyme raw material supply chain, the source of the producing organism, the fermentation media, and the possibility of cross-contamination. A halal certificate from a trusted institution must be included for each enzyme used. In addition, the *halal by design approach* is important in the research and development of food products, by ensuring that all components and processes used have been designed in accordance with sharia principles from the start.

With the increasing demand for halal food products in the global market, the food biotechnology industry is required to be more transparent and proactive in ensuring the halalness of enzymes used in biomass conversion. This is not only a matter of compliance with religious law, but also concerns consumer trust and ethical responsibility in product development. Therefore, synergy between scientists, halal regulators, and industry players is needed to build an innovative biomass production system that remains in accordance with halal values.

12.8 Genetically Engineered Products and Genetically Modified Organisms (GMOs)

The application of genetic engineering in the food sector continues to grow rapidly, enabling the creation of products with higher nutritional value, longer shelf life, and tolerance to extreme environmental conditions. However, in the context of consumption by Muslims, important questions arise regarding the halalness of genetically engineered products or Genetically Modified Organisms (GMOs). Although there is no explicit argument that prohibits genetic engineering technology in general, the main issue lies in the origin of the genes used and their impact on the final properties of the product.

One of the main concerns regarding halal GMOs is the use of genes derived from animals that are forbidden in Islam, such as pigs. If genes from haram organisms are used in the genetic engineering process, even in very small quantities, this can raise doubts about the halalness of the final product. According to the Indonesian Ulema Council (MUI), the principle of *istihalah tammah* (perfect transformation) can be used as a basis for assessment, but caution is still needed in the practice [46].

Halal regulatory agencies in various countries pay special attention to GMOs. For example, the JAKIM states that the use of GMOs is permitted as long as it does not involve haram ingredients as a source of genes and does not cause harm to the health consumer [23]. This assessment is in line with the basic principles in Islam, namely that everything is halal unless there is an argument stating that it is forbidden, as well as the principle of *la darar wa la dirar* (not endangering oneself or others).

Scientific studies also show that GMO technology is basically neutral. The safety and halalness of a product depend on the process and ingredients involved. GMO products such as Bt corn, transgenic soybeans, and pest-resistant tomatoes have been widely used, but still require a halal certification process to ensure that the entire production chain is free from contamination of unclean or forbidden materials [47].

From a contemporary fiqh perspective, the use of modern technology including GMOs is permissible as long as it meets the principles of halal and *thayyib. Halal* means in accordance with Islamic law, while *thayyib* means good and not harmful. Therefore, aspects of food safety, sustainability, and health are also included in the halal considerations of a GMO product [48].

In addition to the gene source, the GMO production process also needs to be closely monitored. Recombinant enzymes used in the fermentation process, for example, must come from halal sources and not from pathogenic microorganisms. Production equipment must also meet hygiene standards and not experience cross-contamination with non-halal products [49]. A thorough audit is required to ensure the halal integrity of the product from the research stage to distribution.

Another important aspect is transparency in labeling. Muslim consumers have the right to know whether a product uses GMO ingredients, and whether the product has obtained halal certification. Honest and accurate labeling not only helps consumers make decisions according to their beliefs, but also increases public trust in the food biotechnology industry [50].

In facing this challenge, synergy between food scientists, sharia experts, and halal authorities is urgently needed. The development of halal guidelines for GMO products is urgent to avoid confusion at the consumer and industry levels. Halal certification institutions are also expected to be able to follow technological developments by conducting comprehensive scientific studies on new genetically engineered products.

Overall, GMO food products have great potential to support global food security, including in Muslim-majority countries. However, in order to be widely accepted, the halal aspect must be scientifically and sharia compliant. Halal certification is an important bridge to ensure that these modern biotechnology products remain in accordance with Islamic values and are acceptable to all levels of Muslim society.

13 Implementation of Good Manufacturing Practice (GMP) and Halal Assurance System

Good Manufacturing Practice (GMP) is a basic guideline in ensuring the quality and safety of food products, including in the context of halal. In the halal assurance system, GMP must be adapted to accommodate the principles of Islamic law such as the separation of unclean and halal materials, cleanliness of production equipment, and the use of additional materials that do not come from haram or *syubhat* (doubtful) elements. The implementation of halal GMP requires the existence of special Critical Control Points (CPCs) for halal, which are not found in conventional GMP [51]. Therefore, the halal-based food biomass industry must align its GMP with the Halal Assurance System (SJH) standards set by authority institutions such as JAKIM or BPJPH.

The traceability system or *halal traceability* is key to ensuring the halal status of a product as a whole from upstream to downstream. This system allows tracking of the origin of raw materials, production processes, and distribution of final products to ensure that there is no cross-contamination with haram or impure elements. Technologies such as blockchain, QR codes, and digital certification have been utilized to strengthen the halal traceability system [52]. In the context of the food biomass industry, which often involves various chemical stages and process engineering, halal traceability provides assurance that every step of production is in accordance with sharia principles and can be audited transparently.

The halal-based food biomass industry faces major challenges in reformulating its production process. Many conventional solvents or fermentation media do not comply with halal standards, such as the use of ethanol from non-halal sources or enzymes from non-halal animals. This reformulation not only impacts product quality but also affects process efficiency and production costs. Therefore, industry players must actively conduct research to find substitute materials and process methods that comply with Islamic law without sacrificing product performance.

Green processing is a technological approach that emphasizes energy efficiency, waste reduction, and the use of environmentally friendly materials. In the context of halal, this approach is very relevant because many green technologies such as supercritical water extraction, microbial enzymes from halal sources, and membrane-based separation actually support clean and free contamination processes [33]. With this, green processing not only supports environmental sustainability, but also provides stronger halal added value to food biomass products.

To ensure the sustainability of halal production in the food biomass industry, a comprehensive risk mitigation strategy is needed. This includes halal training for workers, periodic halal audits, an integrated documentation system, and the use of halal-certified materials from the upstream. In addition, a HACCP-Halal (Hazard Analysis and Critical Control Points)-based approach must be developed to identify vulnerable points in the production process [52]. In this way, the industry can reduce the potential for halal deviations and maintain the trust of Muslim consumers globally.

14 Conclusion

The development of halal food biomass products faces multidimensional challenges, ranging from the availability of halal raw materials, sharia-compliant production processes, to the diversity of cross-country regulations. One of the main strategic solutions is the substitution of raw materials with halal and sustainable sources. For example, the use of microbial enzymes from plant-based fermentation replaces animal enzymes, as well as the use of plant solvents or environmentally friendly extraction technologies such as supercritical CO_2. In addition, the selection of halal culture media such as plant hydrolysates or molasses supports the principles of *halal by design* and sustainability environment.

The second strategy is to strengthen the halal *traceability* and audit system through digital technology such as halal blockchain. This system allows comprehensive tracking from raw materials, fermentation media, to final distribution, thereby increasing transparency and efficiency in the international halal certification process. On the other hand, harmonization of halal certification regulations is also a strategic urgency. Collaboration between world halal certification institutions such as BPJPH, JAKIM, GSO, and ESMA needs to be strengthened by referring to global standards such as MS1500:2019, GSO 2055-1, and Codex Alimentarius to facilitate the export and penetration of halal products into the global market.

In order for the development of sustainable halal biomass, investment in technological innovation and research is essential. Technologies such as *solid-state fermentation*, GMO-free enzymes, and green extraction methods need to be encouraged to produce functional food products that have added health and spiritual value. Halal education for industry players and academics must also be improved through certified training and multi-sector collaboration. Partnerships between universities, industry, and halal certification bodies will strengthen the science and technology-based halal innovation ecosystem. This approach will ensure that the development of halal food biomass is not only compliant with sharia but also globally competitive and relevant to the needs of today's consumers.

References

1. Bauen A, Berndes G, Junginger M, Vuille F, Londo M (2009) Bioenergy—a sustainable and reliable energy source, a review of status and prospects, pp 1–108. http://www.globalbioenergy.org/uploads/media/0912_IEA_Bioenergy_-_MAIN_REPORT_-_Bioenergy_-a_sustainable_and_reliable_energy_source._A_review_of_status_and_prospects.pdf also https://www.researchgate.net/publication/46722642
2. FAO (2017) The future of food and agriculture—trends and challenges. Food Agric Organ. United Nations. Rome
3. WHO (2020) Healthy diet. World Health Organization
4. Tamang JP, Watanabe K, Holzapfel WH (2016) Review: diversity of microorganisms in global fermented foods and beverages. Front Microbiol 7:377. https://doi.org/10.3389/fmicb.2016.00377
5. Ranum P, Peña-Rosas JP, Garcia-Casal MN (2014) Global maize production, utilization, and consumption. Ann NY Acad Sci 1312(1):105–112. https://doi.org/10.1111/nyas.12396
6. IPCC (2019) Climate change and land: an IPCC special report on climate change, desertification, land degradation, sustainable land management, food security, and greenhouse gas fluxes in terrestrial ecosystems. Shukla PR, Skea J, Calvo Buendia E, Masson-Delmotte V, Pörtner H-O, Roberts DC, Zhai P, Slade R, Connors S, van Diemen R, Ferrat M, Haughey E, Luz S, Neogi S, Pathak M, Petzold J, Portugal Pereira J, Vyas P, Huntley E, Kissick K, Belkacemi M, Malley J (eds). In press. https://www.ipcc.ch/site/assets/uploads/2019/11/SRCCL-Full-Report-Compiled-191128.pdf
7. World Bank (2020) Food systems and the role of agriculture. World Bank Group
8. FAO (2021) State of food security and nutrition in the world. Food Agric Organ. United Nations
9. FAO (2021) The future of food and agriculture—microalgae as a sustainable food source. Food Agric Organ
10. Al-Qaradawi Y (2017) Halal and Haram in Islam
11. Azam MSE, Abdullah, MA (2020) Global Halal industry: realities and opportunities. Int J Islam Bus Ethics 5(1):47. https://doi.org/10.30659/ijibe.5.1.47-59
12. Al-Qaradawi Y (2001) The lawful and the prohibited in Islam

13. Malboobi MT, Malboobi MA (2010) Halal concept and products derived from modern biotechnology. In: Saifudeen SM, Salleh SM (eds) International workshop for Islamic scholars on Agribiotechnology: Shariah Compliance. MABIC-ISAAA, Georgetown, Malaysia
14. Karahalil E (2020) Principle of Halal-compliant fermentations: microbial alternatives for the Halal food industry. Trends Food Sci Tech 98:1–9. https://doi.org/10.1016/j.tifs.2020.01.031
15. Ahmad AM, Sipra HM, Tanveer N, Aslam N, Hassan A, Hassan SA (2025) Ensuring Halal food integrity: an overview of modern molecular and technological solutions. Food Biomacromolecules 2:5–22. https://doi.org/10.1002/fob2.12022
16. LPPOM MUI (2023) List of foreign Halal certification institutions recognized by MUI
17. BPJPH (20222) Halal certification procedures for food products
18. Kurniadi M, Frediansyah A (2016) Perspektif Halal Produk Pangan Berbasis Bioproses Mikrobia Reaktor 16 (3):147–160. https://doi.org/10.14710/reaktor.16.3.147-160
19. Nurwulan, Rusdan IH (2023) Halal fermented functional food in Indonesia: a review. Halalsphere 3(2):59–66 J Halal Ind Serv. https://doi.org/10.31436/hs.v3i2.73
20. BPJPH (2021) Government regulation of the republic of Indonesia number 39 of 2021 concerning the implementation of the Halal product assurance sector
21. Harsanto B, Farras JI, Firmansyah EA, Pradana M, Apriliadi A (2024) Digital technology 4.0 on Halal supply chain: a systematic review. Logistics 8:21. https://doi.org/10.3390/logistics8010021
22. Feng H, Wang X, Duan Y, Zhang J, Zhang X (2020) Applying blockchain technology to improve agri-food traceability: a review of development methods, benefits and challenges. J Clean Prod 260:121031. https://doi.org/10.1016/j.jclepro.2020.121031
23. JAKIM (2019) Malaysian Halal certification procedure manual (domestic), 2020 edn
24. LPPOM MUI (2022) Halal assurance system (SJH) guidelines
25. GSO (2020) Halal food guidelines for GCC countries
26. Tieman M, van der Vorst JGAJ, Che Ghazali M (2012) Principles in Halal supply chain management. J Islam Mark 3(3):217–243. https://doi.org/10.1108/17590831211259727
27. Azwanida NN (2015) A review on the extraction methods use in medicinal plants, principle, strength and limitation. Med Aromat Plants 04(03):3–8. https://doi.org/10.4172/2167-0412.1000196
28. Paiva A, Craveiro R, Aroso I, Martins M, Reis RL, Duarte ARC (2014) Natural deep eutectic solvents—solvents for the 21st century. ACS Sustain Chem Eng 2(5):1063–1071. https://doi.org/10.1021/sc500096j
29. Arifin H (2023) Analysis of the Halal certification system for the self-declare category, SINOMIKA. J Publ Ilm Bid Ekon dan Akunt 1(5):1173–1180. https://doi.org/10.54443/sinomika.v1i5.592
30. Majelis Ulama Indonesia (2018) MUI Fatwa number 33 of 2018 concerning Halal products
31. Noordin N, Noor NLM, Samicho Z (2014) Strategic approach to Halal certification system: an ecosystem perspective. Procedia—Soc Behav Sci 121(September):79–95. https://doi.org/10.1016/j.sbspro.2014.01.1110
32. Selamat B, Sahilah J, Jamilah AM (2022) Advances in Halal food science and technology. Springer Nature
33. Abdul Rahman M, Ali R, Ismail ME (2021) Halal bioactive ingredients for functional foods and nutraceuticals. Woodhead Publ
34. JAKIM (2019) Malaysian standard: Halal food—general requirements. Department of Islamic Development Malaysia
35. Che Man, Aida AA, Raha AR, Son R (2007) Identification of pork derivatives in food products by species-specific polymerase chain reaction (PCR) for Halal verification. Food Control 18(7):885–889. https://doi.org/10.1016/j.foodcont.2006.05.004
36. LPPOM MUI (2022) Halal assurance system criteria
37. Codex Alementarius Commission (2004) General guidelines for use of the term "Halal"
38. Wan Sulaiman WSH, Md Zain HH, Othman R, Mohd Latif NH, Mohd Hatta FA, Ramya R (2023) Algae carotenoid pigments as new sources of Halal bioactive ingredients. J Halal Sci Technol 2(1):68–82. https://doi.org/10.59202/jhst.v2i1.669

39. Tieman M (2011) The application of Halal in supply chain management: in-depth interviews. J Islam Mark 2(2):186–195. https://doi.org/10.1108/17590831111139893
40. Kalia S, Dufresne A, Cherian BM, Kaith BS, Averous L, Njuguna J, Nassiopoulos E (2011) Cellulose-based bio- and nanocomposites: a review. Int J Polym Sci 2011. https://doi.org/10.1155/2011/837875
41. Yapias RJM, León NNP, Areche FO, Rivera TJCH, Donayre EMF, Calderon GRL, Huayhua LLA, Huaman JT (2024). Dietary fiber extraction from agricultural residues and agroindustrial waste with emerging technologies: A systematic review. SABRAO J Breed Genet 56(5):2102–2109. https://doi.org/10.54910/sabrao2024.56.5.32
42. Siracusa V, Rocculi P, Romani S, Rosa MD (2008) Biodegradable polymers for food packaging: a review. Trends Food Sci Technol 19(12):634–643. https://doi.org/10.1016/j.tifs.2008.07.003
43. LPPOM MUI (2020) Fatwa and Halal certification guidelines
44. IFANCA (2021) Enzymes and Halal certification. Islamic food and nutrition council of America
45. LPPOM MUI (2020) HAS 23000: Halal guarantee system
46. LPPOM MUI (2021) Halal assurance system (SJH) guidelines
47. Khattak JZK, Mir A, Anwar Z, Wahedi HM, Abbas G, Khattak HZK, Ismatullah H (2011) Concept of Halal food and biotechnology. Adv J Food Sci Technol 3(5):385–389
48. RiazMN, Chaudry MM (2004) Halal food production. CRC Press
49. LPPOM MUI (2022) Audit trail of bakery products and microbial enzymes
50. Bonne K, Verbeke W (2008) Religious values informing Halal meat production and the control and delivery of Halal credential quality. Agric Human Values 25(1):35–47. https://doi.org/10.1007/s10460-007-9076-y
51. Lestari F, Kurniawan R, Arifin J, Yasir M, Muhammad Saleh M, Akbarizan (2023) An integrated framework for the measurement of Halal good manufacturing practices on the case of SMEs in the food sector. J Islam Mark 14(1):82–105. https://doi.org/10.1108/JIMA-04-2021-0105
52. Abd Kadir MH, Raja Mohd Rasi RZ, Omar SS, Abdul Manap ZI (2016) Halal supply chain management streamlined practices: issues and challenges. International Engineering Research and Innovation Symposium (IRIS) IOP Publishing. 2016. IOP Conf Ser.: Mater Sci Eng 160:012070. https://doi.org/10.1088/1757-899X/160/1/012070

Biodiesel Production and Techniques from Various Biomass Feedstocks

Mohammad Aliff Shakir, Mardiana Idayu Ahmad, Mohd Asyraf Kassim, and Fozy Binhweel

Abstract The rising global energy demand and the environmental consequences of fossil fuel consumption have accelerated the pursuit of renewable alternatives. While biodiesel is recognized as a sustainable and cleaner substitute for fossil diesel, the limitation lies in the selection and optimization of suitable biomass feedstocks. The variability in feedstock availability, cost, and conversion efficiency presents a major challenge in scaling up biodiesel production. This chapter addresses the gap by exploring a wide range of biomass feedstocks categorized into animal-based and plant-based sources. Furthermore, it discusses critical extraction techniques of lipids from feedstocks, followed by biodiesel production methods such as pyrolysis, transesterification, microemulsion, and dilution. By evaluating the compatibility of different feedstocks with these techniques, the chapter offers insights into improving biodiesel yield, quality, and economic viability. Ultimately, the chapter positions biodiesel derived from diverse biomass sources as a practical solution to current energy and environmental challenges.

Keywords Biodiesel · Biomass feedstock · Transeterification · Renewable energy · Sustainable

1 Introduction

In recent decades, the global energy landscape has undergone dramatic changes, which have been primarily driven by an ever-increasing demand for fuel across various sectors. This growing energy appetite has placed substantial pressure on conventional, non-renewable sources such as petroleum-based fossil fuels. These energy resources, while once considered abundant and reliable, are now facing rapid depletion due to excessive extraction and widespread consumption [1]. Furthermore, the environmental repercussions of fossil fuel use including greenhouse gas emissions, air pollution, and climate change have become critical concerns on a global

M. A. Shakir · M. I. Ahmad (✉) · M. A. Kassim · F. Binhweel
School of Industrial Technology, Universiti Sains Malaysia, 11800 Penang, Malaysia
e-mail: mardianaidayu@usm.my

M. A. Kassim et al. (eds.), *Sustainable Biorefinery*, Green Energy and Technology,
https://doi.org/10.1007/978-981-95-8181-8_4

scale. Projections by the International Energy Agency (IEA) estimate that by the year 2040, global energy consumption could rise by as much as 30%, highlighting the urgency of developing sustainable alternatives [2]. The continuous reliance on petroleum fuels, despite their finite availability and harmful environmental effects, has prompted extensive discussions and research into renewable substitutes that can support long-term energy stability.

In response to these challenges, the exploration and implementation of renewable energy sources have become a strategic priority for many nations. Options such as solar power, wind energy, hydropower, geothermal systems, biogas, and biofuels are increasingly being integrated into energy frameworks worldwide. Among these alternatives, biodiesel has garnered particular interest due to its renewable nature, biodegradability, and potential to reduce carbon emissions [3]. Derived from biological sources, biodiesel offers a practical and scalable solution to lessen dependency on fossil fuels. Its wide feedstock availability, relative affordability, and environmentally friendly characteristics position biodiesel as a promising candidate in the transition toward a cleaner and more sustainable energy future.

Biodiesel is widely recognized as a promising renewable fuel that aligns with the goals of environmental sustainability and reduced ecological impact. It has gained attention as a cleaner alternative to conventional diesel, particularly for its potential to reduce carbon emissions and reliance on fossil resources. While definitions may vary slightly across research studies. These fatty acids are typically obtained through the chemical conversion of biological lipids sourced from plant oils or animal fats. Some studies emphasize a stricter classification, identifying biodiesel specifically as the product of transesterification involving pure vegetable or animal-based oils [4]. Others adopt a broader view, suggesting that blends of petroleum diesel with oils of biological origin such as straight vegetable oil may still be classified under the biodiesel category, especially when used as fuel [5]. Regardless of the definition, the distinction lies in the renewable nature of its feedstocks and the potential for cleaner combustion.

Biodiesel can be synthesized from a diverse range of lipid-rich materials, including edible and non-edible vegetable oils [6], rendered animal fats [7], waste cooking oils [8], and even non-traditional sources such as macroalgae and microalgae [9]. These varied feedstocks provide significant flexibility and adaptability in biodiesel production across different geographical and economic contexts. With proper conversion methods, the resulting biodiesel can meet standard fuel specifications and function efficiently in existing diesel engines, making it a practical and scalable energy solution. Despite its many advantages, the accelerated expansion of biodiesel production from first-generation sources has raised critical concerns regarding its long-term sustainability [10]. First-generation biodiesel, which is primarily derived from edible oils such as soybean, palm, and rapeseed, has been associated with several socioeconomic and environmental challenges. One of the most pressing issues is the competition between fuel and food, as large-scale biodiesel production can divert agricultural resources away from food crops. This shift has the potential to disrupt food supply chains, increase commodity prices, and exacerbate food insecurity, especially in vulnerable regions.

In response to these limitations, researchers and policymakers have turned their attention to more sustainable alternatives, paving the way for the development of second- and third-generation biodiesel technologies. Second-generation biodiesel sources typically include non-edible oils, agricultural residues, and waste materials which do not compete directly with food production [11, 12]. While this generation offers improved sustainability, challenges related to feedstock availability and conversion efficiency still persist. Third-generation biodiesel, which is produced from algae and other fast-growing microorganisms, has emerged as a particularly promising alternative [13, 14]. It offers high lipid productivity, requires less arable land, and can be cultivated using wastewater or non-potable water. This makes it an environmentally and economically favorable option. With ongoing advancements in cultivation and processing technologies, third-generation biodiesel holds significant promise to enhance national energy security while supporting a more resilient and cost-effective energy infrastructure.

This chapter aims to bridge the knowledge gap surrounding the selection of suitable biomass feedstocks and the techniques used in biodiesel production. While numerous studies have explored individual aspects of biodiesel, this chapter provides a comprehensive comparison of both animal- and plant-based feedstocks, alongside diverse production routes from conventional transesterification to emerging methods. The novelty of this chapter lies in its integrated approach to evaluating feedstock characteristics, processing techniques, and conversion efficiency, which offers a holistic understanding of how different biomass sources influence biodiesel quality and yield. By doing so, the chapter seeks to support researchers, policymakers, and industry stakeholders in identifying viable pathways for sustainable and scalable biodiesel production.

2 Biodiesel Feedstock

The selection of suitable feedstock is a fundamental aspect of biodiesel production, as it directly influences fuel properties, processing efficiency, and overall production cost. Biodiesel can be derived from a wide variety of lipid-rich sources broadly categorized into animal-based and plant-based feedstocks [15, 16]. The type and quality of feedstock used not only determine the yield and composition of the resulting biodiesel but also impact its environmental and economic viability. This section provides an in-depth overview of the major feedstock types, focusing on their characteristics, availability, advantages, and limitations in the context of sustainable biodiesel production.

2.1 Animal-Based Feedstock

Animal-based feedstocks have emerged as an important alternative for biodiesel production in the context of reducing reliance on edible oils and minimizing waste. These feedstocks are typically derived from rendered animal fats, including tallow (from cattle), lard (from pigs), poultry fat, and fish oil [17]. In addition to being abundant in many agro-industrial regions, these by-products are often considered waste materials from the meat processing industry, which makes their use both cost-effective and environmentally favorable. Animal fats are rich in long-chain saturated fatty acids, which contribute to the production of biodiesel with good oxidative stability and high cetane numbers. However, these fats generally have higher melting points and contain more saturated compounds compared to plant-based oils, which may lead to poor cold-flow properties in colder climates. Despite these limitations, with proper pre-treatment and processing techniques such as degumming [18], deacidification [19], and temperature-controlled transesterification [20], animal fats can yield high-quality biodiesel that meets international fuel standards.

Another advantage of using animal-based feedstocks lies in waste valorization and circular economy principles. Converting animal fats into biodiesel not only offers a renewable fuel alternative but also helps in managing waste generated from slaughterhouses and food industries [21]. As shown in Fig. 1, this approach reduces the environmental burden associated with disposal while enhancing the sustainability of biofuel production systems. With ongoing advancements in catalyst development and reaction optimization, the conversion efficiency of animal fats into biodiesel continues to improve. These developments support the viability of animal-based feedstocks as a dependable and scalable source for biodiesel, especially in regions where such materials are plentiful and underutilized.

2.2 Plant-Based Feedstock

Plant-based feedstocks are the most extensively studied and widely utilized sources for biodiesel production due to their relatively high lipid content and ease of processing. These feedstocks include a variety of edible and non-edible oils derived from crops such as soybean, palm, sunflower, rapeseed, jatropha, castor, and camelina [22, 23]. Among these, soybean and palm oil are dominant in commercial biodiesel markets, especially in regions like North and South America, Southeast Asia, and Europe. Edible oils are generally preferred for their low free fatty acid (FFA) content, which simplifies the transesterification process and yields high-quality biodiesel [24]. However, the large-scale use of edible oils has drawn criticism for contributing to food versus fuel conflicts, land use change, and deforestation. These concerns have prompted increased interest in non-edible oil crops such as jatropha, neem, and karanja. These species can grow on marginal lands unsuitable for food crops, reducing competition with agriculture and enhancing sustainability.

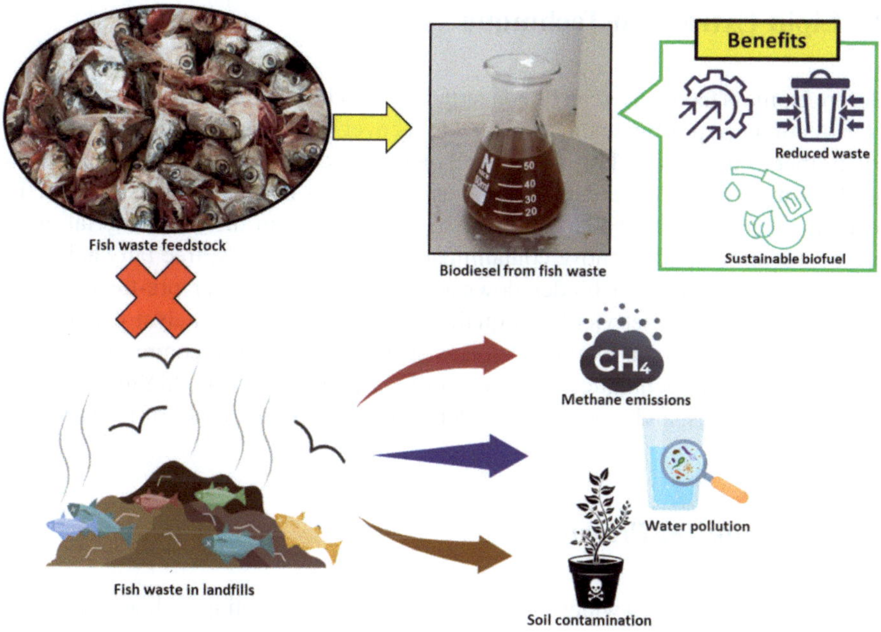

Fig. 1 Utilization concept of potential feedstock from fish waste into biodiesel production

Plant-based oils predominantly contain unsaturated fatty acids, which result in biodiesel with better cold-flow properties but lower oxidative stability compared to animal fats. The fatty acid composition of each oil influences the fuel properties such as viscosity, flash point, cloud point, and cetane number [25]. Therefore, feedstock selection is critical for tailoring biodiesel quality to meet regional climate and engine performance requirements. Additionally, waste vegetable oils and used cooking oils represent an increasingly valuable sub-category of plant-based feedstocks. These materials are often discarded after food preparation, yet they retain significant amounts of residual lipids that can be efficiently converted into biodiesel following proper filtration and pre-treatment [26]. Utilizing these waste oils not only supports waste reduction efforts but also offers a cost-effective and environmentally responsible route to biodiesel production. Due to their diversity and global availability plant-based feedstocks continue to play a central role in biodiesel development. With appropriate technological interventions and sustainable sourcing practices, they offer a reliable foundation for expanding biofuel capacity while addressing environmental and energy security concerns.

3 Lipid Extraction Technique

Before biomass feedstocks can be converted into biodiesel, they must undergo a series of extraction and pre-treatment processes to recover lipids and ensure suitable quality for further conversion. These steps are crucial, especially when dealing with low-grade or high-impurity feedstocks such as animal fats or algal biomass. Extraction techniques are used to isolate lipids or triglycerides from the raw material, while pre-treatment methods remove contaminants like water, solids, free fatty acids, and other impurities that can hinder downstream reactions. Proper pre-treatment not only improves biodiesel yield and quality but also enhances catalyst performance and reduces operational challenges. This section outlines the common extraction and pre-treatment techniques employed in biodiesel production, highlighting their roles, mechanisms, and practical considerations for different types of feedstocks.

3.1 Mechanical Pressing

Mechanical pressing is a widely adopted technique for extracting oil from oleaginous feedstocks in biodiesel production. This method involves applying physical pressure to oil-rich materials such as seeds, nuts, or kernels using equipment like screw presses or hydraulic systems to extract the crude oil [27]. The process is straightforward and does not require chemical solvents. Therefore, this process makes it an attractive option for feedstock preparation in decentralized or small-to-medium-scale biodiesel facilities. The effectiveness of oil recovery through mechanical pressing is influenced by several variables, including the oil content of the biomass, the moisture level, particle size, and the magnitude of pressure applied. Pressing at higher pressure can rupture cellular walls and increase oil release, although excessive pressure may also generate heat, which in turn can affect oil quality. This method is typically optimized for feedstocks with moderate to high oil content, such as sunflower, rapeseed, or Jatropha. However, it generally allows for the processing of only one type of feedstock per cycle, limiting flexibility in multi-feedstock operations.

While mechanical pressing offers several operational advantages including lower cost, minimal chemical usage, and reduced environmental risks it also has limitations. One of the major drawbacks is the relatively high residual oil content left in the seed cake, which reflects lower extraction efficiency compared to solvent-based methods. Additionally, pressing often requires pre-heating of the biomass to reduce oil viscosity and improve yield, which adds an energy input to the process [28]. Furthermore, the equipment must be periodically maintained to ensure continuous pressure efficiency and prevent mechanical wear. Despite these limitations, mechanical pressing remains a preferred technique for oil recovery in regions with restricted access to chemical solvents or where environmental regulations discourage their use. The simplicity of operation combined with a relatively safe working environment and lower setup costs has supported its continued relevance in both rural biodiesel initiatives and preliminary extraction steps prior to chemical upgrading or refining.

3.2 Solvent Extraction

Solvent extraction is a widely employed technique in biodiesel feedstock preparation, particularly when targeting high oil recovery from biomass with relatively low oil content or dense cellular structures. This method relies on the use of organic solvents to dissolve and separate lipids from solid matrices. Among the various extraction systems, Soxhlet extraction as shown in Fig. 2, remains one of the most frequently utilized approaches in laboratory-scale applications due to its ability to perform continuous solvent cycling and achieve high extraction efficiency [29]. In this method, the crushed or powdered biomass is placed within a porous holder known as a thimble made of cellulose or filter paper, which is then positioned in an extraction chamber. A selected solvent such as hexane, petroleum ether, or chloroform is heated in a lower flask, evaporates, condenses in the upper chamber, and repeatedly washes over the sample. Once the chamber reaches a certain solvent level, an automatic siphoning mechanism returns the solvent to the boiling flask, completing a cycle. This process continues through multiple cycles until most of the extractable lipids are transferred into the solvent phase.

While solvent extraction offers higher oil recovery than mechanical pressing, its use is accompanied by a range of practical and environmental challenges. The method is time-intensive, typically requiring 6 to 24 h for full extraction, depending on the biomass and solvent used. It also involves several labor-intensive steps such as

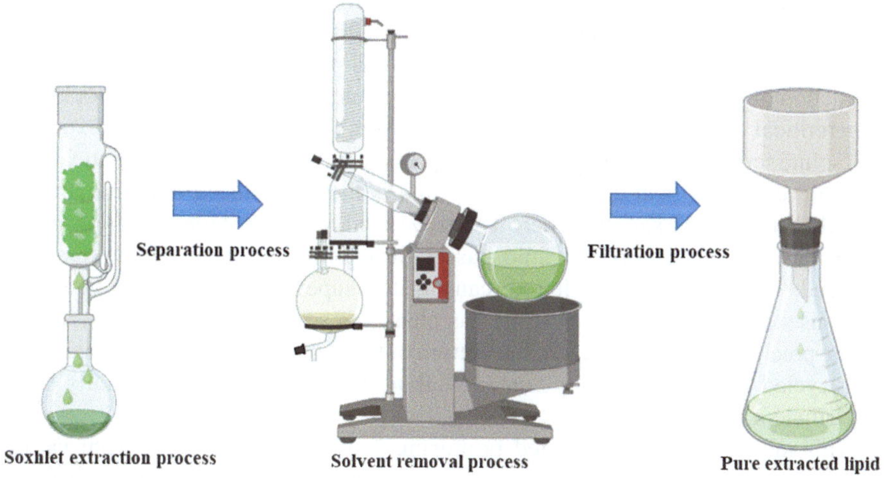

Fig. 2 Soxhlet extraction method

sample loading, equipment assembly, post-extraction filtration, and solvent recovery, often using a rotary evaporator [30]. Furthermore, the reliance on volatile organic solvents presents serious health, safety, and environmental concerns due to their toxicity, flammability, and non-biodegradability. However, solvent extraction remains a valuable tool in analytical and small-scale research applications due to its reproducibility and high oil yield, especially for feedstocks that are difficult to process mechanically. In larger-scale or industrial contexts, the environmental implications and high solvent usage often necessitate alternative technologies or enhanced solvent recovery systems to reduce emissions and improve process sustainability.

3.3 Supercritical Fluid Extraction

Supercritical fluid extraction is an advanced technique that utilizes fluids in their supercritical state, where they exhibit both gas-like and liquid-like properties to efficiently extract target compounds from solid matrices. When a fluid surpasses its critical temperature and pressure, it becomes a supercritical fluid, characterized by low viscosity, high diffusivity, and enhanced solvating power. These unique properties facilitate superior mass transfer rates and significantly reduce extraction time compared to conventional methods [31]. Among various supercritical fluids, carbon dioxide (CO_2) is the most commonly used in food, pharmaceutical, and biodiesel industries. CO_2 is preferred due to its non-toxic, non-flammable, and chemically inert nature, along with its abundance and cost-effectiveness. Moreover, supercritical CO_2 extraction is a low-temperature process, which helps preserve the integrity of heat-sensitive (thermolabile) compounds, thereby improving the quality and functional properties of the extracted oils [32].

To enhance its solvation capacity, especially for polar compounds, CO_2 can be used in combination with co-solvents such as ethanol, methanol, or water. These modifiers increase the polarity of the supercritical phase by enabling more efficient extraction of a broader range of bioactive compounds from complex feedstocks. The supercritical extraction process generally involves two key stages: solute extraction and solute separation. Initially, CO_2 is compressed and heated beyond its critical point to achieve supercritical conditions. The fluid is then introduced into an extraction vessel containing the sample, where it dissolves and carries the target lipids or solutes. After the extraction step, the fluid passes through a separator where pressure is reduced, or temperature is altered, causing the supercritical CO_2 to revert to its gaseous state. As CO_2 returns to its gas phase, its solvating power drops, and the solute precipitates and collects at the bottom of the separator. The CO_2 gas can either be recycled for reuse or safely vented, making the process environmentally sound and efficient.

This method has demonstrated promising results in oil recovery from various biomass sources. Supercritical CO_2 extraction continues to gain traction in sustainable oil recovery due to its clean, solvent-free nature, precise selectivity, and ability to produce high-purity extracts, though it requires high-pressure equipment and greater initial investment compared to simpler techniques. Nonetheless, its benefits in preserving bioactivity and reducing solvent residues make it a valuable option in advanced biodiesel feedstock preparation.

3.4 Microwave-Assisted Extraction

Microwave-assisted extraction (MAE) is an advanced technique that utilizes microwave radiation to heat biomass and accelerate the extraction of target compounds such as lipids for biodiesel production. Microwaves operate within the frequency range of 300 MHz to 300 GHz, and heating occurs as electromagnetic waves are absorbed by the sample material [33]. The energy dissipation in the sample depends on both the electric field intensity and the dielectric properties of the biomass. Unlike conventional heating, the heat is transferred from an external source to the material by microwave heating, which causes direct volumetric heating by generating thermal energy within the sample itself. This is achieved through two primary mechanisms: dipole rotation and ionic conduction [34]. Dipole rotation involves the alignment and realignment of polar molecules in response to the oscillating microwave field. Ionic conduction results from the movement of charged particles. Both mechanisms cause molecular agitation, leading to rapid and uniform heating of the biomass.

One of the major advantages of MAE over traditional methods is its ability to elevate temperature quickly and uniformly, significantly reducing extraction time and improving energy efficiency. In the context of oil extraction, microwave irradiation disrupts cell walls, enhances solvent penetration, and facilitates the release of lipids from the biomass matrix. MAE has been widely explored for the recovery of oils from various feedstocks, including soybeans, rice bran, and algal biomass. Furthermore, the technology has been successfully applied in continuous systems. Due to its high extraction efficiency, short processing time, and compatibility with a wide range of feedstocks, MAE continues to gain traction as a green and energy-efficient alternative in biodiesel feedstock preparation. However, scalability, solvent selection, and microwave absorption properties of the material must be carefully optimized to ensure consistent and safe operation at industrial levels.

3.5 Ultrasonic-Assisted Extraction

Ultrasound-assisted extraction (UAE) is widely recognized as a sustainable and efficient technique that aligns with the principles of green chemistry. This method utilizes high-frequency acoustic waves to generate alternating cycles of compression and rarefaction within a liquid medium [35]. During the rarefaction (low-pressure) phase, micro-bubbles are formed, which then collapse violently during the compression (high-pressure) phase. This phenomenon, known as acoustic cavitation, results in the formation of localized high-energy micro-jets and shockwaves capable of disrupting cellular structures within the feedstock. The mechanical disruption caused by cavitation enhances mass transfer efficiency, facilitating the release of intracellular components such as lipids. The UAE has shown particular effectiveness in the extraction of fats from animal-based tissues, where dense cellular matrices often pose a challenge for traditional methods [36].

Numerous studies have demonstrated that UAE offers significant improvements over conventional lipid extraction techniques in terms of both processing time and extraction efficiency. One of the advantages of this method is the preservation of fatty acid composition. Ultrasonic processing does not appear to alter the chemical structure of the lipids as it maintains a profile comparable to that obtained via standard methods like Soxhlet extraction. This indicates minimal thermal or oxidative degradation during the ultrasound process, making it especially attractive for preserving the integrity of heat-sensitive compounds. In addition to its performance benefits, UAE is considered an environmentally friendly technique due to its reduced reliance on organic solvents and lower energy consumption. The method enables high yields of undistorted fatty acids within a significantly shorter extraction time, making it suitable for both analytical and potential industrial applications. Its combined advantages of efficiency, selectivity, and sustainability position the UAE as a promising method for lipid recovery in biodiesel feedstock preparation.

4 Biodiesel Production

The production of biodiesel involves a range of technological approaches, each designed to accommodate specific feedstock properties, operational conditions, and desired fuel characteristics. Among the various methods developed, four techniques have emerged as the most widely adopted due to their practicality, conversion efficiency, and adaptability, such as pyrolysis, transesterification, microemulsion, and dilution. These processes differ in terms of chemical pathways, equipment requirements, and fuel quality outcomes. This section offers an in-depth overview of each method, discussing their mechanisms, advantages, and inherent limitations, with the aim of providing a comprehensive understanding of their application across different biodiesel production systems.

4.1 Pyrolysis

Pyrolysis is a high-temperature thermochemical technique used to decompose biomass in the absence of oxygen. This process relies on heat to break down complex organic structures within feedstocks, yielding multiple valuable products, most notably bio-oil, along with solid char and combustible gases. Typically operating at temperatures above 500 °C, pyrolysis effectively disintegrates the stable biopolymers present in organic matter [37]. Its fundamental principle lies in the thermal degradation of these substances under oxygen-limited conditions. A familiar example of pyrolysis is the charring of wood into charcoal, where wood thermally decomposes without fully combusting [38]. The bio-oil generated can serve as a precursor for biodiesel production, while the solid and gaseous by-products are often utilized in energy and material applications.

The pyrolysis process is highly dynamic and is influenced by key operational parameters such as temperature, heating rate, reaction time, and biomass particle size. Based on these conditions, pyrolysis is typically classified into three main categories: slow pyrolysis, fast pyrolysis, and flash pyrolysis [39]. Each category results in different product distributions due to variations in the reaction environment. Slow pyrolysis or conventional pyrolysis operates under lower heating rates and extended residence times, favoring higher char yields. On the other hand, fast and flash pyrolysis uses more intense heating conditions to rapidly volatilize organic content and maximize bio-oil output. The decomposition of biomass components usually begins within the temperature range of 350–550 °C and can extend up to 800 °C in the absence of oxygen. The three main outputs, such as liquid (bio-oil), solid (char), and gas, are produced simultaneously with their proportions dependent on the specific pyrolysis setup [40].

During pyrolysis, the long molecular chains composed of carbon, hydrogen, and oxygen atoms are broken into smaller fragments, producing a complex mix of syngas (H_2, CO, CO_2), char, and condensable vapors that include oils and tars [41]. The resulting liquid, when properly processed, may be used as a renewable fuel alternative. Pyrolysis is praised for its simplicity and relatively low waste generation, but it is not without challenges. The requirement for high-temperature, corrosion-resistant equipment leads to substantial capital costs. Moreover, the bio-oil produced often contains a high percentage of oxygenated compounds and impurities, resulting in low purity and suboptimal fuel properties. Thus, these factors limit its direct use as biodiesel on an industrial scale.

Beyond traditional biomass, pyrolysis has been effectively applied to the conversion of waste frying oil into biodiesel. In this approach, waste cooking oil is subjected to thermal cracking at elevated temperatures in an oxygen-free environment, leading to the generation of bio-oil that comprises a complex mixture of hydrocarbons, oxygenates, and other organic molecules [42]. The nature of this bio-oil differs significantly from its feedstock. Waste frying oil, primarily composed of triglycerides along with degradation by-products from cooking, differs in chemical profile from bio-oil. While waste oil retains a simpler composition centered on fats, pyrolysis-derived bio-oil contains a wider variety of compounds due to its more severe decomposition conditions [43]. These differences in chemical structure and origin affect their respective suitability for fuel production and determine the processing routes required for each.

The pathway for converting waste frying oil into usable biodiesel via pyrolysis involves several key steps. Initially, the oil is heated under controlled conditions to produce bio-oil, which is then subjected to a series of upgrading processes. One critical step is hydrodeoxygenation (HDO), which is a catalytic process that removes oxygenated species to enhance the energy density and stability of the fuel [44]. Following HDO, stabilization treatments are employed to eliminate volatile fractions and reduce reactivity. The stabilized oil is then separated into light and heavy fractions. The heavier fraction is defined by a boiling point above 350 °C. Then, it will be directed to hydrocracking, a process that further breaks down complex molecules into lighter hydrocarbons. The biodiesel produced from this refining sequence can be utilized in diesel and gasoline engines. This multifaceted conversion strategy illustrates the potential of combining pyrolysis with advanced upgrading technologies to valorize waste oils into biodiesel and other valuable fuel products.

4.2 Transesterification

Transesterification is one of the most well-established and widely implemented chemical routes for converting lipid-based feedstocks into biodiesel. As shown in Fig. 3, this process involves the transformation of triglycerides as the primary molecular constituents of fats and oils into fatty acid methyl esters (FAMEs), which are commonly recognized as biodiesel [45, 46]. A secondary by-product of this reaction is glycerol, a compound that can be further processed for use in various industries such as pharmaceuticals and cosmetics. The transesterification reaction typically occurs when triglycerides are reacted with a short-chain alcohol, such as methanol or ethanol, in the presence of a catalyst. The role of the catalyst is vital as it significantly influences the rate of reaction and the overall yield of biodiesel. This reaction is reversible and must be carefully controlled to ensure a complete conversion of triglycerides into methyl esters. Parameters such as temperature, alcohol-to-oil molar ratio, catalyst concentration, and reaction time must be optimized to maximize efficiency.

Catalysts used in transesterification are typically classified into three major categories: alkaline (basic), acidic, and enzymatic. Among these, alkaline catalysts such as sodium hydroxide ($NaOH$) and potassium hydroxide (KOH) are the most commonly used due to their high reactivity and cost-effectiveness. However, their use is mostly limited to feedstocks with low free fatty acid (FFA) content. If the FFA content exceeds 2 mg KOH/g, these catalysts may lead to soap formation, which reduces biodiesel yield and complicates the purification process. Waste frying oils are a common feedstock for biodiesel that usually contain elevated FFA levels due to prolonged exposure to heat and air during cooking. For such feedstocks, an initial esterification step is necessary to reduce the FFA content before undergoing the base-catalyzed transesterification.

Esterification is typically performed using methanol or ethanol in the presence of an acid catalyst, most often sulfuric acid. This pre-treatment step involves heating the oil to reduce its viscosity and improve mixing, after which the alcohol–catalyst mixture is introduced to initiate the conversion of free fatty acids into esters. This process effectively minimizes soap formation and facilitates a more efficient transesterification reaction. This two-step process involving acid-catalyzed esterification

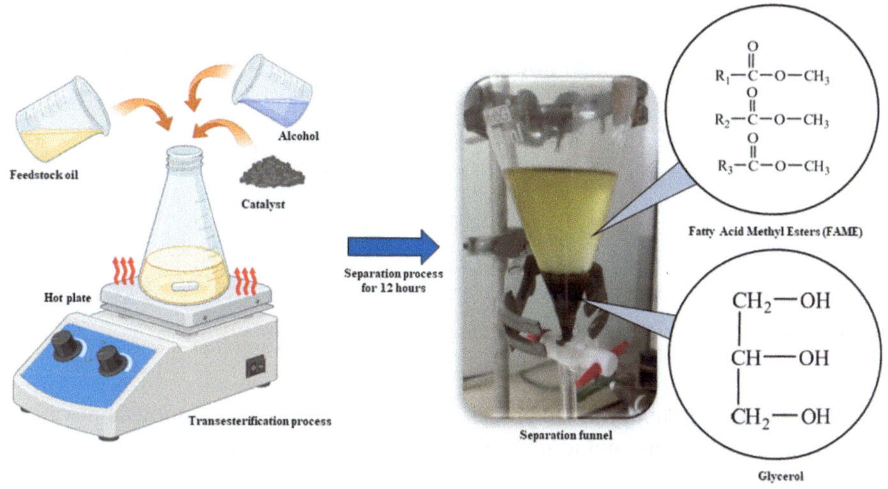

Fig. 3 Biodiesel production using transesterification process

followed by base-catalyzed transesterification has proven effective in processing low-quality oils into high-quality biodiesel [47, 48].

Once the reaction is complete, the product mixture undergoes a series of separation and purification steps. These include the removal of glycerol, residual alcohol, excess catalyst, and impurities. The biodiesel is then typically washed with water to eliminate remaining contaminants and dried to remove moisture. The purified biodiesel must meet standardized specifications such as ASTM D6751 or EN 14214 before it can be used in engines or blended with petro-diesel. Transesterification remains a cornerstone technology in biodiesel production due to its high conversion efficiency, moderate operational conditions, and scalability. It is applicable to a broad spectrum of feedstocks, including virgin oils, waste oils, and animal fats.

The method is simple and has relatively low infrastructure requirement make it suitable for both small-scale and industrial-scale biodiesel production facilities. Catalyst selection continues to be a focal point in optimizing transesterification performance. Homogeneous catalysts, which dissolve in the reaction medium, include alkaline bases (NaOH, KOH) and acidic types such as sulfuric acid and sulfonic acid derivatives [49]. These offer fast reaction times but introduce difficulties in catalyst recovery, product separation, and wastewater generation. In contrast, heterogeneous catalysts operate in a different phase than the reactants and are thus easier to separate and reuse. These include basic solids like calcium oxide (CaO) and acid-functionalized mesoporous materials, some of which are derived from industrial by-products like fly ash or biomass residues.

Despite its strengths, transesterification is not without drawbacks. One notable challenge is the purification of biodiesel post-reaction, which can be labor-intensive and may require multiple washing cycles, as shown in Fig. 4. Moreover, glycerol removal, emulsion formation, and energy input for heating and mixing contribute to

Fig. 4 Washing process of biodiesel after transesterification process

the operational complexity of the process [50]. Nevertheless, continuous innovations in catalyst development, including bifunctional catalysts capable of performing both esterification and transesterification, as well as enzyme-based systems, are helping to mitigate these challenges. These advancements are paving the way for more sustainable, cost-effective, and environmentally friendly biodiesel production, particularly from non-edible and waste-derived feedstocks.

4.3 Micro-Emulsification

Micro-emulsification has emerged as a notable alternative technique in the field of biofuel production due to its ability to overcome some of the limitations associated with high-viscosity oils. At its essence, a microemulsion is a thermodynamically stable and optically isotropic fluid system [51]. It is distinct from conventional emulsions that are only kinetically stable and tend to separate into oil and water phases over time. This thermodynamic stability allows micro-emulsions to remain uniform without phase separation, making them attractive for fuel formulations that require consistent performance. Micro-emulsions typically consist of nano-sized structures ranging from 1 to 150 nm and form spontaneously when immiscible liquids such as vegetable oil and a short-chain alcohol, such as methanol, ethanol, or butanol, are mixed in the presence of surfactants [52]. These surfactants play a vital role by creating a dispersed system of fine oil droplets stabilized in a continuous alcohol phase. The formation of micro-emulsions can significantly reduce the viscosity of vegetable oils, improving their flow properties, atomization behavior, and combustion efficiency in diesel engines.

One of the advantages of micro-emulsification in biodiesel production is the absence of a chemical reaction. Unlike transesterification or pyrolysis, this method does not involve molecular transformation of the oil. Instead, it blends the feedstock with alcohol and surfactants to create a fuel mixture that mimics diesel in behavior and engine compatibility. This not only simplifies the process but also eliminates the generation of by-products such as glycerol, which are common in other biodiesel methods. The resulting biofuel demonstrates improved spray characteristics and combustion quality, contributing to lower NOx emissions and better fuel–air mixing [53]. In practical applications, the micro-emulsification technique has shown promise in addressing the high viscosity of vegetable oils like soybean, sunflower, and waste frying oils. Viscous fuels can cause issues such as poor atomization, incomplete combustion, injector needle sticking, and excessive carbon build-up in engine components. By converting these oils into micro-emulsion-based fuels, their fluidity is enhanced, reducing stress on fuel systems and supporting more efficient combustion.

However, despite its benefits, the technique is not without limitations. One notable concern is the long-term stability of the micro-emulsified fuel under varying storage and operating conditions. The physical and chemical properties of the fuel may change over time, leading to phase separation or instability, which can compromise engine performance [54]. Additionally, while the process avoids complex chemical conversions, it can be energy-intensive and may require relatively high concentrations of alcohol and surfactants, which may raise economic and environmental concerns if not optimally managed. Nevertheless, micro-emulsification continues to be actively researched as a low-residue, fuel-blending alternative, especially for waste-derived and non-edible oils. As research progresses, improvements in surfactant design and process optimization may further enhance the feasibility and sustainability of this technique in future biodiesel applications.

4.4 Dilution

Dilution represents a straightforward, non-reactive method in biodiesel processing, distinguished by its physical rather than chemical nature. The technique involves reducing the viscosity and density of vegetable oils or waste frying oils by blending them directly with solvents, most commonly diesel or ethanol, without inducing any molecular transformation [55]. The dilution process aims to improve the fuel characteristics of raw oils by simply lowering their concentration through solvent addition. Unsaturated oils such as those derived from sunflower and cottonseed are typically preferred for this method due to their relatively favorable fluidity profiles. One of the primary advantages of dilution lies in its simplicity, low operational cost, and minimal equipment requirement. The absence of chemical reactions means there are no catalysts or by-products involved, making the process relatively clean and energy-efficient. Moreover, the resulting fuel blend demonstrates lower viscosity, which facilitates easier injection, better atomization, and smoother combustion in diesel engines.

Despite these benefits, the dilution method is not without significant limitations. Fuels produced through direct dilution often suffer from poor volatility, reduced thermal stability, and incomplete combustion. All of these factors can negatively impact engine performance. Prolonged usage may lead to carbon build-up, injector nozzle clogging, and frequent maintenance requirements [56]. Furthermore, diluted fuels, especially those with high proportions of raw oil, frequently fail to comply with established fuel standards such as ASTM D6751 or EN 14214 [57]. These deficiencies limit the widespread adoption of the dilution method in commercial biodiesel applications.

In practical terms, a 20% volume blend (B20) of biodiesel with diesel is widely regarded as a safe and effective ratio for use in conventional diesel engines without requiring major modifications. Exceeding this ratio can begin to replicate the behavior and drawbacks of using pure biodiesel (B100), including cold-flow issues and oxidative instability [58]. The chosen dilution ratio plays a pivotal role in determining key fuel properties such as viscosity, calorific value, ignition delay, and combustion efficiency. In many cases, achieving acceptable performance may still require additional treatments or blending strategies to enhance fuel quality. Overall, while the dilution approach offers a low-cost entry point for utilizing waste oils as fuels, it often serves as a preliminary or supplemental step rather than a stand-alone solution.

5 Sustainability Perspectives in Biodiesel Production

The pursuit of sustainable energy alternatives has positioned biodiesel as a promising candidate in addressing global concerns related to energy security, environmental degradation, and the depletion of fossil fuels. As highlighted in this chapter, biodiesel production from various biomass feedstocks, including plant-derived oils and animal fats, offers a viable pathway toward circular economy practices. Recently, insects such as *Hermetia illucens,* known as the Black Soldier Fly (BSF), have attracted attention as an emerging feedstock for biodiesel production [59]. This is due to their high lipid content and ability to grow on organic waste substrates. Their use supports circular economy principles and contributes to Sustainable Development Goal (SDG) 12 by converting waste materials into renewable fuel precursors. The utilization of these feedstocks not only reduces dependency on finite petroleum resources but also promotes the valorization of waste materials that would otherwise contribute to pollution or landfill accumulation [60]. By integrating waste management with fuel generation, biodiesel contributes to a more resource-efficient and environmentally responsible energy framework. Life Cycle Assessment (LCA) is increasingly applied to evaluate the overall sustainability of biodiesel production by quantifying environmental impacts across its entire value chain from feedstock cultivation to fuel utilization. Through LCA, key indicators such as greenhouse gas emissions, energy balance, and resource efficiency can be systematically compared among different feedstocks and processing routes. Incorporating LCA into biodiesel evaluation strengthens decision-making toward more sustainable and low-impact production systems.

Each biodiesel production method discussed pyrolysis, transesterification, microemulsification, and dilution presents unique sustainability benefits and limitations. Transesterification, particularly when applied to non-edible and waste oils, remains one of the most efficient and widely adopted techniques due to its high conversion rates and adaptability. Pyrolysis allows for broader feedstock flexibility, including lignocellulosic biomass and used cooking oil, but requires energy-intensive conditions and advanced upgrading processes. Micro-emulsification and dilution, while simpler and less resource-intensive, face challenges related to fuel stability and engine compatibility. Thus, sustainable biodiesel production relies not only on feedstock availability but also on the careful selection of technology based on environmental impact, energy balance, and local resource conditions.

From a broader sustainability perspective, the transition to biodiesel aligns with global goals such as the United Nations Sustainable Development Goals (SDGs), particularly those related to affordable and clean energy (SDG 7), responsible consumption and production (SDG 12), and climate action (SDG 13). By leveraging waste-based and renewable biomass feedstocks, biodiesel production reduces greenhouse gas emissions, promotes energy diversification, and supports rural development. Furthermore, the continued advancement of green catalysts, energy-efficient processing technologies, and integrated biorefinery systems is essential for minimizing the environmental footprint of biodiesel production. Ultimately, sustainable biodiesel practices represent a critical step toward creating a resilient and low-carbon energy future.

6 Conclusion

This chapter has provided a comprehensive overview of biodiesel production using diverse biomass feedstocks by emphasizing their role in fostering a more sustainable and circular bioeconomy. The discussion began with an overview of biodiesel feedstock classifications, such as vegetative and animal-based, followed by extraction techniques essential for enhancing lipid extraction efficiency. Waste-derived feedstocks offer the greatest potential due to their abundance, low cost, and alignment with waste valorization efforts. Various extraction methods, such as mechanical pressing, solvent extraction, supercritical CO_2, microwave-assisted, and ultrasonic-assisted techniques, were examined. It highlights their respective strengths in processing efficiency, energy use, and sustainability. Emerging non-conventional techniques like microwave- and ultrasonic-assisted extraction were found to improve lipid recovery while reducing processing time and environmental impact.

The chapter also detailed four major biodiesel production pathways, such as pyrolysis, transesterification, micro-emulsification, and dilution. Each method was critically evaluated for its operational characteristics, feedstock compatibility, yield potential, and limitations. Transesterification emerged as the most widely used and effective method, particularly when preceded by esterification for high-FFA feedstocks. Importantly, the sustainability dimension was woven throughout the discussion, affirming that biodiesel production is not merely a technical process but also a strategic approach to environmental stewardship. By converting waste feedstocks into valuable biodiesel, this approach helps in reducing greenhouse gas emissions and promoting resource recovery. Biodiesel aligns with global sustainability goals.

As research advances in green processing and integrated biorefineries, biodiesel is poised to play a vital role in the global shift toward low-carbon and renewable energy systems.

References

1. Shakir MA et al (2025) Advancing insights into building electrical energy consumption: a comprehensive review of conventional and artificial intelligence methods. Adv Build Energy Res 1–44
2. Ghasemian S et al (2020) An overview of global energy scenarios by 2040: identifying the driving forces using cross-impact analysis method. Int J Environ Sci Technol 1–24
3. Binhweel F, Ahmad MI, Shakir MA (2025) Harnessing waste lipids for biodiesel: optimization and thermodynamic investigation of beef tallow transesterification. Int J Green Energy 1–13
4. Vasileiadis V et al (2026) Advancing the bioeconomy through biodiesel: a comparative evaluation of plant-based, animal-based, and mixed feedstocks for sustainable use of heating diesel. Biofuels 1–17
5. Aguado-Deblas L et al (2021) Evaluation of dimethyl carbonate as alternative biofuel. Performance and smoke emissions of a diesel engine fueled with diesel/dimethyl carbonate/straight vegetable oil triple blends. Sustainability 13(4):1749
6. Emmanouilidou E et al (2024) A comparative study on biodiesel production from edible and non-edible biomasses. J Mol Struct 1306:137870
7. Shalfoh E et al (2025) Biomass to biofuel: optimizing sustainable biodiesel production from fish waste and thermodynamic-kinetic analysis. Biofuels Bioprod Biorefin 19(3):654–677
8. Senusi W et al (2024) Biodiesel production and characteristics from waste frying oils: sources, challenges, and circular economic perspective. Environ Sci Pollut Res 31(23):33239–33258
9. Senusi W et al (2024) Comparative assessment for biodiesel production from low-cost feedstocks of third oil generation. Renew Energy 236:121369
10. Khan N, Sudhakar K, Mamat R (2021) Role of biofuels in energy transition, green economy and carbon neutrality. Sustainability 13(22):12374
11. Shakir MA et al (2024) Biomass to biofuel: Palm kernel shells as catalyst supports for enhanced biodiesel production. Biofuels Bioprod Biorefin 18(6):2038–2052
12. Fasanya OO, Osigbesan AA, Avbenake OP (2021) Biodiesel production from non-edible and waste lipid sources. In: Biodiesel technology and applications, pp 389–427
13. Shakir MA, Ahmad MI (2025) Biodiesel production from Ulva lactuca oil extracted by supercritical carbon dioxide using a biogenic activated carbon catalyst. Bioresour Technol Rep 32:102393
14. Binhweel F et al (2025) Biodiesel production from Ulva lactuca: insights into kinetic and thermodynamic evaluation of lipid transesterification. Biocatal Agric Biotechnol 70:103856
15. Aniza R et al (2023) A review of biowaste remediation and valorization for environmental sustainability: artificial intelligence approach. Environ Pollut 324:121363
16. Shakir MA, Ahmad MI (2025) Biodiesel production from Ulva lactuca oil extracted by supercritical carbon dioxide using a biogenic activated carbon catalyst. Bioresour Technol Rep 102393
17. Binhweel F, Hossain MS, Ahmad MI (2023) Recent trends, potentials, and challenges of biodiesel production from discarded animal fats: a comprehensive review. BioEnergy Res 16(2):778–800
18. Passos RMd et al (2024) Ethyl biodiesel production from crude soybean oil using enzymatic degumming-transesterification associated process. Ind Crop Prod 222:119930
19. Bai H et al (2022) Process optimization of esterification for deacidification in waste cooking oil: RSM approach and for biodiesel production assisted with ultrasonic and solvent. Fuel 318:123697

20. Shakir MA et al (2024) Effect of activation parameters on palm kernel shell-based activated carbon as catalyst support for biodiesel from discarded beef tallow. In: IOP Conference Series: Earth and Environmental Science. IOP Publishing
21. Singh SK, Chauhan A, Sarkar B (2024) Resilience of sustainability for a smart production system to produce biodiesel from waste animal fat. J Clean Prod 452:142047
22. Stamenković OS et al (2023) Biodiesel production from camelina oil: present status and future perspectives. Food Energy Secur 12(1):e340
23. Esmi F et al (2023) Tabatabaei M, Nizami A-S (eds) Chapter 3—Biodiesel feedstocks: location, location, and location, in sustainable biodiesel. Academic Press, pp 71–90
24. Suzihaque MUH et al (2022) Biodiesel production from waste cooking oil: a brief review. Mater Today: Proc 63:S490–S495
25. Kumar S, Singhal MK, Sharma MP (2022) Predictability of biodiesel fuel properties from the fatty acid composition of the feedstock oils. Arab J Sci Eng 47(5):5671–5691
26. Hosseinzadeh-Bandbafha H et al (2022) Environmental life cycle assessment of biodiesel production from waste cooking oil: a systematic review. Renew Sustain Energy Rev 161:112411
27. Chew CL et al (2021) Improving sustainability of palm oil production by increasing oil extraction rate: a review. Food Bioprocess Technol 14(4):573–586
28. Zulqarnain et al (2021) Solvent extraction and performance analysis of residual palm oil for biodiesel production: experimental and simulation study. J Environ Chem Eng 9(4):105519
29. Shakir MA, Yhaya MF, Ahmad MI (2017) Physical effects of pre-hydrolysis and alkali treatment of empty fruit bunch (EFB) fibers. Int J Chem Eng Appl 8(6):387–391
30. Lee SY et al (2021) Techniques of lipid extraction from microalgae for biofuel production: a review. Environ Chem Lett 19(1):231–251
31. Wetterwald L et al (2023) Supercritical CO2 extraction of neutral lipids from dry and wet Chlorella vulgaris NIES 227 microalgae for biodiesel production. J Environ Chem Eng 11(5):110628
32. Shalfoh E et al (2024) Fish waste oil extraction using supercritical CO2 extraction for biodiesel production: mathematical, and kinetic modeling. Renew Energy 220:119659
33. Aparamarta HW et al (2022) Optimization and kinetic study of biodiesel production from nyamplung oil with microwave-assisted extraction (MAE) technique. Heliyon 8(8)
34. Rezaei Motlagh S et al (2021) Ionic liquid-based microwave-assisted extraction of lipid and eicosapentaenoic acid from Nannochloropsis oceanica biomass: experimental optimization approach. J Appl Phycol 33(4):2015–2029
35. Tabtimmuang A et al (2024) Ultrasonic-assisted synthesis of mono- and diacylglycerols and purification of crude glycerol derived from biodiesel production. Ind Crops Prod 208:117891
36. Ahmad MA, Samsuri S (2021) Biodiesel purification via ultrasonic-assisted solvent-aided crystallization. Crystals 11(2):212
37. Kartal F, Dalbudak Y, Özveren U (2023) Prediction of thermal degradation of biopolymers in biomass under pyrolysis atmosphere by means of machine learning. Renew Energy 204:774–787
38. Supriyanto, Ylitervo P, Richards T (2024) Fast co-pyrolysis of wood and plastic: evaluation of the primary gaseous products. Energy Convers Manag: X 22:100613
39. Adelawon BO et al (2022) Comparison of the slow, fast, and flash pyrolysis of recycled maize-cob biomass waste, box-benhken process optimization and characterization studies for the thermal fast pyrolysis production of bio-energy. Chem Eng Commun 209(9):1246–1276
40. Brindhadevi K et al (2021) Effect of reaction temperature on the conversion of algal biomass to bio-oil and biochar through pyrolysis and hydrothermal liquefaction. Fuel 285:119106
41. Mariyam S et al (2022) A critical review on co-gasification and co-pyrolysis for gas production. Renew Sustain Energy Rev 161:112349
42. Gad MS et al (2021) Effect of oil blends derived from catalytic pyrolysis of waste cooking oil on diesel engine performance, emissions and combustion characteristics. Energy 223:120019
43. Su G et al (2022) Pyrolysis of waste oils for the production of biofuels: a critical review. J Hazard Mater 424:127396

44. Qu L et al (2021) A review of hydrodeoxygenation of bio-oil: model compounds, catalysts, and equipment. Green Chem 23
45. Ahmad MI et al (2024) Fueling the future: extraction and production technological development of marine macroalgae-seaweed biodiesel. In: Rafatullah M et al (eds) Multidisciplinary applications of marine resources: a step towards green and sustainable future. Springer Nature Singapore, Singapore, pp 285–302
46. Naseef HH, Tulaimat RH (2025) Transesterification and esterification for biodiesel production: a comprehensive review of catalysts and palm oil feedstocks. Energy Convers Manag: X 26:100931
47. Yatish KV et al (2024) Demonstrating photocatalytic esterification as a potential strategy to improve the properties of feedstock oil derived from dairy waste scum for biodiesel production. Energy Convers Manag 309:118463
48. Alsaadi S, Ahmad MI, Shakir MA (2025) Kinetics and thermodynamics study of lipid extraction from Camelus dromedarius fat waste and feedstock application in biodiesel production. Discov Sustain 6(1):1102
49. Hasan N, Ratnam MV (2022) Biodiesel production from waste animal fat by transesterification using H2SO4 and KOH catalysts: a study of physiochemical properties. Int J Chem Eng 2022(1):6932320
50. Orege JI et al (2022) Recent advances in heterogeneous catalysis for green biodiesel production by transesterification. Energy Convers Manag 258:115406
51. Murray R, King G, Wyse-Mason R (2021) Micro-emulsification vs. transesterification: an investigation of the efficacy of methanol use in improving vegetable oil engine performance. Biofuels 12(9):1165–1174
52. Kaur P, Singh G, Kaur K (2025) Nanoemulsions of essential oils for sustainable agriculture. In: Baskar C, Ramakrishna S, Rosa ADL (eds) Encyclopedia of green materials. Springer Nature Singapore, Singapore, pp 1439–1453
53. Deepak B, Mohamed Ibrahim M (2024) A critical review on emulsion fuel formulation and its applicability in compression ignition engine. Biofuels 15(5):555–573
54. El Nagy HA, Mohamed MAE-A (2024) Stable diesel microemulsion using diammonium ionic liquids and their effects on fuel properties, particle size characteristics and combustion calculations. Sci Rep 14(1):7728
55. Brahma S et al (2022) Biodiesel production from mixed oils: a sustainable approach towards industrial biofuel production. Chem Eng J Adv 10:100284
56. Xu M et al (2021) Evaluation of ignition process and NOx reduction of coal under moderate and intensive low-oxygen dilution combustion by implementing fuel-rich/lean technology. Fuel 296:120657
57. Bitire SO, Jen TC (2022) Performance and emission analysis of a CI engine fueled with parsley biodiesel-diesel blend 11(2):143–153
58. Ndiaye M et al (2020) Reuse of waste animal fat in biodiesel: biorefining heavily-degraded contaminant-rich waste animal fat and formulation as diesel fuel additive. Renew Energy 145:1073–1079
59. Mangindaan D, Kaburuan ER, Meindrawan B (2022) Black soldier fly larvae (Hermetia Illucens) for biodiesel and/or animal feed as a solution for waste-food-energy nexus: bibliometric analysis. Sustainability 14(21):13993
60. Shakir MA (2025) et al Development and characterization of fiber-reinforced biopolymer composites from rice husk waste with Rhizopus oligosporus as a bio-binder. Express Polym Lett 19(9)

Role of Biomass as Carbon Neutral

**Mohd Asyraf Kassim, Mardiana Idayu Ahmad,
and R. R. Dirgarini Julia Nurlianti Subagyono**

Abstract This chapter aims to review on the role of biomass as carbon neutral for sustainable environment. Several approaches to capture such as carbon capture and storage (CCS) and carbon capture storage and utilization (CCSU) are discussed in this chapter. In addition, new concept to integrate with other technologies are also reviewed. CCSU approach with additional bioproduct such as biochar and bioplastic are discussed. Finally, closed-loop carbon capture is also discussed.

Keywords Greenhouse mitigation · Bioproduct · Biocircular economy · Bioprocess

1 Status of Global CO_2 Emission

Carbon dioxide (CO_2) emissions are one of the major contributors of climate change caused by human activities such as burning fossil fuel, industrial processes, and deforestation. Accumulation of CO_2 in atmosphere has significantly contributed to greenhouse effect which this gas along with other greenhouse gases will trap heat in the atmosphere resulting to the increment of planet temperature. Current global warming data indicated that global mean near-surface temperatures are likely to exceed 1.5 °C above pre-industrial levels. It is projected that global temperature will exceed the 1.5 °C threshold within 5 years. In 2023, the top five countries with the highest CO_2 emissions were: China with the largest emitter, producing 11.9 billion metric tons of CO_2, followed by the U.S. with 4.9 billion metric tons. Other countries such as India, Russia, and Japan generated 2.69 and 1.91 billion metric tons respectively, while Japan emitted 944 million metric tons of CO_2 in 2023 from combustion and industrial activities.

M. A. Kassim (✉) · M. I. Ahmad
School of Industrial Technology, Universiti Sains Malaysia (USM), Gelugor, Malaysia
e-mail: asyrafkassim@usm.my

R. R. D. J. N. Subagyono
Chemistry Department, Faculty of Mathematics and Natural Sciences, Mulawarman University, Samarinda, Indonesia

M. A. Kassim et al. (eds.), *Sustainable Biorefinery*, Green Energy and Technology,
https://doi.org/10.1007/978-981-95-8181-8_5

Table 5.1 CO_2 capture and utilization technology in various industries

Technology	Description	Application/product
CarbonCure	Injects captured CO_2 into fresh concrete mix	Stronger concrete, less cement needed
LanzaTech Gas Fermentation	Microbes ferment captured CO_2/CO into fuels	Ethanol, jet fuel, chemicals
Covestro Polyols	CO_2 converted into polyols for polyurethanes	Mattresses, insulation, adhesives
Newlight Technologies—AirCarbon	CO_2 + methane converted into PHB bioplastic	Biodegradable plastics (cutlery, bags)
BASF Dry Reforming	Converts $CO_2 + CH_4$ to syngas using catalysts	Used in chemical manufacturing
Mineralization (e.g., Blue Planet)	CO_2 turned into calcium carbonate	Synthetic limestone for construction

Several carbon capture and storage, utilization have been introduced in order to reduce the accumulation of CO_2 in the atmosphere [1]. This approach is mainly a process that captures carbon dioxide emissions from sources like coal-fired power plants and either reuses or stores it so it will not enter the atmosphere. Carbon capture approaches can be classified into Carbon capture and storage (CCS) and utilization (CCUS) [2, 3]. CCS—technology refers to the process of separating CO_2 from relevant emission sources, then transporting it to the storage site and isolating CO_2 from the atmosphere for a long time. Generally, CCS can remove the existing CO_2 in the atmosphere on a large scale, which is a relatively good method to alleviate the greenhouse effect. On the other hand, CCUS is an emerging CO_2 emission reduction technology with great potential, and it is also a new development trend of CCS technology. The CO_2 captured in the production process is put into a new production process and can be converted and utilized in different ways. This process is not simply storage, but effectively realizes the recycling and reuse of CO_2 CCUS technology can recycle CO_2 and generate economic benefits, which is more practical. Usage (or utilization) refers to using the captured CO_2 to produce commercially marketable products or services. Table 5.1 shows the current CCUS technology that utilize CO_2 as feedstock for production of various value-added products.

2 Carbon Neutral by Biomass

Biomass can play a significant role in Carbon Capture, Utilization, and Storage (CCUS) by acting both as a carbon sink and as a feedstock for carbon-based products. The biomass cycle begins with photosynthesis, where plants absorb carbon dioxide (CO_2) from the atmosphere and convert it into organic matter using sunlight (Eq. 5.1). This natural process effectively captures carbon and stores it in the form of biomass, such as trees, crops, or algae. From a CCUS standpoint, this is a crucial first step, as it represents a natural and renewable method of drawing CO_2 out of the atmosphere and temporarily storing it in plant material.

$$6CO_2 + 6H_2O + \text{light energy} \rightarrow C_6H_{12}O_6 + 6O_2 \tag{5.1}$$

Figure 5.1 illustrates a closed-loop carbon capture and utilization (CCU) cycle involving biomass. This concept encompasses a series of interconnected natural and engineered processes, beginning with the capture of atmospheric CO_2 through photosynthesis. In this stage, plants absorb CO_2 and convert it into organic carbon compounds, forming the biomass. This process is a vital natural mechanism for carbon capture, temporarily storing atmospheric carbon in renewable biological forms such as trees, crops, and other vegetation. These natural systems function as carbon sinks, contributing significantly to the reduction of atmospheric CO_2 and supporting sustainable carbon management strategies.

Once the biomass is harvested, it can be processed into a variety of renewable products and energy sources. This represents the utilization phase of the cycle, where fixed carbon is transformed into value-added outputs such as biofuels, bioplastics, construction materials, or direct heat and electricity. Unlike fossil fuels, which introduce new carbon into the atmosphere, biomass-based systems recycle previously captured carbon, aligning with circular economy principles and lowering overall carbon intensity. Eventually, biomass may undergo combustion or decomposition, releasing CO_2 back into the atmosphere. However, because this carbon originated from atmospheric CO_2 recently absorbed via photosynthesis, it constitutes part of a short-term, renewable carbon loop. When coupled with sustainable land practices such as replanting and land restoration and emerging technologies like biochar production or Bioenergy with Carbon Capture and Storage (BECCS), this system

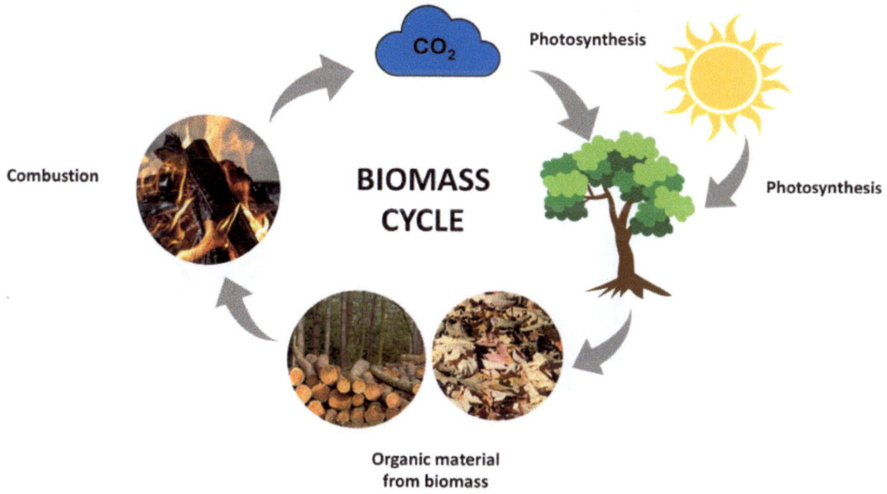

Fig. 5.1 Closed-loop carbon capture and utilization by biomass

can become carbon-negative. BECCS enables the capture and permanent storage of CO_2 released during biomass combustion, effectively removing more CO_2 from the atmosphere than is emitted, and offering a scalable solution for both clean energy generation and long-term climate mitigation.

3 Bioenergy with Carbon Capture and Storage (BECCS)

Bioenergy with carbon capture and storage (BECCS) concept as a carbon removal technology that involves integration of bioenergy generation from biomass with carbon capture and storage system. In this approach, bioenergy that is commonly used in fermentation, anaerobic digestion, combustion and gasification is produced using biomass as a feedstock, and the CO_2 generated from these processes are captured and stored underground (Fig. 5.2).

Bioenergy from biomass can be in various forms either solid or liquid (Table 5.2). Most biomass that contains a high calorific value are used to produce solid fuel which is usually used in the combustion and gasification processes.

Among the main processes for producing energy and CO_2 is through combustion and gasification. During the combustion of biomass, several important chemical and physical processes occur that convert organic matter into energy, gas, and ash [4]. During combustion, biomass consisting of several main chemical elements such as Carbon (C), hydrogen (H), and oxygen (O) will undergo oxidation in the presence of

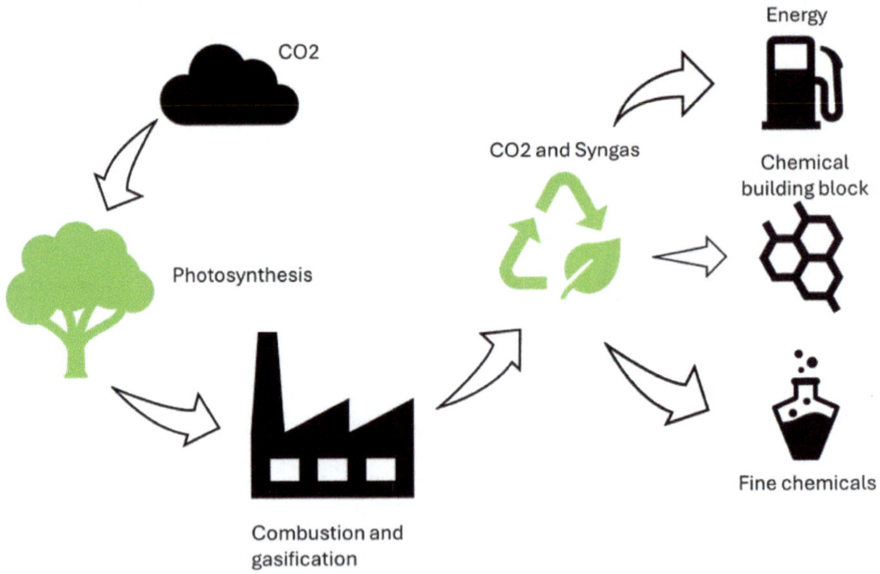

Fig. 5.2 Carbon capture approach via bioenergy with carbon capture storage

Table 5.2 Bioenergy produced from different biomass and conversion methods

Biomass type	Source examples	Conversion method	Bioenergy product	Notes
Woody biomass	Forest residues, wood chips, sawdust	Combustion, gasification	Heat, electricity, syngas	High energy content; widely used in power plants and heating
Agricultural residues	Straw, husks, corn stover, rice husk	Combustion, anaerobic digestion	Biogas, heat, electricity	Abundant and cheap; seasonal availability; ash may require management
Energy crops	Switchgrass, miscanthus, sugarcane	Fermentation, gasification	Bioethanol, biogas, syngas	Grown specifically for energy; may compete with food crops
Animal manure	Livestock waste	Anaerobic digestion	Biogas (CH_4 + CO_2)	Reduces methane emissions from waste; suitable for farm-scale use
Algae	Microalgae, macroalgae	Transesterification, fermentation	Biodiesel, bioethanol	High productivity; can use wastewater and CO_2; still under development
Food waste	Leftover food, expired products	Anaerobic digestion, fermentation	Biogas, bioethanol	Helps reduce landfill emissions; efficient when source-separated
Industrial organic waste	Glycerol, distillery slops, pulp waste	Anaerobic digestion, gasification	Biogas, syngas, bio-oil	Adds value to industrial waste; requires pre-treatment in some cases
Municipal solid waste (MSW)	Organic fraction of MSW	Incineration, gasification	Heat, electricity, syngas	High potential in urban areas; needs sorting and air pollution controls

oxygen. During this process, exothermic chemical reactions will occur and energy is produced as the main product in the reaction. Usually, the energy or heat produced is used to generate electricity and energy for industrial activities. In addition, the CO_2 produced will be captured through the CCS process where if the biomass used is from biogenic materials, then the combustion and CO_2 capture process produces an overall carbon-negative process (Eqs. 5.2 and 5.3)

$$\text{Carbon} + O_2 \rightarrow CO_2 \quad (5.2)$$

$$H_2 + {}^1\!/_2O_2 \rightarrow H_2O\ (\text{water}) \quad (5.3)$$

In contrast to the gasification reaction, during biomass gasification, a thermochemical process occurs where organic matter (biomass) is converted into synthetic gas (syngas), rather than being burned directly as in combustion. This process occurs in limited oxygen conditions (not full combustion) and involves high temperatures between 700 and 1000 °C. During the gasification process, biomass will undergo several phases of change where biomass will be converted into the main product, which is syngas. This gas is a combination of several main gases such as hydrogen, carbon monoxide and also methane (CH_4).

The initial stage of gasification involves a heating and drying process where the water (moisture) present in the biomass is removed without chemical reaction. The main purpose of this process is to make the biomass dry and suitable for use in the next chemical reaction. Next, the reaction continues with the pyrolysis reaction where the chemical reaction occurs at high temperatures between 300 and 500 °C in the absence of oxygen. During this reaction, the biomass will decompose and produce several main products such as gas (CO, CH_4, H_2, and tar), biochar and also bio-oil (a heavy liquid containing complex hydrocarbons). The reaction continues with the combustion of combustible materials such as carbon and partially burned gas as described in Eqs. (5.4) and (5.5)

$$\text{Carbon} + {}^1\!/_2O_2 \rightarrow CO \quad (5.4)$$

$$H_2 + {}^1\!/_2O_2 \rightarrow H_2O \quad (5.5)$$

Finally, the resulting gas will undergo a reaction between carbon, water vapor, and CO_2 to produce more gas as in Eqs. (5.6) and (5.7). Through gasification + CCS can be used to capture CO_2 from syngas or during the conversion process to other fuels. The sorption-enhanced water gas shift (SEWGS) process removes CO_2 from a high-temperature and pressure syngas stream. This is achieved by passing it through multiple vessels filled with a catalytically active CO_2 adsorbent. The result is an essentially carbon-free, hydrogen-rich product stream.

$$\text{Carbon} + H_2O \rightarrow CO + H_2 \quad (5.6)$$

$$CO + H_2O \rightarrow CO_2 + H_2 \quad (5.7)$$

4 Carbon Utilization via Biomass Conversion

Apart from capture CO_2, biomass also can be converted into carbon-rich based product that can be partially replaced fossil-based products alternative (Fig. 5.3). Biomass that poses wide range of chemical such as cellulose, hemicellulose, lignin, carbohydrate, lipid, and protein can be converted into various products including biochar, bioplastic, and other building blocks (Fig. 5.4).

4.1 Carbon Capture by Biochar from Biomass

Biochar, a charcoal produced from biomass pyrolysis, is a promising and cost-effective adsorbent for capturing CO_2 from the atmosphere and other sources. Biochar from biomass is produced through a process called pyrolysis, which involves heating organic material in the absence or limited presence of oxygen [5]. Table 5.3 shows the biochar production from pyrolysis of different type of biomass. It can be seen that pyrolysis of biomass at different temperature could significantly affect the biochar yield. In addition, biochar prepared from different raw materials and pyrolysis conditions of biomass generally exhibits different physicochemical properties in terms of pore structures, specific surface areas, surface oxygen-containing functional groups, and elemental compositions [6]. This biochar works by utilizing its porous structure and unique surface properties to capture CO_2 molecules through physical and chemical adsorption. This process not only removes CO_2 from the environment but also can sequester carbon within the biochar for a long period.

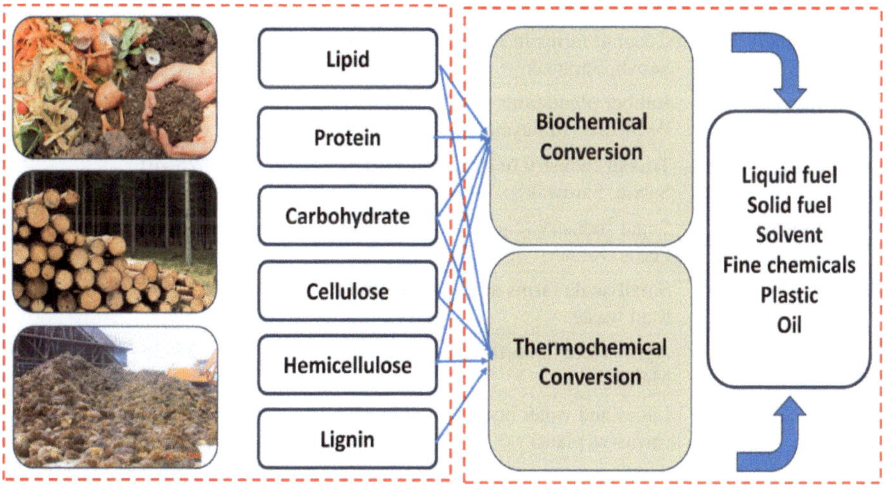

Fig. 5.3 Biomass valorization to various value added products

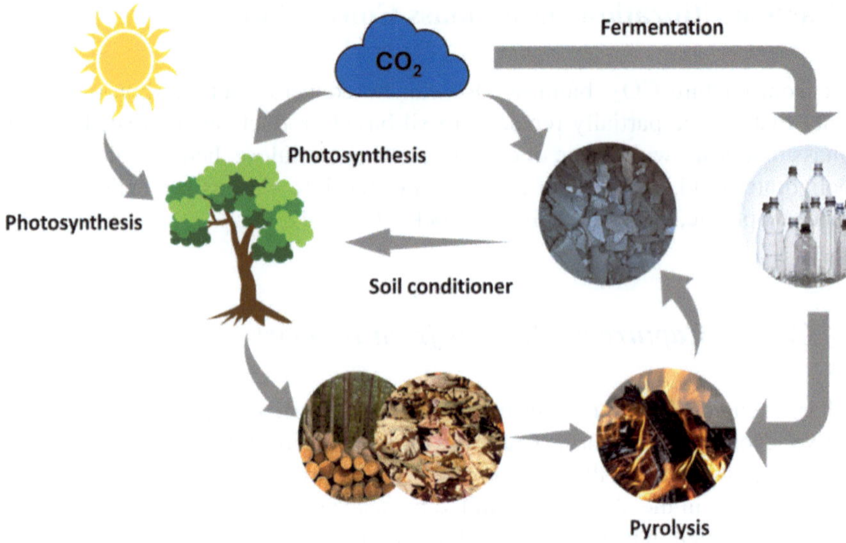

Fig. 5.4 CO_2 capture utilization via biochar and bioplastic production

Table 5.3 Biochar production from pyrolysis of different biomass feedstocks

Biomass type	Common sources in Malaysia	Pyrolysis temperature (°C)	Biochar yield (% wt)
Oil palm biomass	Empty fruit bunch (EFB), palm kernel shell (PKS), palm fronds	300–600	25–45
Rice husk	Rice mills in Kedah, Perlis, Perak	400–600	30–40
Coconut shell	Coconut farms in Johor, Sabah, Sarawak	450–700	25–35
Rubber wood	Rubber plantations in Peninsular Malaysia	350–600	20–30
Sawdust	Timber industry (e.g., Sabah, Sarawak)	300–550	25–40
Sugarcane bagasse	Sugar industry (e.g., Perlis, Kedah)	350–500	30–45
Banana pseudostem	Small-scale farms and food waste	300–500	20–30
Corn cob	Agro-waste in northern Malaysia	350–600	25–35
Water hyacinth	Lakes and water bodies (invasive plant)	300–450	35–50

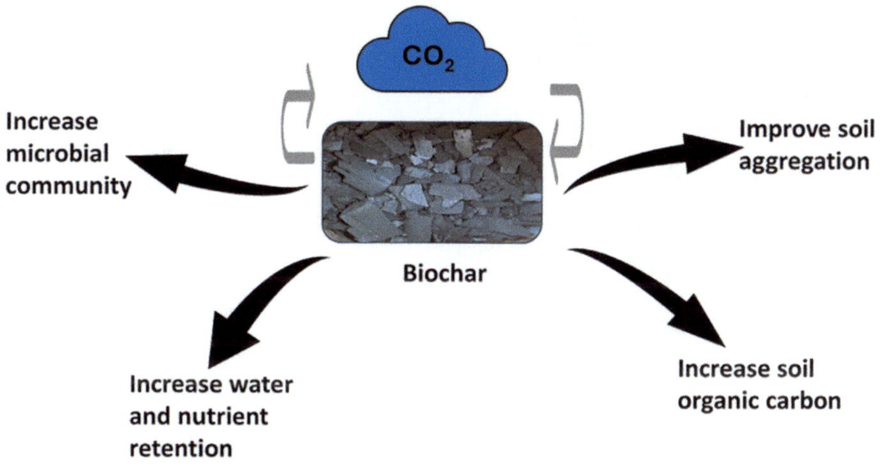

Fig. 5.5 Carbon dioxide capture by biochar from biomass mechanism

Biochar captures carbon dioxide (CO_2) primary mechanism involves physically adsorb CO_2 to its porous structure and large surface area, allowing CO_2 molecules to adhere to internal pore surfaces through Van der Waals forces [7]. Additionally, chemically active sites on biochar, particularly in alkaline or mineral-rich variants, can bind CO_2 through chemisorption, forming stable carbonates with elements such as calcium or magnesium. Beyond these direct mechanisms, biochar also contributes indirectly to CO_2 capture by enhancing soil fertility, improving water retention, and supporting microbial activity, which in turn promotes plant growth and increased atmospheric CO_2 uptake through photosynthesis. Furthermore, biochar-amended soils often emit less nitrous oxide (N_2O) and methane (CH_4), both potent greenhouse gases, thereby contributing to overall greenhouse gas reduction. Collectively, these mechanisms demonstrate biochar's multifaceted role in capturing and stabilizing atmospheric carbon within environmental systems (Fig. 5.5).

4.2 *CO_2 to Bioplastic Technology*

CO_2-to-bioplastic technology is another approach that has been introduced to capture and utilize CO_2. This technology refers to an innovative process that converts carbon dioxide, a major greenhouse gas, into sustainable bioplastics using microbial or catalytic pathways (Fig. 5.6). The conversion of CO_2 to bioplastics involves three main technologies: microbial fermentation, photosynthetic biofixation, and catalytic polymerization.

Conversion of CO_2 to plastic typically involves the use of various types of microorganisms including engineered microorganisms such as cyanobacteria, algae, or bacteria. Generally these microorganisms can fix CO_2 through photosynthesis

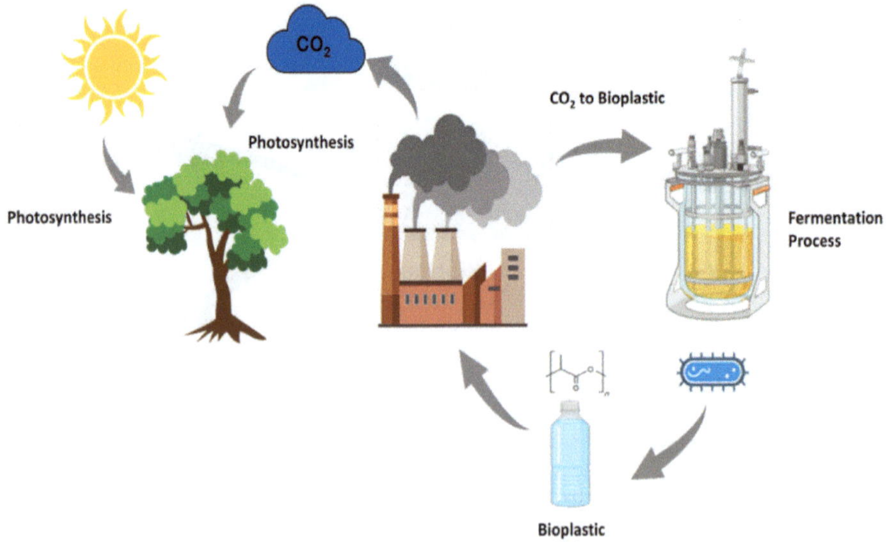

Fig. 5.6 Conversion of CO_2 to bioplastic via fermentation process

or chemoautotrophic metabolism and convert it into organic compounds [8]. In microbial fermentation, specialized bacteria such as *Cupriavidus necator* utilize CO_2 (often with hydrogen as an energy source) to produce polyhydroxyalkanoates (PHAs), a class of biodegradable plastics. Other types of microorganisms that has been reported able to convert CO_2 into bioplastic or PHA including *Bacillus cereus*, *Rastonia eustropa*, *Synechocystis* sp. and *Synecococcus* sp. [8–11]. Other studies have also engineered or optimized the microbes to produce biopolymers like polyhydroxyalkanoates (PHAs) or polylactic acid (PLA), which are biodegradable and can serve as alternatives to petroleum-based plastics [12]. Photosynthetic biofixation, on the other hand, employs algae or cyanobacteria to capture CO_2 directly through photosynthesis and convert it into biomass or precursors like sugars and organic acids, which are further processed into bioplastics such as polylactic acid (PLA). Table 5.4 summaries the production of bioplastic by different types of microbes using CO_2 as carbon sources. According to this summary, it shows that the production of PHA from microbes is influenced by several factors such as the cultivation method, CO_2 concentration and also the type of microbe used.

In parallel, chemical catalysis approaches are also being developed, where CO_2 is combined with epoxides or other substrates under specific conditions to produce polymers such as polycarbonates. These technologies not only provide a circular approach to plastic production by utilizing waste CO_2 but also reduce dependence on fossil fuels and help mitigate environmental pollution. Though still in the development and scale-up stages, CO_2-to-bioplastic systems are gaining attention as part of the broader carbon capture and utilization (CCU) strategy for a more sustainable and low-carbon economy.

Table 5.4 Bioplastic production from CO_2 via biological approach

Microbes	Substrate	PHA (mol%)	PHA (productivity g/L/d)	References
Cupriavidus necator B-10646	CO_2:O_2:H_2	85	0.45	[12]
*Ralstonia eutro*pa B5786	CO_2	51	NA	[13]
Ralstonia eutropa	CO_2	38	NA	[14]
Bacillus cereus	CO_2	55	NA	[15]
Wautersia eutropa H16	CO_2 + valeric acid	10	6.1	[16]
Synetococcus sp. PCC7942	CO_2 + acetic acid	35	NA	[17]

5 Integration with Industrial CCUS

Integrating CO_2 capture with Carbon Capture, Utilization, and Storage (CCUS) technologies is another approach introduced to reduce CO_2 in atmosphere. This technology offers a pathway to both reduce CO_2 emissions and create valuable products from the captured gas. This approach, often referred to as Integrated CO_2 Capture and Conversion (ICCC), can be more efficient and economical than traditional CCUS, which focuses solely on storage. This approach offer several benefits as it can improve energy efficiency by converting captured CO_2 directly into chemicals or fuels. This approach could reduce the energy intensive steps of compression and transportation for underground storage.

Capturing carbon dioxide (CO_2) from flue gases and utilizing it as a feedstock for the production of renewable chemicals and fuels has been identified as an essential step toward mitigating climate change. The conversion of CO_2 into other products such as methane, methanol, ethanol, and other chemicals will create new bio-product that results to enhance economic viability of this technology [18, 19].

A new concept presented here combines carbon capture with power-to-methane: Bio-integrated carbon capture and utilization (BICCU) [20, 21]. It exploits the robustness of microbial catalysts to simultaneously desorb the CO_2 from an amine-based absorbent and convert the captured CO_2 to synthetic natural gas (CH_4) in a single step using renewable hydrogen. Robust microbe is important to ensure the stability of CO_2 capture process. The presence of high CO_2 will alter pH condition resulting to lower microbe growth and CO_2 capture process. Thus, selection of the microbes that have capability to adapt to acidic condition is needed to ensure the CO_2 capture efficiency. Study by Kassim et al. [22]—compared the capability of two microalgae strain *Chlorella* sp. and *Tetraselmis suecica*—indicated that different strains exhibited different CO_2 capture capability. Another study also indicated that utilization of acidophilic microalgae also offer a great potential to be applied in this system as changes of pH environment not significantly affect the microalgae metabolism [23]. Implementation of ICCUS using microalgae also offer more advantages as microalgae biomass generated from the system can be further converted into value added product and support biocircular economy (Fig. 5.7). The microalgae biomass generated can be converted either low value product such as animal feed or biofertilizer and high value product such as biofuel.

Currently, the ICCUS concept has been implemented by several industries. Referring to Table 5.5, CO_2 produced by flue gas or other gas sources can be captured and made into an additional product. With this method, it is indirectly able to reduce CO_2 emissions into the atmosphere and add value by producing new products.

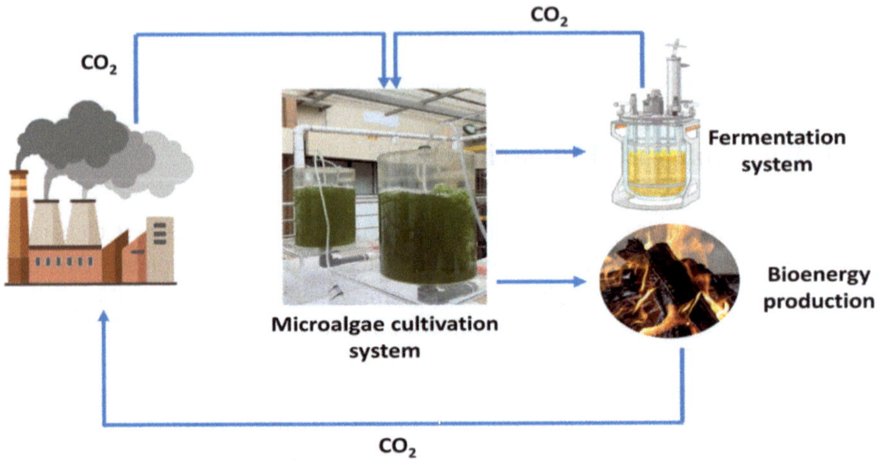

Fig. 5.7 Integrated microalgae-based carbon capture and utilization for value added product production

Table 5.5 Current established ICCU technology for various end products

Company/project	Location	CO_2 source	Utilization	End product
Carbon Clean + Tata Chemicals	India	Flue gas	Sodium bicarbonate	Food-grade $NaHCO_3$
LanzaTech	USA/China	Industrial off-gas (steel, refinery)	Fermentation by microbes	Ethanol, chemicals
CarbonCure	Canada/USA	Captured CO_2 from various emitters	Injected into concrete during mixing	Stronger, low-carbon concrete
SkyNRG + Climeworks + partners	Netherlands (in development)	Direct air capture	Synthetic jet fuel	Sustainable aviation fuel

6 Conclusion

Uncontrolled carbon dioxide emissions into the atmosphere have caused many negative impacts on the environment. Emphasis on methods or initiatives to reduce CO_2 content in the air is very important. Several control methods have been introduced to reduce CO_2 content in the air. Among them is using biomass as an agent to trap CO_2 and convert it into other value-added products. Although the ability of biomass to trap CO_2 has been identified, until now not many technologies have been adapted for large scale. Several emphases need to be taken into account such as net-carbon emissions and appropriate harvesting methods to balance the production and reduction of CO_2 in the air. In addition, the use of robust microbes that are capable of converting CO_2 into biobased products also needs to be considered for the feasibility process. Several genetic and road site modifications to improve the production of bioproducts from CO_2 are also needed.

References

1. Fu L, Ren Z, Si W, Ma Q, Huang W, Liao K, Huang Z, Wang Y, Li J, Xu P (2022) Research progress on CO_2 capture and utilization technology. J CO_2 Util 66:102260
2. Tardin-Coelho R, Bharadwaj BM, Ashworth P (2025) Carbon capture utilisation and storage (CCUS) and public perceptions: a systematic literature review. Int J Greenh Gas Control 145:1045393
3. Garifullina ChA., Klimov DS (2024) CCU technologies as a tool to achieve scope and ESG goals. In: Proceeding E3S web of conferences, vol 498, p 01015, 1–10
4. Marangwanda GT, Madyira DM, Babarinde TO (2020) Combustion models for biomass: a review. Energy Rep 6:664–672
5. Yaashika PR, Kumar PS, Varjani S, Saravanan A (2020) A critical review on the biochar production techniques, characterization, stability and applications for circular bioeconomy. Biotechnol Rep 28:e00570
6. Torchia DFO, Zonta E, Andrade AM, Garcia CG (2022) Production and characterization of biochar obtained from different biomass and pyrolysis temperature, vol 39, pp 415–427
7. Guo S, Li Y, Wang Y, Wang L, Sun Y, Liu L (2022) Recent advances in biochar-based adsorbents for CO_2 capture. In: Carbon capture science and technology, vol 4, p 100059
8. Almomani F, Abdelbar A, Ghanimeh S (2023) A review of the recent advancement of bioconversion of carbon dioxide to added value products: a state of the art, vol 15, no 1
9. Maheswari N, Kumar M, Thakur IS, Srivastava S (2018) Production, process optimization and molecular characterization of polyhydroxyalkanoate (PHA) by CO_2 sequestering B. cereus SS105, pp 75–82
10. Young-Cheol C, Reddy MV, Mawatari Y, Sarkar O (2025) Enhanced polyhydroxyalkanoate biosynthesis by Cupriavidus sp. CY-1 utilizing CO_2 under controlled non-explosive conditions, vol 373, p 144181
11. Barla RJ, Raghuvanshi S, Gupta S (2024) Reforming CO_2 bio-mitigation utilizing Bacillus cereus from hypersaline realms in pilot-scale bubble column bioreactor. Sci Rep 6354
12. Carpine R, Olivieri G, Hellingwerf KJ, Pollio A, Marzocchella A (2020) Industrial production of poly-β-hydroxybutyrate from CO_2: can Cyanobacteria meet this challenge?, vol 8, p 323
13. Ghysel S, Mozumder MSI, Wever HD, Volcke EIP, Garcia-Gonzalez L (2018) Targeted poly(3-hydroxybutyrate-co-3-hydroxyvalerate) bioplastic production from carbon dioxide. Bioresour Technol 249:858–868

14. Volvona TG, Vainu NA (2003) Kinetic parameters of a culture of the hydrogen-oxidizing bacterium Ralstonia eutropha grown under conditions favoring polyhydroxybutyrate biosynthesis. Appl Biochem Microbiol 39:166–170. https://doi.org/10.1023/A:1022585912873
15. Khosravi-Darani K, Mokhtari ZB, Amai T, Tanaka K (2013) Microbial production of poly(hydroxybutyrate) from C1 carbon sources. Appl Microbiol Biotechnol 97:1407–1424. https://doi.org/10.1007/s00253-012-4649-0
16. Maheshwari N, Kumar M, Thakur IS, Srivastava S (2018) Production, process optimization and molecular characterization of polyhydroxyalkanoate (PHA) by CO_2 sequestering B. cereus SS105. Bioresour Technol 254:75–82. https://doi.org/10.1016/j.biortech.2018.01.002
17. Volova TG, Kalacheva GS, Steinbüchel A (2008) Biosynthesis of multicomponent polyhydroxyalkanoates by the bacterium Wautersia eutropha. Macromol Symp 269:1–7. https://doi.org/10.1002/masy.200850901
18. Stegman S, Melkonian C, Tamarit D, Huang-Lin E, Lebrero R, Cantera S (2025) Prospective bioconversion of CO_2 and CO into fine chemicals via halophilic purple phototrophic bacteria, vol 24, pp 29–41
19. Arunachalam R, Chinnaraja E, Subramaniam PS, Efficient homogeneous catalysts for conversion of CO_2 to fine chemicals. In: Catalysis for cleaning energy and environmental sustainability, pp 599–641
20. Sieborg MU, Nielsen AKH, Ottosen LDM, Daasbjerg K, Kofoed MVW (2024) Bio-integrated carbon capture and utilization: at the interface between capture chemistry and archaeal CO_2 reduction, vol 15, pp 1–11
21. Arunachalam R, Chinnaraja E, Subramanian PS, Efficient homogeneous catalysts for conversion of CO_2 to fine chemicals. In: Catalysis for clean energy and environmental sustainability, pp 599–641
22. Kassim MA, Meng TK et al (2017) Carbon dioxide (CO_2) biofixation by microalgae and its potential for biorefinery and biofuel production. Sci Total Environ 584:1121–1129
23. Meng TK, Kassim MA (2020) Growth, carbohydrate productivity, and growth kinetic study of Halochlorella rubescens cultivated under CO_2-rich conditions. Malays Appl Biol 49:1–11

Addressing Pollution Through Green Processes for Environmental Mitigation and Remediation

Norli Ismail and Amir Talebi

Abstract The increasing prevalence of synthetic polymers, hazardous dyes, and toxic solvents in industrial processes necessitates urgent transitions toward greener, safer, and more circular environmental management technologies. This chapter synthesizes two major green process innovations: the reinforcement of sago starch biodegradable films using microbial bio-nanofillers, and the extraction of methyl violet dye using waste vegetable oil-based organic solvents. Reinforcement using Bacillus spp. enhances tensile strength, structural compactness, thermal stability, and moisture resistance of sago starch films, demonstrating their potential as sustainable alternatives to petroleum-derived polymers. Simultaneously, the valorization of waste cooking oil as a liquid–liquid extraction medium achieves high removal efficiency of methyl violet while replacing toxic commercial solvents. This chapter consolidates experimental findings, mechanistic insights to contextualize the environmental, industrial, and circular economy relevance of both technologies.

Keywords Microbial composite films · Bacterial nanofillers · Green solvent extraction · Biodegradable polymers · Dye wastewater remediation

1 Introduction

Industrialization and global economic expansion have intensified environmental pressures, particularly due to the extensive use of synthetic plastics, dyes, and hazardous solvents. These materials persist in the environment, disrupt ecological balance, and impose long-term risks on human and environmental health. Current pollution control technologies often rely on chemical-intensive or energy-intensive processes that paradoxically generate additional waste streams. As a result, contemporary environmental engineering is undergoing a paradigm shift toward sustainable

N. Ismail (✉) · A. Talebi
Environmental Technology Division, School of Industrial Technology, Universiti Sains Malaysia, Penang, Malaysia
e-mail: norlii@usm.my

M. A. Kassim et al. (eds.), *Sustainable Biorefinery*, Green Energy and Technology,
https://doi.org/10.1007/978-981-95-8181-8_6

solutions grounded in green chemistry, biological reinforcement, and waste valorization. The field has seen significant growth in the application of renewable polymers, biogenic nanomaterials, and non-toxic solvents for pollution mitigation [1, 2].

Green process engineering integrates sustainability principles with material innovation, emphasizing renewable feedstocks, low-carbon processes, and circular resource flows. Within this framework, two research domains have attracted considerable scientific attention. First, the biodegradable polymer films reinforced with biological fillers addresses the urgent global need to replace petrochemical plastics. Second, the transition to green solvents derived from waste biomass supports safer industrial separation processes and reduces chemical toxicity. Recent studies have demonstrated the feasibility of microbial systems and waste oils as functional components in these innovations [3–5].

In this chapter, the focus is placed on two complementary green process strategies: reinforcement of sago starch films using *Bacillus* spp., and the removal of methyl violet dye using waste vegetable oil–based liquid–liquid extraction (LLE). Both represent practical, scalable pathways toward environmentally responsible technology, aligning with global sustainability priorities and circular economy objectives.

Environmental pollution resulting from the extensive use of plastics, synthetic dyes, and hazardous industrial chemicals continues to pose a major global threat, with far-reaching ecological and public health implications. Conventional plastics remain persistent in terrestrial and aquatic ecosystems, contributing to microplastic formation and long-term environmental loading due to their resistance to biodegradation [5]. Similarly, synthetic dyes such as methyl violet (MV) represent a persistent class of pollutants; they are known to be toxic, mutagenic, and capable of bioaccumulating in aquatic organisms, complicating wastewater treatment and posing substantial risks to human and environmental health [6, 7]. These challenges are intensified in industrial sectors where toxic organic solvents and petrochemical-based materials are frequently used, generating complex waste streams that conventional technologies struggle to treat effectively.

In response, green process engineering has gained significant attention in the last six years as industries and researchers prioritize solutions that minimize environmental impact while maintaining functional performance. The principles of green chemistry emphasize pollution prevention, safer chemical synthesis, renewable resource utilization, and waste minimization—an approach that has guided the development of greener materials and processes within industrial and environmental engineering [8]. Recent advancements have underscored the potential of biodegradable polymers, bio-nanocomposites, and waste-derived functional materials to address contemporary pollution challenges while supporting circular economy objectives [9, 10]. These innovations not only reduce dependence on fossil-derived materials but also promote resource recovery from industrial and agricultural waste streams.

Within this broader sustainability context, the research presented in this chapter explores two promising green technologies developed in recent years: (i) microbial reinforcement of sago starch films, and (ii) the extraction of methyl violet dye using waste vegetable oil–based solvents. The microbial reinforcement approach enhances the mechanical, barrier, and thermal properties of biodegradable starch films through the incorporation of Bacillus spp. as bio-nanofillers. Recent studies have demonstrated that bacterial cells and their nanoscale structural components improve polymer packing density, reduce microvoids, and enhance film strength, positioning microbial fillers as viable, environmentally benign alternatives to synthetic nanomaterials. Bacterial cellulose nanofibers (BCNF) and bacterial cellulose nanocrystals (BCNCs) have been shown to enhance the tensile strength, Young's modulus, and overall mechanical properties of polymer films. For instance, BCNF-containing films exhibited a 49% increase in tensile strength compared to control samples [11]. Similarly, BCNCs improved the tensile strength and Young's modulus of films based on gelation, pectin, and hydroxypropylmethyl cellulose (HPMC) [12]. The incorporation of bacterial cellulose into polymer matrices, such as poly(L-lactic acid) (PLLA) and poly(3-hydroxybutyrate-co-3-hydroxyvalerate) (PHBV), has been shown to improve film strength and reduce microvoids. The use of waste vegetable oil (WVO) as a diluent in liquid–liquid extraction offers an eco-friendly solution for dye removal, replacing toxic petrochemical solvents with renewable, low-toxicity waste oils that possess strong hydrophobic extraction capabilities. These approaches leverage readily available and underutilized waste resources, thereby aligning with global goals for waste valorization and circularity. Waste Cooking Oil (WCO) as diluent is considered a low-cost substance, non-toxic organic solvent in a green emulsion liquid membrane (GELM) process for the removal of Methyl violet 2B dye. The process achieved a maximum extraction efficiency of 99.1% under optimized conditions [13]. Another study explored the use of WCO in the emulsion liquid membrane technique to extract Rhodamine B from wastewater, achieving a dye removal efficiency of 95% [14].

Collectively, these advancements reflect the growing academic and industrial emphasis on sustainable materials and processes. They align with global agendas such as the United Nations Sustainable Development Goals (SDGs), particularly those addressing clean water and sanitation (SDG 6) and responsible consumption and production (SDG 12). The integration of biodegradable polymer technologies with waste-derived green extraction systems highlights a multidisciplinary direction in pollution mitigation one that combines material innovation, biotechnology, and circular resource management. The subsequent sections examine in detail the scientific basis, methodologies, findings, and wider implications of these emerging environmentally friendly technologies.

2 Reinforcement Mechanism of Sago Starch Films by Bacteria

2.1 *Bacterial Cellulose as Organic Filler*

Biodegradable polymers have gained prominence due to intensifying concerns regarding the environmental persistence of conventional plastics. Sago starch is particularly promising as a biodegradable matrix because of its wide availability in Southeast Asia, high amylopectin content, and film-forming characteristics. However, its practical use remains constrained by poor mechanical performance and hydrophilicity. Since 2020, research has shifted toward bio-nanocomposite reinforcement using biological fillers that enhance structural, mechanical, and barrier properties while maintaining sustainability. Biodegradable films made from seaweed polysaccharides have garnered significant attention due to their potential to replace conventional plastic packaging, which poses environmental hazards. Seaweed polysaccharides such as carrageenan, alginate, and agar are particularly promising due to their renewability, biodegradability, and biocompatibility [15]. Understanding the effects of processing routes on the antibacterial activity and biocompatibility of biodegradable polymer composites remains a challenge [16].

Microbial fillers offer distinct advantages over inorganic nanomaterials such as carbon nanotubes or metal oxides. Bacteria, particularly *Bacillus subtilis*, possess nanoscale structural components including flagella and robust peptidoglycan-rich cell walls that mimic the fiber-like reinforcement behavior seen in synthetic nanofillers [4]. Their integration into polymer matrices represents a novel approach in the design of biogenic nanocomposite films.

Recent literature consistently demonstrates that bacterial cellulose (BC) is one of the most effective reinforcing agents for starch-based biopolymer films, including those derived from sago starch. The incorporation of bacterial cellulose (BC) nanofibers into a starch-chitosan matrix has been shown to significantly enhance the mechanical properties of the resulting biocomposite films. This improvement is primarily attributed to the formation of a percolating nanofibrillar network and enhanced hydrogen bonding at the filler-matrix interface. The addition of BC nanofibers to starch-chitosan matrices results in notable increases in tensile strength. For instance, incorporating 20% BC nanofibers into chitosan films increased the tensile strength by 39.60 MPa [17]. Similarly, the addition of BC nanofibers to tapioca starch/chitosan films resulted in a maximum tensile strength of 4.7 MPa [18].

Incorporating BC nanofibers into a starch-chitosan matrix produces notable improvements in tensile strength due to the formation of a percolating nanofibrillar network and enhanced hydrogen bonding at the filler-matrix interface. The mechanical enhancements are largely due to the formation of a percolating nanofibrillar network within the matrix. This network facilitates better load transfer and stress distribution across the composite, leading to improved mechanical properties [18, 19]. Another study highlighted nanocellulose-induced hydrogen bonding, reduced polymer chain mobility, and tortuous-path mechanisms as dominant factors contributing to mechanical and barrier improvements in starch composites [20]. The interaction between BC nanofibers and the starch-chitosan matrix is strengthened by hydrogen bonding. This enhanced bonding at the filler-matrix interface contributes to the increased tensile strength and overall mechanical performance of the films [17, 21]. The synergistic effects of combining BC nanofibers with chitosan and starch are evident in the significant improvements in mechanical properties. For example, Young's modulus and tensile strength of films with 20% BC nanofibers showed increments of up to 15 and 30 MPa, respectively [19]. Besides mechanical improvements, the incorporation of BC nanofibers also enhances other properties such as thermal stability, moisture resistance, and water vapor barrier properties. For instance, the thermal resistance of tapioca starch/chitosan films increased with the addition of BC nanofibers [18].

A growing number of studies have also explored bacterial cellulose (BC) derived directly from sago-processing waste streams. Sago liquid waste can serve as an efficient substrate for BC production, and the resulting BC significantly enhances tensile and moisture-related characteristics when incorporated into starch-based films. Sago liquid waste has been identified as an effective substrate for BC. Studies have shown that using sago liquid waste as a nutrient source can yield significant amounts of BC. For instance, one study reported a BC yield of 12.37 g/L, which increased to 14.52 g/L with a 10% sugar concentration in the sago liquid waste medium after a 14-day fermentation period. The mechanical properties of BC produced from sago liquid waste include a tensile strength of 44.2–87.3 MPa, elongation at break of 4.8–5.8%, and Young's modulus of 0.86–1.64 GPa, indicating its robustness and potential for various applications [22]. Bacterial cellulose improves film strength up to an optimal loading, beyond which fibril aggregation can diminish performance. Their SEM micrographs further confirm efficient stress transfer from matrix to fibrils, denote the importance of dispersion quality.

The incorporation of BC into starch-based films significantly enhances their mechanical and moisture-related properties. For instance, the addition of BC to tapioca starch/chitosan-based films increased the tensile strength to 4.7 MPa and improved moisture resistance by 8.9% and water vapor barrier properties by 27% [18]. Another study demonstrated that the inclusion of cellulose nanofibers in thermoplastic starch (TPS) films improved tensile strength and water vapor barrier properties, making the films more suitable for food packaging applications [23]. Sago liquid waste is a viable and efficient substrate for producing bacterial cellulose, which, when incorporated into starch-based films, significantly enhances their mechanical and moisture-related characteristics. This makes BC-starch composite films promising candidates for environmentally friendly and effective food packaging solutions. The consistent findings across multiple studies emphasize the potential of this approach in industrial applications.

Complementary studies using alternative polymeric additives support the importance of matrix compatibility when designing sago starch composites. Reported study on sago starch gelatin films found that hydrogen bonding and plasticizer content strongly influence flexibility and mechanical strength, suggesting parallels to BC starch interactions in terms of chain mobility and interfacial adhesion [24, 25]. Research on sago starch films reinforced with microfibrillated cellulose from plant sources provides additional mechanistic support. [26] further enhanced crystallinity and tensile performance in such composites, demonstrating similar behavior to BC reinforcement and validating the general role of cellulose nanofibrils in stiffness enhancement and load transfer. Broader reviews of starch-based edible films, further emphasize nanocellulose as a promising reinforcement due to its ability to strengthen hydrogen bonding networks and create tortuous diffusion pathways for improved barrier properties [27]. Overall, these studies collectively establish that BC whether produced from purified cultures or generated using sago by-products acts as a highly effective reinforcement phase for sago starch films, improving mechanical performance, interfacial adhesion, and barrier properties through interconnected nanofibrillar networks and strong hydrogen-bonding interactions.

2.2 Bacterial Isolation and Characterization Composite Film Preparation

In this research, *Bacillus* species were isolated from palm kernel cake (PKC) and palm oil mill effluent (POME), both of which contain nutrient-rich substrates supporting diverse microbial populations. Recent studies have confirmed that POME harbors thermophilic and mesophilic *Bacillus* spp. with robust cell morphologies suitable for reinforcement applications [28, 29]. Microscopically, the isolates demonstrated diameters between 0.7 and 2.3 μm, consistent with previously reported environmental *Bacillus* dimensions [3].

Characterization included morphological screening, freeze-drying for structural preservation, and quantification of cell dimensions. The presence of flagella was of particular interest because flagellar filaments act as natural nanofibers capable of enhancing interfacial adhesion, trapping microvoids, and improving load transfer within polymer matrices [4].

The isolation and characterization of bacteria for use as reinforcement agents in biodegradable composite films have become increasingly prominent in green material science. Recent studies highlight that bacteria particularly Bacillus spp. possess structural features such as rigid peptidoglycan cell walls, proteinaceous surface layers, and nanoscale flagella that can serve as natural bio-nanofillers within polymer matrices. Researchers have focused on isolating such bacteria from agro-industrial waste streams including palm kernel cake, soil, and palm oil mill effluent (POME), which contain diverse microbial communities capable of producing robust cell morphologies suitable for material applications. Ho et al. [3] demonstrated POME chamber thermophilic and mesophilic Bacillus species with unique cell-wall architectures and high mechanical resilience, further supporting their potential use in composite films. Bacteria isolated from POME have been used as fillers to reinforce calcium chloride cured sodium alginate films. The inclusion of bacteria improved the morphological, physical, and mechanical properties of the composite films. Specifically, the addition of 0.3 g and 0.4 g of bacteria resulted in less curling on the film surface and increased tensile strength by approximately 22.70% (from 36.10 ± 1.94 MPa to 44.29 ± 0.60 MPa). However, the bacteria fillers did not significantly enhance the elongation properties of the films, with only a 0.6% increase observed.

The microbial community in POME includes various bacteria such as Bacillus, Pseudomonas, and other species that can degrade organic matter and potentially contribute to the mechanical properties of composite films [30, 31]. Mixed cultures of bacteria have shown more extensive degradation of organic load (COD and BOD) than pure single cultures, suggesting that a diverse microbial community might enhance the mechanical resilience of composite films by improving the uniformity and integrity of the biofilm [31]. Microbial sources from POME play a significant role in enhancing the mechanical resilience of composite films by improving tensile strength and promoting biofilm formation, while also contributing to sustainable waste management practices.

Key to successful bacterial reinforcement is the ability to preserve nanoscale structures during isolation and characterization. Freeze-drying is widely adopted to maintain the integrity of bacterial cells and flagella, preventing morphological collapse during incorporation into polymer matrices. Freeze-dried microbial fillers exhibit improved stability and dispersion when integrated into biopolymer films compared to wet biomass [32]. Creating a synthetic polymer network within bacterial cells through intracellular gelation can prevent cell division and maintain the structural integrity of the cells during incorporation into polymer matrices [33]. A UV-crosslinking and plasma treatment can enhance bacterial adhesion and prevent morphological collapse by ensuring strong interactions between bacterial cells and the polymer matrix [34, 35]. By employing these strategies, researchers can effectively reinforce bacterial cells within polymer matrices while maintaining their structural integrity and preventing morphological collapse. Figure 1 illustrates various methods in maintaining the integrity of bacterial cells and flagella and mechanisms to prevent the morphological collapse of both cells.

Recent research emphasizes the significance of microscopic and spectroscopic characterization in understanding bacterial contributions to composite material performance. SEM and TEM imaging allow visualization of bacterial size, morphology, and surface texture, which directly influence reinforcement efficiency. Rod-shaped bacteria, such as those used in the study of Bifidobacterium infantis, can aggregate and form continuous networks within polymer matrices. This aggregation helps in occupying microvoids and creating stress-transfer pathways, which enhances the mechanical properties of the composite material [36]. FTIR is used to analyze the chemical interactions between the bacterial fillers and the polymer matrix. It helps in identifying the presence of functional groups and the nature of chemical bonding. For example, FTIR analysis revealed good interaction among PMMA, fillers, and GO nanosheets, indicating compatibility and effective dispersion within the polymer matrix [37]. The hydrophobic nature and chemical bonding of natural fibers in composites were also assessed using FTIR, which is crucial for

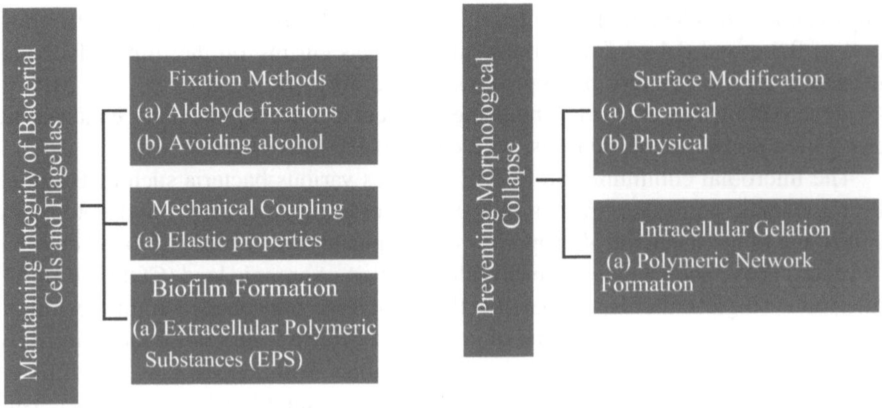

Fig. 1 Methods in maintaining the integrity of bacterial cells and flagella and mechanisms to prevent the morphological collapse of both cells

understanding the compatibility and interaction between the fillers and the polymer matrix [38].

The importance of targeted bacterial isolation, robust preservation techniques, and precise characterization in optimizing microbial reinforcement of biodegradable films. These studies affirm that bacteria-derived fillers offer a sustainable alternative to synthetic nanomaterials, contributing not only to improved film performance but also to broader circular economy strategies through the valorization of biological waste streams.

2.3 *Structural and Morphological Changes, Mechanical and Thermal Performance*

SEM micrographs revealed marked differences between bacterial-reinforced and control films. Control films displayed prominent voids, cleavages, and microcracks, reflecting the intrinsic brittleness and weak polymer chain entanglement characteristic of unmodified starch films. Conversely, the microbial composite films demonstrated smoother surfaces, reduced porosity, and tightly packed cross-sectional structures. The embedded bacteria appeared to fill microvoids and interact intimately with starch matrices, supporting the hypothesis that bacterial fillers act both as physical reinforcements and as nucleation sites for more ordered polymer alignment.

The significant potential of nanofillers and plasticizers in advancing the field of starch-based bioplastics in line with the current challenges and leveraging new technologies, these materials can play a crucial role in promoting sustainable and eco-friendly alternatives to conventional plastics [39]. These findings align with recent work demonstrating that bio-based fillers including bacterial cellulose, yeast biomass, and microbial nanofibers improve the homogeneity and mechanical function of biodegradable films.

Mechanical testing confirmed significant improvements in tensile strength and elongation properties of the bacterial-reinforced films. The enhanced strength arises from stronger filler–matrix interactions, decreased defect density, and improved load transfer mechanisms. Moisture absorption tests also revealed lower water uptake due to reduced porosity and partial hydrophobic shielding conferred by bacterial structures. These results correspond closely to findings reported in Ho et al. [3] where bacterial reinforcement improved the mechanical and barrier properties of alginate films.

Thermal stability and polymer degradation with bio-nanofillers study consistently shown that the incorporation of bio-nanofillers derived from microorganisms significantly enhances the thermal stability and delays the onset of degradation in biodegradable polymer matrices. This enhancement is primarily attributed to the presence of bacterial cell-wall proteins, peptidoglycan networks, and flagellar components, which restrict polymer chain movement, thereby contributing to heat resistance. Thermal analysis indicated that bacterial inclusion increased thermal stability and delayed the onset of polymer degradation. The presence of bacterial cell-wall proteins, peptidoglycan networks, and flagellar components restricts polymer chain movement, contributing to heat resistance [40].

The integration of bio-nanofillers into biodegradable polymer matrices significantly enhances thermal stability and delays degradation. This is achieved through mechanisms such as the formation of dense networks that restrict polymer chain movement. However, the impact on biodegradability varies depending on the type and content of the nanofiller used. These findings alter the potential of bio-nanofillers in developing high-performance, biodegradable materials for various applications.

2.4 Reinforcement Mechanism

The reinforcement mechanism is structured. First, bacterial rods and flagella act as natural nanofibers that integrate into the polymer matrix to form continuous stress-transfer pathways. Second, hydrogen bonding between starch hydroxyl groups and bacterial cell-wall functional groups increases interfacial adhesion. Third, the bacteria serve as microstructural fillers that occupy voids, reducing matrix defects and increasing packing density. These synergistic effects collectively enhance mechanical strength and functional performance, confirming the viability of microbial nanofillers as environmentally benign reinforcement agents.

Microorganisms can interact with polymer matrices both directly and indirectly. Direct interactions include the colonization of polymer surfaces by microorganisms, which can lead to physical changes in the polymer structure. Indirect interactions involve the transformation of the surrounding media by microorganisms, which can affect the polymer's properties [41]. The formation of biofilms on polymer surfaces can significantly influence degradation rates. For instance, biofilms can create a physical barrier that slows down the degradation process by protecting the polymer from direct microbial attack [42, 43]. Reinforcement mechanisms in biofilm composites are primarily driven by the composition and interactions within the EPS matrix, the influence of environmental factors, and the potential for engineering biofilm composites with enhanced mechanical properties. Understanding these mechanisms can aid in developing strategies for controlling biofilm formation and leveraging their properties for various applications.

Structural and Functional Adaptations of the biofilm composite are influence by a few factors. To begin with a cell wall biosynthesis which the biosynthesis of cell wall components such as peptidoglycan, wall teichoic acids, and teichuronic acids is crucial for biofilm stability and resistance. Inhibiting these pathways can disrupt biofilm formation [44]. Next the antioxidant mechanisms ensure the biofilm cells activate reactive oxygen species (ROS) detoxification pathways to protect against oxidative stress, which is essential for biofilm maintenance [45]. External factors include environmental and nutritional reveals the biofilm formation is influenced by the availability of nutrients, which can trigger metabolic adaptations to optimize resource utilization while the initial attachment of cells to surfaces is mediated by pili and other adhesion molecules, which are critical for the establishment of biofilms [46, 47].

The reinforcement mechanism is governed by three primary interactions are illustrated in Fig. 2 as reported in [4].

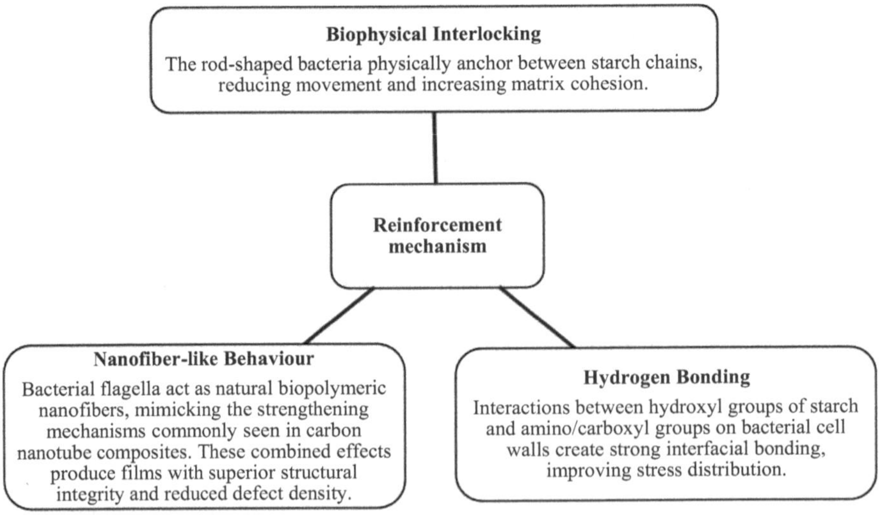

Fig. 2 The main interactions of the reinforcement mechanism

3 Replacing Volatile Organic Solvents with Bio-derived and Waste-Derived Alternatives

Synthetic dyes such as methyl violet remain persistent in wastewater streams due to their aromatic structures and resistance to oxidation and biodegradation. The shift toward greener, bio-based solvents is driven by the need to reduce the environmental and health impacts associated with traditional volatile organic solvents (VOCs). Waste vegetable oils have emerged as promising candidates due to their low toxicity, widespread availability, and hydrophobic properties, making them suitable for various extraction applications. The global movement toward green solvent systems have accelerated with increased emphasis on replacing volatile organic solvents with bio-derived and waste-derived alternatives. Waste vegetable oils are non-toxic and environmentally friendly, making them ideal for replacing petroleum-based solvents [48]. Waste vegetable oils have emerged as promising candidates due to their low toxicity, widespread availability, and hydrophobic properties suitable for extraction applications. The hydrophobic nature of vegetable oils makes them effective in extracting non-polar compounds, which is beneficial for applications such as pollution remediation and biodiesel production [49].

Vegetable oils have been successfully used to extract bioactive compounds from natural resources. For instance, soybean oil has shown high efficiency in extracting phenolic compounds from rosemary leaves [50]. Additionally, the use of vegetable oils in ultrasound-assisted extraction has proven effective in obtaining phenolic content from Crocus sativus L. waste [51]. Waste cooking oils can be converted into biodiesel using mixed ionic liquid systems, which enhance the extraction and conversion of free fatty acids to methyl esters [49]. This process not only utilizes waste oils but also improves the efficiency of biodiesel production. Waste vegetable oils have been used to treat contaminated soils by mobilizing polycyclic aromatic hydrocarbons (PAHs). Although slightly less efficient than unheated oils, waste oils still significantly reduce the non-bioaccessible PAH fraction [52]. Waste vegetable oils present a viable and sustainable alternative to traditional VOCs, offering environmental benefits and effective extraction capabilities. Their application spans various fields, including biodiesel production, soil remediation, and the extraction of bioactive compounds. However, optimizing extraction processes and addressing challenges such as viscosity and deacidification are crucial for maximizing their potential.

3.1 Extraction Performance

Under the optimized conditions of 125 mM D2EHPA, 50 mM TBP, and pH 4.0, the system achieves up to 95% extraction of methyl violet. This high extraction efficiency can be attributed to several factors. The concentration of the carrier (D2EHPA) plays a significant role. Higher concentrations of D2EHPA in the organic phase have been shown to increase extraction efficiency [53, 54]. For example, a study demonstrated that increasing D2EHPA concentration in the organic phase led to higher extraction rates of vanadium [53]. Similar systems and parameters that influence extraction efficiency suggest that optimizing pH, carrier concentration, and the use of mixed extractant systems can significantly enhance extraction efficiency, potentially achieving results comparable to or surpassing conventional solvent-based systems. The pH of the feed phase is crucial. For instance, a pH of 4 ± 0.2 was found optimal for the extraction of methyl violet using D2EHPA as a carrier [55]. This efficiency marks several conventional solvent-based systems reported in recent dye removal studies.

The mechanism of extraction is rooted in ion-pair formation between methyl violet and D2EHPA, followed by hydrophobic partitioning into the waste oil phase. The presence of TBP ensures the stability of the organic phase by reducing interfacial rigidity of such phase modifiers in modern green extraction systems. Waste frying palm oil methyl ester (WPOME) has been successfully used to produce biodiesel with a high yield of 95.7%. The biodiesel properties such as kinematic viscosity, density, and cetane number are higher than those of conventional diesel, although the calorific value is lower [56]. Waste-oil-based extraction not only reduces solvent toxicity but also supports circular economy principles by converting a common household and industrial waste into a functional extraction solvent.

WCO is effectively used in the ELM technique to remove dyes such as Rhodamine B and Methyl Violet 2B from wastewater. This method involves creating a stable emulsion with WCO, which acts as the organic phase, and optimizing parameters like surfactant concentration and phase volume ratios to enhance dye extraction efficiency [13, 14, 57].

Under optimized conditions, the waste-oil solvent system is able to achieve extraction efficiencies exceeding 95%. Waste palm cooking oil (WPCO) also demonstrated strong solvency behavior, achieving efficiencies around 90%, consistent with recent work on waste vegetable valorization (WVO). The hydrophobic interactions of WVO support the incorporation of cationic dye-extractant complexes. The use of synergistic extractants formulated in cooking oil can enhance the extraction efficiency of dyes. For example, combining Aliquot 336 and D2EHPA in cooking palm oil significantly improves the extraction of Orange 3R dye [58]. Waste cooking oil can replace petroleum-based solvents in LLE processes. It has been shown to achieve high extraction and back-extraction efficiencies for dyes like methylene blue (MB) under optimized conditions for instance equilibrium pH of 8, aqueous to organic phase ratio of 1:1 [59]. Optimal pH levels vary depending on the type of oil and dye, with alkaline conditions generally favoring higher removal efficiencies [59, 60] (Table 1).

Table 1 Dye wastewater extraction using waste vegetable oil and palm-oil-based solvents

Parameter	Waste vegetable oil	Palm-oil-based solvents
Dyes studied	Methyl Violet 2B (MV) [14], Rhodamine B [13]	Reactive Red 3BS [61], Orange 3R [58], Methyl Orange [62]
Solvent type	Waste cooking oil [13, 14]	Palm oil [58, 61, 62]
Surfactants used	Span 80 [13, 14]	Span 80 [61], Saponin [62]
Stripping agents	Hydrochloric acid [14], NaOH [13]	$NaHCO_3$ [61], NaOH [58]
Extraction efficiency	MV: 99.1% [14], Rhodamine B: 95% [13]	- Reactive Red 3BS: 90% [61] - Orange 3R: 90% [58] - Methyl Orange: Optimized [62]

3.2 Integrating Green Process Approaches for Sustainable Pollution Control

The extraction mechanism involves the formation of an ion-pair complex between methyl violet and D2EHPA, which partitions into the hydrophobic waste oil matrix. TBP plays a critical role in preventing emulsification, a common challenge in waste-oil-based extraction systems validate the critical function of phase modifiers in stabilizing green solvent extraction systems.

The use of waste vegetable oil aligns strongly with the principles of circular economy, as it diverts a major waste stream from landfills and reintroduces it into industrial processes. Furthermore, replacing toxic organic solvents with biodegradable waste oils decreases the environmental footprint of the extraction process. These benefits position waste-oil-based liquid–liquid extraction (LLE) systems as viable candidates for large-scale industrial dye removal. The reinforcement of starch-based films and the use of waste-oil-based solvent extraction represent two distinct yet synergistic approaches to sustainable pollution mitigation. Both technologies capitalize on renewable and waste-derived materials, reduce reliance on harmful chemicals, and enhance the functionality of existing environmental treatment systems. Their integration into industrial practice aligns with global agendas promoting cleaner production, resource efficiency, and environmentally responsible materials.

The dual contributions of microbial reinforcement of biodegradable films and waste-oil-based extraction highlight the versatility and potential of green process engineering in addressing varied environmental challenges. Both technologies reflect recent scientific emphasis on reducing lifecycle emissions and creating closed-loop material systems. Reinforced sago starch films offer alternatives to single-use plastics by enhancing durability without compromising degradability, while green extraction solvents derived from waste streams mitigate hazardous waste generation in liquid separation processes.

4 Conclusion

This chapter highlights two significant advancements in green process engineering: microbial reinforcement of biodegradable starch films and waste-oil-based extraction of synthetic dyes. Both technologies demonstrate how waste biomass and biological systems can be harnessed to replace environmentally harmful materials. The research findings, supported by the potential of these approaches to contribute meaningfully to sustainable pollution control, circular resource loops, and safer industrial processes.

This expanded chapter presents two interlinked green technologies that is microbial composite starch films and waste-oil-based solvent extraction of dyes demonstrating their potential to achieve high-performance environmental solutions while maintaining sustainability. Reinforcement of sago starch films with Bacillus spp. improves mechanical, thermal, and barrier properties, while waste vegetable oil-based extraction replaces hazardous solvents without compromising efficiency. Together, they showcase the practical and scientific viability of integrating bio-based materials and waste valorisation to address pollution challenges. Continued research should focus on scale-up evaluation, long-term biodegradation analysis, and techno-economic feasibility to accelerate adoption at the industrial level.

References

1. Prasad J, Kumar N, Pratibha, Jaiswal R, Yadav A, Sharma SP, Fawole OA, Kaushik N (2025) Biopolymer based composite packaging: a sustainable approach for fruits and vegetables preservation. Appl Food Res 5(2):101211. https://doi.org/10.1016/j.afres.2025.101211
2. Peramune D, Manatunga DC, Dassanayake RS, Premalal V, Liyanage RN, Gunathilake C, Abidi N (2022) Recent advances in biopolymer-based advanced oxidation processes for dye removal applications: a review. Environ Res 215:114242. https://doi.org/10.1016/j.envres.2022.114242
3. Ho BKX, Azahari B, Yhaya MF, Talebi A, Ng CWC, Tajarudin HA, Ismail N (2020) Green technology approach for reinforcement of calcium chloride cured sodium alginate films by isolated bacteria from palm oil mill effluent (POME)'. Sustainability 12(22):9468. https://doi.org/10.3390/su12229468
4. Chun CNW, Tajarudin HA, Ismail N, Azahari B, Makhtar MMZ, Yan LK (2020) Bacterial flagellum versus carbon nanotube: a review article on the potential of bacterial flagellum as a sustainable and green substance for the synthesis of nanotubes'. Sustainability 13(1):21. https://doi.org/10.3390/su13010021
5. Andrady L (2011) Microplastics in the marine environment. Mar Pollut Bull 62(8):1596–1605. https://doi.org/10.1016/j.marpolbul.2011.05.030
6. Yaseen M, Scholz U (2019) Textile dye wastewater characteristics and constituents of synthetic effluents: a critical review. Int J Environ Sci Technol 16:1193–1226. https://doi.org/10.1007/s13762-018-2130-z
7. Khattab TA, Abdelrahman MS, Rehan M (2019) Textile dyeing industry: environmental impacts and remediation. Environ Sci Pollut Res 27(4):3803–3818. https://doi.org/10.1007/s11356-019-07137-z
8. Anastas P, Eghbali N (2010) Green chemistry: principles and practice. Chem Soc Rev 39(1):301–312. https://doi.org/10.1039/b918763b

9. Geissdoerfer M, Savaget P, Bocken NMP, Hultink EJ (2017) The circular economy—a new sustainability paradigm? J Clean Prod 143:757–768. https://doi.org/10.1016/j.jclepro.2016.12.048
10. Ghisellini P, Cialani C, Ulgiati S (2016) A review on circular economy: the expected transition to a balanced interplay of environmental and economic systems. J Clean Prod 114:11–32. https://doi.org/10.1016/j.jclepro.2015.09.007
11. Shabanpour B, Kazemi M, Ojagh SM, Pourashouri P (2018) Bacterial cellulose nanofibers as reinforce in edible fish myofibrillar protein nanocomposite films. Int J Biol Macromol 117:742–751. https://doi.org/10.1016/j.ijbiomac.2018.05.038
12. da Costa FM, Melo PTS, Nishimoto PHK, Lorevice MV, Aouada FA, de Moura MR (2025) Percolation threshold of bacterial nanocrystals in biopolymeric matrices to build up strengthened biobased food packaging. Foods 14(7):1123. https://doi.org/10.3390/foods14071123
13. Shokri, PD, Zereshki S (2020) Water decolorization using waste cooking oil: An optimized green emulsion liquid membrane by RSM. J Water Process Eng 33:101021. https://doi.org/10.1016/j.jwpe.2019.101021
14. Purnama H, Dewi DE, Jannah NM, Kamaratih AA, Hidayati N (2025) Liquid waste treatment containing Rhodamine B using emulsion liquid membrane from waste cooking oil. Res Eng 25:104066. https://doi.org/10.1016/j.rineng.2025.104066
15. Khalil HP, Tye YY, Saurabh CK, Leh CP, Lai TK, Chong EW, Fazita MN, Hafiidz JM, Banerjee A, Syakir MI (2017) Biodegradable polymer films from seaweed polysaccharides: a review on cellulose as a reinforcement material. Express Polym Lett 11(4):244–265. https://doi.org/10.3144/expresspolymlett.2017.26
16. Smola-Dmochowska A, Lewicka K, Macyk A, Rychter P, Pamuła E, Dobrzyński P (2023) Biodegradable polymers and polymer composites with antibacterial properties. IJMS 24(8):7473. https://doi.org/10.3390/ijms24087473
17. Che Aziz NN, Mat Yasin NH, Adam F, Abu Samah R (2025) Eco-friendly synthesis of cellulose nanofibrils via mechanical-assisted ultrasonication treatment and film-forming characteristics for sustainable packaging materials. Polym Eng Sci 65(8):4440–4455. https://doi.org/10.1002/pen.70015
18. Abral H, Pratama AB, Handayani D, Mahardika M, Aminah I, Sandrawati N, Sugiarti E, Muslimin AN, Sapuan SM, Ilyas RA (2021) Antimicrobial edible film prepared from bacterial cellulose nanofibers/starch/chitosan for a food packaging alternative. Int J Polym Sci 2021:1–11. https://doi.org/10.1155/2021/6641284
19. Tomé LC, Fernandes SC, Perez DS, Sadocco P, Silvestre AJ, Neto CP, Marrucho IM, Freire CS (2013) The role of nanocellulose fibers, starch and chitosan on multipolysaccharide based films. Cellulose 20(4):1807–1818. https://doi.org/10.1007/s10570-013-9959-6
20. Muñoz-Gimena PF, Oliver-Cuenca V, Peponi L, López D (2023) A review on reinforcements and additives in starch-based composites for food packaging. Polymers 15(13):2972. https://doi.org/10.3390/polym15132972
21. Petrova VA, Gofman IV, Dubashynskaya NV, Golovkin AS, Mishanin AI, Ivan'kova EM, Romanov DP, Khripunov AK, Vlasova EN, Migunova AV, Baranchikov AE (2023) Chitosan composites with bacterial cellulose nanofibers doped with nanosized cerium oxide: characterization and cytocompatibility evaluation. IJMS 24(6):5415. https://doi.org/10.3390/ijms24065415
22. Yanti NA, Ahmad SW, Muhiddin NH, Ramadhan LO, Walhidayah T (2021) Characterization of bacterial cellulose produced by Acetobacter xylinum Strain LKN6 using sago liquid waste as nutrient source. Pakistan J Biol Sci 24(3):335–344. https://doi.org/10.3923/pjbs.2021.335.344
23. Fazeli M, Simão RA (2018) The effect of cellulose nanofibers on the properties of starch biopolymer. Macromol Symp 380(1). https://doi.org/10.1002/masy.201800110
24. Al-Hassan A, Norziah MH (2012) Starch–gelatin edible films: water vapor permeability and mechanical properties as affected by plasticizers. Food Hydrocolloids 26(1):108–117. https://doi.org/10.1016/j.foodhyd.2011.04.015

25. Ahmad M, Hani NM, Nirmal NP, Fazial FF, Mohtar NF, Romli SR (2015) Optical and thermo-mechanical properties of composite films based on fish gelatin/rice flour fabricated by casting technique. Prog Org Coat 84:115–127. https://doi.org/10.1016/j.porgcoat.2015.02.016
26. Nasution H, Harahap H, Fath MA, Afandy Y (2019) Tensile properties of sago starch biocomposites reinforced with nanocrystalline cellulose from rattan biomass. IOP Conf Ser: Mater Sci Eng 505(1):012118. https://doi.org/10.1088/1757-899x/505/1/012118
27. Karnwal A, Rauf A, Jassim AY, Selvaraj M, Al-Tawaha AR, Kashyap P, Kumar D, Malik T (2025) Advanced starch-based films for food packaging: Innovations in sustainability and functional properties. Food Chem: X 29:102662. https://doi.org/10.1016/j.fochx.2025.102662
28. Aqel H, Farah H, Al-Hunaiti A (2024) Ecological versatility and biotechnological promise: comprehensive characterization of the isolated thermophilic Bacillus strains. PLoS ONE 19(4):e0297217. https://doi.org/10.1371/journal.pone.0297217
29. Bala JD, Lalung J, Saeed Al-Gheethi AA, Hossain K, Ismail N (2018) Microbiota of palm oil mill wastewater in Malaysia. TLSR 29(2):131–163. https://doi.org/10.21315/tlsr2018.29.2.10
30. Bala JD, Lalung J, Ismail N (2014) Studies on the reduction of organic load from palm oil mill effluent (POME) by bacterial strains. Int J Recycl Org Waste Agricult 4(1):1–10. https://doi.org/10.1007/s40093-014-0079-6
31. Nur MMA, Faizah H, Tutik Muji S, Tunjung Wahyu W, Sri Wahyu M, Raden Roro Endang A, Mahreni M, Ira Nurhayati D, Hadiyanto H (2025) Integrated microalgae-based biophotovoltaic systems using palm oil mill effluent for electricity generation, biogas purification, and bioproduct valorization. Waste Biomass Valor. https://doi.org/10.1007/s12649-025-03308-x
32. Khoshkava V, Kamal MR (2014) Effect of drying conditions on cellulose nanocrystal (CNC) agglomerate porosity and dispersibility in polymer nanocomposites. Powder Technol 261:288–298. https://doi.org/10.1016/j.powtec.2014.04.016
33. Yan J-H et al (2024) Intracellular gelation-mediated living bacteria for advanced biotherapeutics in mouse models. Langmuir 40(31):16605–16614. https://doi.org/10.1021/acs.langmuir.4c02215
34. Choi H et al (2007) Improvement of bacterial tethering using both physical and chemical surface modification for flagella spin actuators. Sens Actuators, B Chem 123(1):269–276. https://doi.org/10.1016/j.snb.2006.08.019
35. Yoo H, Lee S, Cho D (2014) Motility control of bacteria-actuated biodegradable polymeric microstructures by selective adhesion methods. Micromachines 5(4):1287–1295. https://doi.org/10.3390/mi5041287
36. Zheng J, Yeung ES (2003) Mechanism of microbial aggregation during capillary electrophoresis. Anal Chem 75(4):818–824. https://doi.org/10.1021/ac026146t
37. Hassan K, Hamidinezhad H, Al-Bermany E (2024) Antibacterial activity and optical behavior for restoration of micro and nano dental fillers using functional graphene nanosheets with polymethyl methacrylate. Nano Biomedicine and Engineering 16(4):652–664. https://doi.org/10.26599/nbe.2024.9290075
38. Raja T, Oleyan Al-Otibi F, Alharbi RI, Mohanavel V, Velmurugan P, Karthikeyan S, Perumal M, Basavegowda N (2023) A novel study of biological and structural analysis on Cissus quadrangularis fiber-reinforced CaO particulates epoxy composite for biomedical application. J Mater Res Technol 27:692–702. https://doi.org/10.1016/j.jmrt.2023.09.302
39. Tan SX, Andriyana A, Ong HC, Lim S, Pang YL, Ngoh GC (2022) A comprehensive review on the emerging roles of nanofillers and plasticizers towards sustainable starch-based bioplastic fabrication. Polymers 14(4):664. https://doi.org/10.3390/polym14040664
40. Wen X, Zhang K, Wang Y, Han L, Han C, Zhang H, Chen S, Dong L (2010) Study of the thermal stabilization mechanism of biodegradable polypoly(L-lactide)/silica nanocomposites. Polym Int 60(2):202–210. https://doi.org/10.1002/pi.2927
41. Senko O, Stepanov N, Maslova O, Efremenko E (2025) What can contribute to weakening of Poly(Vinyl Alcohol) cryogels used for cell (Self)Immobilization? Appl Microbiol 5(3):97. https://doi.org/10.3390/applmicrobiol5030097
42. Goodwin DG Jr et al (2017) Biodegradation of carbon nanotube/polymer nanocomposites using a monoculture. Environ Sci Technol 52(1):40–51. https://doi.org/10.1021/acs.est.7b02062

43. Abriat C, Gazil O, Heuzey M-C, Daigle F, Virgilio N (2021) The polymeric matrix composition of vibrio cholerae biofilms modulate resistance to silver nanoparticles prepared by hydrothermal synthesis. ACS Appl Mater Interfaces 13(30):35356–35364. https://doi.org/10.1021/acsami.1c07455
44. Bucher T, Oppenheimer-Shaanan Y, Savidor A, Bloom-Ackermann Z, Kolodkin-Gal I (2015) Disturbance of the bacterial cell wall specifically interferes with biofilm formation. Environ Microbiol Rep 7(6):990–1004. https://doi.org/10.1111/1758-2229.12346
45. Caro-Astorga J et al (2020) Biofilm formation displays intrinsic offensive and defensive features of Bacillus cereus. NPJ Biofilms Microbiomes 6(1). https://doi.org/10.1038/s41522-019-0112-7
46. Zhao Y, Zhang J, Ni M, Pan Y, Li L, Ding Y (2024) Cultivation of phosphate-accumulating biofilm: Study of the effects of acyl-homoserine lactones (AHLs) and cyclic dimeric guanosine monophosphate (c-di-GMP) on the formation of biofilm and the enhancement of phosphate metabolism capacity. Sci Total Environ 928:172408. https://doi.org/10.1016/j.scitotenv.2024.172408
47. Dutta B, Lahiri D, Nag M, Ray RR (2023) Functional metabolomics approaches in determining the inhibition of biofilm. In: Microbial biofilms. Elsevier, pp 235–251. https://doi.org/10.1016/b978-0-323-95715-1.00004-2
48. Noppawan P, Sangon S, Supanchaiyamat N, Hunt AJ (2021) Vegetable oil as a highly effective 100% bio-based alternative solvent for the one-pot multicomponent Biginelli reaction. Green Chem 23(16):5766–5774. https://doi.org/10.1039/d1gc00872b
49. Grimes SM, Kewcharoenwong P (2017) Dual-functionality ionic liquid mix for extraction and esterification of fatty acids as a step towards increasing the efficiency of conversion of waste cooking oils to biodiesel. J Chem Tech Biotech 92(8):2098–2105. https://doi.org/10.1002/jctb.5207
50. Li Y et al (2019) Towards a zero-waste biorefinery using edible oils as solvents for the green extraction of volatile and non-volatile bioactive compounds from rosemary. Antioxidants 8(5):140. https://doi.org/10.3390/antiox8050140
51. Slimani C et al (2024) Optimization of ultrasound-assisted extraction of phenols from Crocus sativus by-products using sunflower oil as a sustainable solvent alternative. Food Chem: X 23:101579. https://doi.org/10.1016/j.fochx.2024.101579
52. Scherr KE, Hasinger M, Loibner AP (2012) Increasing PAH (bio)accessibility in historically PAH-contaminated soils: waste versus food-grade oils. J Chem Tech Biotech 87(9):1229–1236. https://doi.org/10.1002/jctb.3823
53. Li X, Wei C, Deng Z, Li M, Li C, Fan G (2011) Selective solvent extraction of vanadium over iron from a stone coal/black shale acid leach solution by D2EHPA/TBP. Hydrometallurgy 105(3–4):359–363. https://doi.org/10.1016/j.hydromet.2010.10.006
54. Chang Z, Zhou X, Bai H, Li D, Zhang L (2023) Intensified reactive extraction of 4-hydroxypyridine with di(2-ethylhexyl) phosphoric acid in 1-octanol by using tributyl phosphate. Chin J Chem Eng 54:199–205. https://doi.org/10.1016/j.cjche.2022.03.021
55. Hajarabeevi N, Mohammed Bilal I, Easwaramoorthy D, Palanivelu K (2009) Facilitated transport of cationic dyes through a supported liquid membrane with D2EHPA as carrier. Desalination 245(1–3):19–27. https://doi.org/10.1016/j.desal.2008.06.009
56. Yogeeswara T, Devendra U, Kalaisselvane A (2020) Physical and chemical characterization of waste frying palm oil biodiesel and its blends with diesel. In: AIP conference proceedings, vol 2225. AIP Publishing, p 030003. https://doi.org/10.1063/5.0005584
57. Zereshki S, Daraei P, Shokri A (2018) Application of edible paraffin oil for cationic dye removal from water using emulsion liquid membrane. J Hazard Mater 356:1–8. https://doi.org/10.1016/j.jhazmat.2018.05.037
58. Rahman HA, Jusoh N, Othman N, Rosly MB, Sulaiman RNR, Noah NFM (2019) Green formulation for synthetic dye extraction using synergistic mixture of acid-base extractant. Sep Purif Technol 209:293–300. https://doi.org/10.1016/j.seppur.2018.07.053
59. Sipi KM, Chang SH (2020) Extraction of methylene blue from aqueous solutions by waste cooking oil and its back-extraction. IOP Conf Ser: Mater Sci Eng 811(1):012001. https://doi.org/10.1088/1757-899x/811/1/012001

60. Mahmoud S, Ghaly AE, Brooks MS (2007) Removal of dye from textile wastewater using plant oils under different pH and temperature conditions. Am J Environ Sci 3(4):205–218. https://doi.org/10.3844/ajessp.2007.205.218
61. Othman N, Raja Sulaiman RN, Rahman HA, Noah NFM, Jusoh N, Idroas M (2018) Simultaneous extraction and enrichment of reactive dye using green emulsion liquid membrane system. Environ Technol 40(11):1476–1484. https://doi.org/10.1080/09593330.2018.1424258
62. Sowmya R, Sivarajasekar N, Manimaran DR, Sivamani S (2020) Biosurfactant assisted liquid-membrane extraction of methyl-orange from aqueous solutions. In: AIP conference proceedings, vol 2291. AIP Publishing, p 020008. https://doi.org/10.1063/5.0019525

Fruits and Vegetable Waste: Harnessing Potential Toward Sustainable Bioenergy Production

Fozy Binhweel, Mardiana Idayu Ahmad, Hassan Pyar, Mohd Asyraf Kassim, and Mohammad Aliff Shakir

Abstract The massive generation of fruit and vegetable waste (FVW) presents a significant challenge in managing this type of organic waste. Conventional disposal methods such as landfilling and incineration pose environmental risks by potentially polluting water, soil, and air. However, emerging conversion technologies offer promising solutions by transforming FVW into valuable bio-based products. FVW is considered a cost-effective feedstock for the bioenergy industry and aligns well with the waste-to-energy concept. This chapter explores the sources of FVW generation and highlights its distinctive characteristics. More importantly, it provides an analysis of thermochemical, biochemical, and physical conversion technologies used to convert FVW into various forms of biofuels such as bio-oil, biodiesel, biogas, syngas, biochar, and compacted energy-dense units. These biofuels can be utilized in a range of applications including residential heating, cooking, electricity generation, and transportation which reduces dependence on fossil fuels and lowering greenhouse gas (GHG) emissions. Ultimately, the development of conversion technologies and supportive policies for utilizing FVW in bioenergy production fosters a circular economy and promotes sustainable practices.

Keywords Fruit and vegetable waste · Bioenergy · Biofuel · Conversion technologies

1 Introduction

Food waste is a serious and urgent global challenge. Humanity wastes around 1.3 billion tons of fresh and processed food annually, which accounts for one-third of total global food production [1]. Food is wasted at various stages, including production,

F. Binhweel · M. I. Ahmad (✉) · M. A. Kassim · M. A. Shakir
School of Industrial Technology, Universiti Sains Malaysia, 11800 Penang, Malaysia
e-mail: mardianaidayu@usm.my

F. Binhweel · H. Pyar
Faculty of Environmental Science and Marine Biology, Hadhramout University, Hahdramout, Yemen

M. A. Kassim et al. (eds.), *Sustainable Biorefinery*, Green Energy and Technology,
https://doi.org/10.1007/978-981-95-8181-8_7

cultivation, harvesting, storage, transportation, processing, distribution, marketing, and consumption. This problem is further exacerbated by the tremendous amounts of water, arable land, and energy used during food cultivation. It is estimated that the food sector consumes 30% of global energy and is responsible for 25% of global greenhouse gas (GHG) emissions [2]. Food waste is characterized by high moisture and organic content, which promotes intense microbial activity. Improper or delayed management of such waste complicates treatment processes and leads to numerous environmental and health issues, including toxicity, soil contamination, water and air pollution, and unpleasant odors [3].

Fruits and vegetables experienced high demand recently due to dramatic changes in human dietary habits. They are supplied either as raw products or processed for long-term use. However, tremendous amounts of fruit and vegetable waste (FVW) are generated in the form of peels, seeds, rotten cores, and pulps. This waste comprises both the edible and inedible organic parts of fruits and vegetables discarded throughout the food supply chain from production to consumption [4]. Increased processing leads to greater losses in fruit and vegetable products. FVW accounts for 25–30% of the waste produced by the food processing industry [5] and 14.8% of total food waste [6]. Notably, FVW is rich in carbohydrates, cellulose, starch, minerals, saccharides, and organic acids, exceeding those concentrations found in the edible pulp in some cases. Therefore, proper management is essential to optimize the utilization of FVW for the synthesis of valuable products such as bioactive compounds and biofuels [1].

Several sustainable and unsustainable practices are employed to dispose FVW. Landfilling is the most common method, as it is a simple and straightforward approach capable of handling the large volumes of FVW generated daily. However, this practice raises concerns due to the significant amounts of methane gas produced in landfills, which account for about one-third of human-caused methane emissions in the atmosphere. Additionally, landfilling leads to the formation of leachate, which is a complex liquid containing organic and inorganic materials, salts, heavy metals, microcontaminants, and microcontaminants that can pollute soil and water resources. Incineration is another widely used method that relies on thermal treatment. Despite its ability to reduce large quantities of FVW, it is limited by the high energy consumption required for combustion and the emission of gases that pollute the atmosphere. Alternative practices, both traditional and valorized, for managing FVW include using it as animal feed, composting, extracting bioactive compounds, and synthesizing biofuels [1].

Accordingly, this chapter reviews the potential utilization of FVW as a cost-effective feedstock for producing various types of biofuels. It discusses the generation of FVW from different sources and presents relevant statistics. An overview of the composition and characteristics of FVW is also provided. More importantly, the chapter explains comprehensively the chemical, physical, and biological technologies employed to convert FVW into valuable biofuels such as biogas, bioethanol, bio-oil, and biochar. The environmental impacts of using FVW as a feedstock for the production of bioenergy are evaluated, highlighting its role in promoting sustainable energy and reducing the accumulating waste worldwide. Economic feasibility is considered for FVW based biofuel, emphasizing the alignment of this approach with the waste-to-energy concept. Lastly, the chapter explores future prospects for improving the utilization of FVW in bioenergy production, including advancements in conversion technologies and strategies for enhancing efficiency and sustainability.

2 Generation of FVW

FVW is generated from the unconsumed and unused parts of fruits and vegetables. It comprises a wide range of parts, including seeds, roots, tubers, stems, peels, rinds, shells, leaves, husks, and skins. Determination of waste from fruits and vegetables remains a subject of debate. While some scholars consider all losses occurring throughout the supply chain from production to final disposal as waste, others limit the definition to waste generated at the consumption stage, as it is easier to identify and quantify. Unsold fruits and vegetables also contribute significantly to the total volume of FVW. In addition, improper handling practices, such as excessive trimming, can lead to substantial waste generation. Furthermore, the morphology of certain fruits plays a key role in determining the proportion of inedible parts as in coconut husks and palm shells which are largely unconsumable [1]. Table 1 presents examples of potential FVW components and their estimated losses.

Table 1 The estimated loss from fruits and vegetables products. Adapted from [7]

Fruits and vegetables	Waste part	Loss percentages (%)
Durian	Rind, seeds	60–70
Jackfruit	Rind, seeds	50–70
Mangosteen	Rind, seeds	60–75
Rambutan	Rind, seeds	50–65
Passion fruit	Rind, seeds	45–50
Pineapple	Rind, core	33
Mango	Peel, stone	45
Banana	Peel	35
Grape	Stem, skin, seeds	20
Guava	Peel, seeds	10
Papaya	Rind, seeds	10–20
Pea	Shell	40
Potato	Peel	15
Tomato	Peel, seeds	20
Onion	Peel	–
Garlic	Peel	–

The generation of waste from fruit and vegetable products occurs at various stages along the supply chain, beginning with harvesting, followed by transportation, storage, processing, and finally reaching the consumption phase. However, the amount of waste generated at each stage varies significantly depending on the country and the level of infrastructure and technology dedicated to handling agricultural products. In developing countries, a substantial portion of FVW is produced during the harvesting and processing stages due to inadequate storage facilities, inefficient harvesting techniques, and poor transportation systems. In contrast, developed countries typically have well-established infrastructure that helps preserve the quality of produce through advanced storage and transportation methods. Despite this, a considerable amount of waste in these regions is generated at the consumption stage, largely due to strict aesthetic standards and consumer preferences. For instance, the United States records an annual loss of approximately 45 million tons of FVW, costing about 1 billion USD, whereas India generates around 5.60 million tons per year [4].

3 FVW Properties

The characteristics of FVW are of great importance, as they play a crucial role in determining the most suitable and effective waste management and conversion strategies. FVW is primarily composed of cellulose, hemicellulose, and lignin which are key structural carbohydrates that contribute to its fibrous nature. Additionally, it contains significant amounts of organic matter, total solids, volatile solids, and various essential minerals. The proportions of these components can vary considerably depending on the type of waste and the specific fruit or vegetable involved. One critical parameter is the carbon-to-nitrogen (C/N) ratio, which typically ranges from 10 to 21% in vegetable waste but is notably higher in fruit waste, reaching up to 19–53%. Likewise, the total solids content is generally greater in fruit waste (7.5–23%) compared to vegetable waste (3–11%). Volatile solids follow a similar pattern, with fruit waste containing 5–12%, while vegetable waste contains only 2–9%. Accurate identification and analysis of these components are essential for selecting the appropriate technology for biofuel production and other valorization processes [4].

FVW exhibits several common physical and chemical properties. Moisture content is one of the most distinctive characteristics and largely depends on the specific type of fruit or vegetable. The moisture content in FVW ranges from 23.50 to 90.40% [8], with fruit waste generally exhibiting higher moisture content than vegetable waste. Texture is another key attribute of FVW. It is significantly influenced by the type of produce, its moisture content, the method of processing, and the degree of degradation. The texture of FVW can vary from soft and watery to fibrous and tough. In terms of acidity, FVW tends to be acidic due to the presence of natural organic acids and the effects of microbial decomposition, which lower its pH scale. Additionally, FVW is rich in nutrients and bioactive compounds, including vitamins, minerals, antioxidants, and enzymes. However, it may also contain trace amounts of herbicides and pesticides applied during cultivation [9].

4 Conversion Technologies of FVW into Bioenergy

Utilizing FVW for bioenergy production is considered one of the most valorized approaches to waste management, as it represents a direct application of the waste-to-energy concept. FVW can be converted into various forms of bioenergy, including heat, gas, liquid, and even solid fuels. Specific types of FVW are suitable for producing particular biofuels such as syngas, biogas, bioethanol, biodiesel, and biochar. The conversion of FVW into these valuable biofuels involves physical, chemical, and biological pathways. Selecting the appropriate conversion technology depends on several factors, including the nature of the FVW, its moisture content, texture, and the targeted type of biofuel [10]. Figure 1 illustrates the main conversion technologies employed to produce biofuels from FVW.

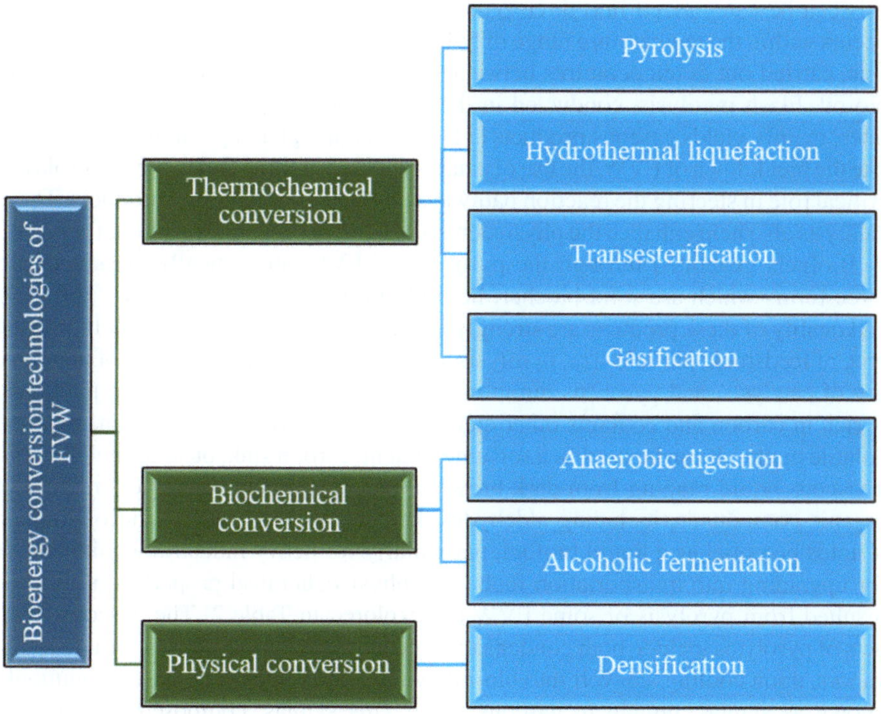

Fig. 1 The main conversion technologies employed for bioenergy production from FVW

4.1 Thermochemical Conversion

Thermochemical conversion refers to technological methods that rely on temperature to alter the chemical composition of FVW into useful forms of biofuels. A wide range of temperatures is employed, from mild conditions, as in transesterification, to high and extreme temperatures, as seen in pyrolysis and gasification. These methods produce various types of biofuels in gaseous, liquid, and solid forms. In this context, pyrolysis, hydrothermal liquefaction, transesterification, and gasification are discussed as thermochemical technologies used to convert FVW into biofuels.

4.1.1 Pyrolysis

Pyrolysis is a thermochemical process that involves the thermal decomposition of FVW in an environment devoid of oxygen. This process occurs at elevated temperatures, which are sufficient to break down the complex organic compounds of cellulose, hemicellulose, and lignin into simpler, smaller molecular compounds, including gases, bio-oil, and char. Based on the operating temperature and heating rate, pyrolysis can be categorized into three types. Conventional or slow pyrolysis typically occurs within the temperature range of 550–900 °C, producing more char. Fast pyrolysis, carried out at temperatures between 850–1250 °C, favors liquid products like bio-oil. Flash pyrolysis, conducted in the 1050–1300 °C range with rapid heating rates, mainly yields gaseous products [11]. Although high temperatures alone facilitate the breakdown of FVW, the use of catalysts such as zeolite and ferric oxide plays a critical role in steering the reaction pathways toward specific biofuel products. These catalysts also help enhance the physicochemical properties of the resulting fuels [12].

Biofuels generated through the pyrolysis of FVW are typically classified into three forms which are solid biochar, liquid bio-oil, and gaseous syngas. The yield and quality of these products are strongly influenced by several factors, including the type of feedstock, particle size, pyrolysis temperature, type and flow rate of inert gas, use of catalysts, and pyrolysis duration. Biochar, the solid residue from pyrolysis, is rich in carbon and contains other elements such as hydrogen, nitrogen, ash, and volatile matter. It can be used as a soil amendment, carbon sink, or solid fuel. Bio-oil, a viscous liquid ranging from dark brown to black, contains a complex mixture of organic compounds, including aldehydes, ketones, esters, amines, and oligomers. It holds potential as a liquid biofuel, particularly for heavy industrial equipment or for upgrading into transportation fuels. The physicochemical properties of bio-oils resulted from pyrolysis of some FVW are explored in Table 2. The gaseous products of pyrolysis include both condensable and non-condensable gases. Condensable vapors, upon cooling, convert into bio-oil, while non-condensable gases, commonly known as syngas, remain in gaseous form. This syngas, composed primarily of hydrogen, carbon monoxide, and methane, can be combusted directly for heating or utilized in power plants for electricity generation [13].

Table 2 Physicochemical properties of bio-oils produced from pyrolyzing some FVW

FVW	Temp. (°C)	Density (kg/m^3)	Moisture content (wt%)	Heating value (MJ/kg)	Acid value (mg KOH/g)	Ref.
Corn stover	450	1080	54.7	2.66	85.8	[14]
Sweetgum	450	1160	38.3	2.65	119.2	
Pine nuts	550	1090	9.36	19.31	–	[15]
Walnut shell	550	950	23.29	–	–	[16]
Rice husk	600	1150	15.82	22.99	–	[17]

4.1.2 Hydrothermal Liquefaction

Hydrothermal liquefaction is a thermochemical process that uses hot and pressurized water to convert FVW into bio-crude oil. The process typically operates at temperatures ranging from 250 to 500 °C and pressures between 5 and 35 MPa, over a duration of 5–60 min [12]. In hydrothermal liquefaction, water acts as a solvent, enabling the breakdown of complex organic matter under subcritical or near-critical conditions. The process begins by preparing the FVW, which involves grinding the feedstock into a smooth and fine texture. The ground material is then mixed with water to form a slurry, which is fed into high temperature and pressure reactor. Inside the reactor, a series of thermochemical reactions occur including decarboxylation, thermal cracking, and pyrolysis. These reactions break down the complex organic compounds in the FVW into simpler and energy-dense molecules, resulting in the formation of bio-crude oil and other byproducts. In some cases, organic solvents may be used instead of water, but water is generally preferred due to its non-toxic nature, favorable solvent properties under hydrothermal liquefaction conditions, and lower cost [18].

The primary product of hydrothermal liquefaction is bio-crude oil that serves as a promising alternative to fossil fuels. Before it can be utilized as a biofuel or refined for other applications, the bio-crude must be purified to remove residual solids and contaminants inherited from the feedstock. One of the distinguishing features of bio-crude oil is its relatively high oxygen content, typically ranging between 10–20% on a wet weight basis, which influences its chemical stability and upgrading requirements. It also possesses a notable higher heating value typically between 30–35 MJ/kg, making it more energy-dense than pyrolysis-derived oils. In addition to bio-crude, hydrothermal liquefaction generates valuable byproducts, including water-soluble organics, gases, and solid residue char. The yield and composition of these products depend on various factors, such as the type of feedstock, feedstock-to-water ratio, reaction temperature and pressure, residence time, and use of catalysts [12, 19].

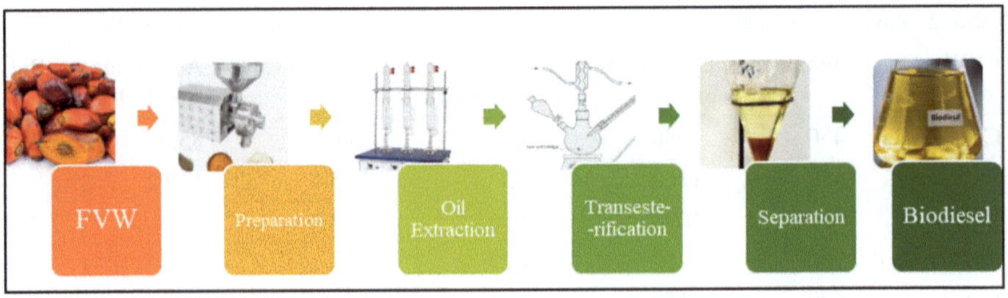

Fig. 2 Sequential steps of transesterification of FVW for biodiesel production

Transesterification

Transesterification is a chemical reaction in which fatty acids from oils react with alcohol in the presence of a base catalyst to produce alkyl esters, commonly known as biodiesel, with glycerol as a byproduct. FVW rich in oil content is a viable feedstock for biodiesel production through transesterification. The process begins with the preparation of FVW for oil extraction, which involves drying, cutting, and grinding it into smaller pieces. The ground FVW is then subjected to oil extraction using mechanical, solvent-based, or instrument-assisted methods. Once the oil is extracted, it undergoes a transesterification reaction, typically involving methanol and a base catalyst such as sodium hydroxide, under specific temperature and time conditions. At the end of the reaction, the mixture is allowed to settle, forming two distinct layers which are biodiesel at the upper layer and glycerol at the lower layer. Finally, the biodiesel is separated and purified for further use [20]. Figure 2 illustrates the sequential steps of transesterification for biodiesel production from FVW.

Biodiesel is the main product obtained from the transesterification of oils extracted from FVW. It consists of alkyl esters of long-chain fatty acids and serves as an eco-friendly, renewable, and sustainable alternative to conventional fossil diesel. Due to its similar physicochemical properties, biodiesel can be used directly in diesel engines without requiring significant modifications. It possesses a high energy content, typically ranging between 20 and 45 MJ/kg, making it suitable for various energy applications. Biodiesel can be used as a pure fuel or blended with petroleum diesel in different ratios. Beyond its use as a transportation fuel, biodiesel is also applicable in power generation and residential or industrial heating systems. When burned, biodiesel produces emissions lower and more biodegradable than fossil fuel diesel, contributing to reduced environmental impact. In addition to biodiesel, transesterification also yields glycerol as a byproduct, which is widely used in the pharmaceutical, cosmetic, and detergent industries [21].

Gasification

Gasification is a thermochemical process in which high temperatures are used to convert FVW into syngas in the presence of gasifying agent such as oxygen, carbon dioxide, steam or a mixture of them. There are two types of gasification processes which are catalytic and non-catalytic. Catalytic gasification employs catalysts to convert FVW into syngas at lower temperatures. In contrast, non-catalytic gasification requires higher temperatures, often exceeding 800 °C, to decompose the carbonaceous components of FVW into syngas [9, 12]. The thermochemical conversion via gasification typically begins with drying the FVW feedstock at temperatures between 100 and 200 °C, which removes approximately 30–60% of the feedstock moisture content. The dried feedstock then undergoes pyrolysis, producing volatile compounds and residual char. In the presence of a gasifying agent at around 700 °C, these residuals are oxidized into carbon monoxide and carbon dioxide through exothermic reactions. Finally, at temperatures exceeding 800 °C, carbon and other volatiles undergo endothermic reactions with the gasifying agents, typically carbon dioxide, resulting in the formation of syngas [12, 22].

Gasification is a thermochemical conversion process that takes place in a gasifier, an apparatus used to convert FVW into syngas. A gasifier typically consists of a feedstock inlet, gasification agent supply, reactor chamber, and gas outlet. Gasifiers are generally categorized into three types which are fixed-bed, fluidized-bed, and entrained-flow. The fixed-bed gasifier is suitable for small quantities of FVW and operates at moderate temperatures. However, higher temperatures can be applied to handle rigid and inorganic components. The fluidized-bed gasifier is more appropriate for large quantities of biomass or pretreated FVW, depending on specific size and feedstock characteristics. This type of gasifier is further divided into bubbling fluidized-bed gasifiers, used mainly for pretreated waste, and circulating fluidized-bed gasifiers, which are suitable for general biomass. Entrained-flow gasifiers are designed for feedstock with very fine particle sizes, typically in the micrometer range. They operate under moderate pressures, ranging from 20 to 80 bar. However, this type of gasifier is generally unsuitable for FVW and is more commonly used with coal-based feedstocks [23].

Syngas is the primary product of the gasification process. This syngas typically consists of carbon monoxide, carbon dioxide, hydrogen, and methane, with its composition varying depending on several operational parameters as presented in Table 3. The quality and thermal energy content of the produced syngas are significantly influenced by various factors including the type and efficiency of the gasifier used, the nature of the feedstock, the temperature and pressure conditions applied during the process, and the type and ratio of the gasifying agent. Additionally, the moisture and ash content of the feedstock also play a critical role in determining the overall performance and calorific value of the resulting syngas. Once produced, syngas can be utilized directly for applications such as heating and electricity generation. Alternatively, it can serve as a precursor for the synthesis of other valuable biofuels and chemicals, including methanol and ethanol, contributing to sustainable energy development [24].

Table 3 Syngas composition from gasification of FVW. Adapted from [12]

Input			Output				
FVW	Temp. (°C)	Gasifying agent	CO	CO_2	H_2	CH_4	Heating value (MJ/Nm^3)
Olive kernel	950	Air	15–20 wt%	40–55 wt%	20–30% wt%	10–12 wt%	8.8–10.4
Palm oil wastes	800	Steam	14–33 v%	14–26 v%	47–58 v%	3–6 v%	8.7–12
Empty fruit bunch	850	Air	32–45 v%	16.6–36 v%	18.3–27.4 v%	12–15 v%	12.3–15.3
Bamboo	400–600	Air	23.5–30.6% m/m	59–63% m/m	6.6–8.1% m/m	4–5% m/m	1.6–1.9

4.2 Biochemical Conversion

Biochemical conversion utilizes both biological agents, such as microorganisms, and chemical factors, like enzymes, to transform FVW into useful forms of biofuels. This method is considered one of the simplest and most environmentally friendly conversion pathways, operating efficiently under ambient temperature and atmospheric pressure conditions. A key advantage of biochemical conversion is its high tolerance to the natural moisture content present in FVW, eliminating the need for dehydration processes. Additionally, it offers high selectivity and productivity, ensuring that the desired biofuel products are generated with minimal byproducts. This enhances the overall efficiency and cost-effectiveness of the process. Common biofuels derived through biochemical pathways include biogas, bioethanol, and biohydrogen. Core biochemical conversion technologies encompass processes such as anaerobic digestion, enzymatic hydrolysis and alcoholic fermentation [12].

4.2.1 Anaerobic Digestion

Anaerobic digestion is a biochemical process in which FVW is converted into biogas and other valuable byproducts. Biochemically, it is a complex process that involves various groups of microorganisms working together to decompose FVW into simpler organic compounds through multiple sequential steps. However, it is widely regarded as a cost-effective and sustainable method of FVW management, as it upcycles organic waste into useful forms of biofuels. The anaerobic digestion of FVW occurs in four main phases which are hydrolysis, acidogenesis, acetogenesis, and methanogenesis as illustrated in Fig. 3. The process begins with hydrolysis, where complex organic compounds such as carbohydrates, lipids, and proteins are broken down into simpler, soluble molecules like simple sugars, fatty acids, and amino acids. This step is facilitated by enzymes produced by hydrolytic bacteria. In the acidogenesis phase, these hydrolysis products are further converted by fermentative bacteria into volatile fatty acids, alcohols, hydrogen, and carbon dioxide. During acetogenesis, the products of acidogenesis are converted by homoacetogenic bacteria into acetate, along with additional hydrogen and carbon dioxide. Finally, in the methanogenesis phase, methanogenic archaea convert acetate and carbon dioxide into methane through two main pathways of acetoclastic or hydrogenotrophic methanogenesis. This step results in the production of biogas [12].

Biogas resulting from the anaerobic digestion of FVW is primarily composed of methane, ranging from 50 to 75%, and carbon dioxide, ranging from 30 to 50%. It also contains trace amounts of other gases, including water vapor (~6%), nitrogen (~3%), oxygen (~1%), hydrogen, hydrogen sulfide, and ammonia [12]. This biogas can either be used directly for heating and electricity generation through direct combustion or purified into biomethane by removing carbon dioxide and other trace gases. The purified biomethane can then be utilized as a transportation biofuel or injected into local gas grids for use in cooking and heating. Thus, biogas from FVW serves as a

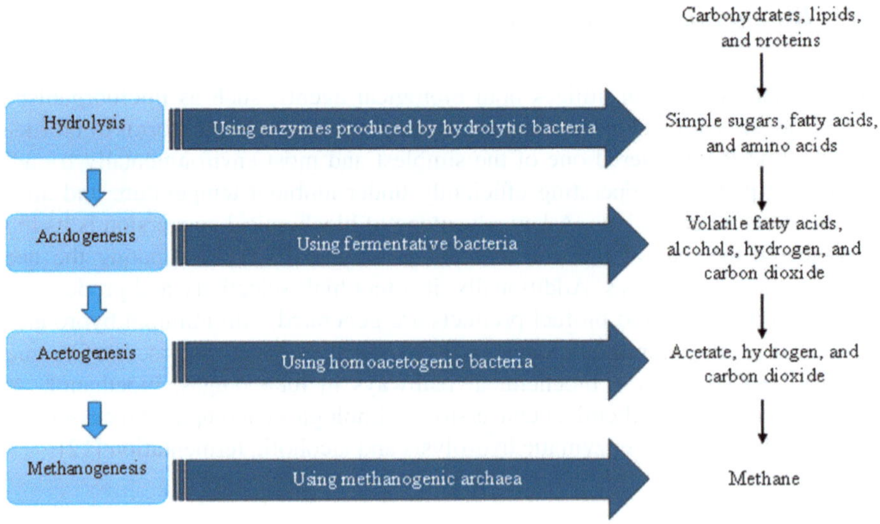

Fig. 3 The four phases of anaerobic digestion for FVW

sustainable biofuel that reduces reliance on fossil fuels and mitigates environmental impact [9].

The application of anaerobic digestion for FVW offers several advantages. In addition to being an efficient and cost-effective waste management approach that reduces GHG emissions, it produces biogas as a sustainable biofuel that can substitute fossil fuel-based gas. Furthermore, the digestate, a byproduct of FVW after anaerobic digestion, is valorized as a nutrient-rich fertilizer. However, this process is limited by rapid acidification, which hinders methanogenesis and produces volatile fatty acids. This behavior is mainly due to the natural acidity of FVW. Additional challenges include the availability of FVW in sufficient quantities, feedstock contamination, and the need for appropriate infrastructure. Nevertheless, careful monitoring and the implementation of suitable pretreatment techniques can help overcome these challenges associated with the anaerobic digestion of FVW [25].

4.2.2 Alcoholic Fermentation

Alcoholic fermentation is a biological process in which sugar-rich materials contained in FVW are converted into alcohol, organic acids, and gases, typically in the absence of oxygen. There are several types of fermentation, depending on the microorganisms involved and the final products. However, the type of fermentation used for biofuel production from FVW is called alcoholic fermentation, as it is carried out by yeast to produce bioethanol. The conversion of FVW into bioethanol generally involves three sequential phases which are pretreatment, enzymatic hydrolysis, and fermentation as presented in Fig. 4. The process begins with pretreatment, which may involve physical grinding, chemical oxidation, biological decomposition, or a combination of these methods. The goal of pretreatment is to break down complex components and disrupt the cell walls of FVW to facilitate the subsequent steps. Next, enzymatic hydrolysis breaks down the pretreated material into simpler molecules using water and specific enzymes. For instance, cellulase enzymes convert cellulose into glucose. Finally, the resulting simple sugars, such as glucose, are fermented into alcohol bioethanol through the activity of yeast [26].

Although alcoholic fermentation is a useful strategy for managing waste and producing bioethanol, it faces several challenges that may limit its application to FVW for biofuel production. One major issue is the low sugar content in certain types of FVW, such as lignocellulosic biomass, which hinders the fermentation process and reduces bioethanol yield. Moreover, the success of fermentation depends on the use of specific microbial strains, which are often necessary to achieve optimal yields. Recently, genetically modified strains of microorganisms and yeasts have been developed to improve fermentation outcomes. In addition to the sugar content of FVW and microbial performance, the economic feasibility remains a concern due to the high resource demands of fermentation, distillation, and purification processes. Despite these limitations, ongoing research and development efforts aim to optimize

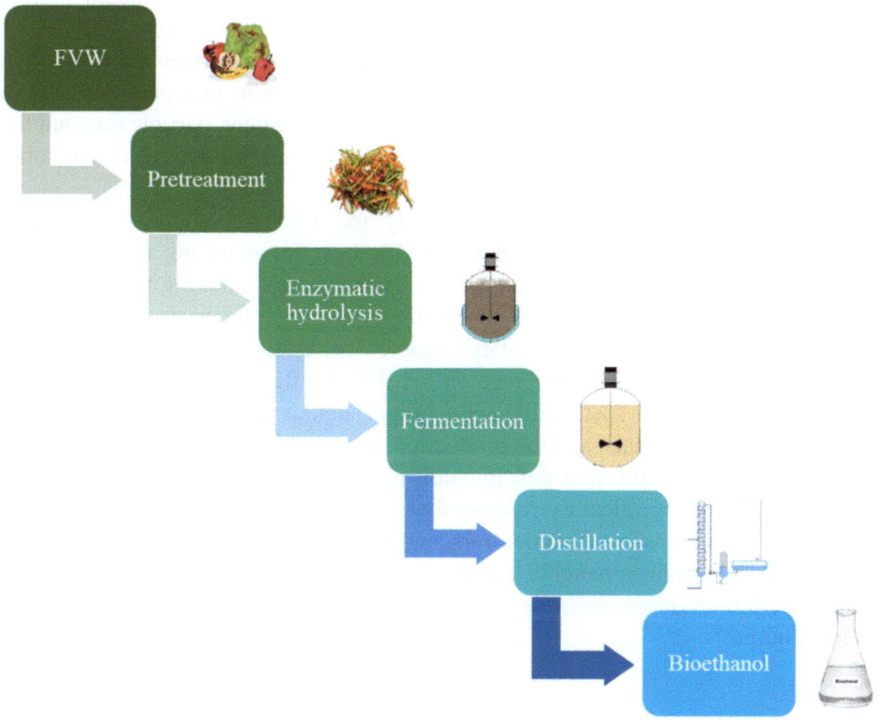

Fig. 4 The sequential steps of alcoholic fermentation of FVW

alcoholic fermentation, reduce production costs, and enhance the overall viability of biofuel products as a sustainable fuel source [9].

4.2.3 Physical Conversion

Densification

The physical conversion of FVW is achieved through densification, a process in which the FVW feedstock is compressed to greatly reduce its volume and significantly increase its density. This is purely a physical transformation, altering the size of the FVW without affecting its chemical composition. Densification of FVW can be carried out through three main approaches which are briquetting, extrusion, and pelletizing. Briquetting involves compressing FVW into compacted briquettes, typically using ground or shredded material. In contrast, extrusion forces the FVW feedstock through narrow openings under heat and pressure, softening the material and forming continuous strand shapes that are later cut into smaller units for easier handling and transport. Finally, pelletizing applies heat and pressure to finely ground FVW feedstock, producing uniform, cylindrical, dense pellets. To enhance the stability of the densified products, binders are often added. These binders improve cohesion and are used across all forms of briquettes, extruded shapes, and pellets. They can be derived from the FVW itself or from external materials such as lignin, starch, and molasses. Achieving high-quality densification products depends on several factors including the type of FVW, its moisture content, particle size, applied pressure and temperature, binder type and ratio, and the duration of the densification process [9].

Densified FVW offers several eco-friendly and biodegradable biofuel applications. It can be directly combusted to provide heat for warming residential and commercial buildings. Additionally, it serves as a practical cooking fuel, especially in remote or off-grid regions where conventional energy sources are limited. Moreover, densified FVW can be utilized as a supplementary fuel in power plants to help generate electricity, reducing reliance on fossil fuels. Beyond bioenergy, densified FVW finds use in other sustainable applications such as animal feed, soil enhancement, packaging materials, and the production of bio-based chemicals. The physical compaction achieved through densification plays a crucial role in effective waste management, as it significantly reduces the volume of accumulated organic waste. This makes it much easier to handle, store, and transport, contributing to a cleaner and more sustainable environment [27]. Table 4 compares thermochemical, biochemical, and physical conversion methods along with the advantages and disadvantages of each technique.

Table 4 Comparisons of main conversion technologies employed for bioenergy production from FVW

Conversion technologies		Description	Advantages	Disadvantages
Thermochemical conversion	Pyrolysis	Thermal decomposition of FVW in the absence of oxygen to produce bio-oil, syngas, and biochar	– It produces multiple biofuels – It reduces bulky waste	– Controlling pyrolysis conditions are required – Further processing of bio-oil is required
	Hydrothermal liquefaction	Converting wet FVW into bio-crude oil under high pressure and moderate temperature in presence of water	– It deals with FVW with high moisture content – It produces bio-crude with high density	– Requirement of high pressure equipment – Separation process is required
	Transesterification	Converting lipids from FVW into biodiesel using alcohol and a base catalyst	– It produces biodiesel can be used in diesel engines without significant modifications – Valuable byproduct of glycerol	– it requires lipid extraction process – Further purification is required
	Gasification	Converting FVW into syngas at high temperatures in the presence of gasifying agent	– Versatile biofuels – It reduces waste mass significantly	– Energy consuming – It requires dry feedstock
Biochemical conversion	Anaerobic digestion	Microbial breakdown of organic matter in FVW in the absence of oxygen to produce biogas	– It is suitable for wet FVW – Affordable procedures – It produces valuable digestate as a biofertilizer	– Extremely slow process – Sensitive operating conditions

(continued)

Table 4 (continued)

Conversion technologies		Description	Advantages	Disadvantages
	Alcoholic fermentation	Converting sugars from FVW into ethanol using yeast and bacteria	– It is simple and scalable process – It produces valuable liquid fuel of bioethanol	– It works with sugary feedstock of FVW
Physical conversion	Densification	Compressing FVW to reduce volume and increase density, forming solid biofuels for combustion	– Tremendous reduction of FVW volume – Easy handling, storing, and transportation	– Binders are required – Additional combustion steps are required

5 Environmental Impact

Utilizing FVW for bioenergy production is increasingly recognized as an effective and sustainable waste management strategy. Globally, massive quantities of food waste are generated on a daily basis, placing significant pressure on environmental systems. It is estimated that approximately 1.3 billion tons of fresh and processed food are wasted each year [1], with around 14.8% of this total consisting of FVW [6]. Managing such enormous amounts of waste presents a major global challenge. In many regions, the predominant methods of disposal, landfilling and incineration, are not environmentally sustainable. These practices contribute to pollution and the degradation of natural resources such as soil, water, and air. In contrast, converting FVW into biofuels provides a cleaner and more responsible alternative. It supports the "waste-to-energy" concept by turning organic waste into valuable energy resources. Moreover, this approach aligns with the principles of a circular economy by minimizing waste and promoting the reuse of resources, ultimately contributing to environmental sustainability and energy security [28].

In addition to effective waste management, converting FVW into bioenergy products plays a vital role in reducing GHG emissions. GHG emissions are a major contributor to global environmental issues such as global warming and climate change. Two of the most significant and growing sources of GHG emissions are landfills and the excessive combustion of fossil fuels for energy. Landfilling organic waste like FVW results in the release of methane, a potent GHG, while burning fossil fuels emits large amounts of carbon dioxide. By diverting FVW from landfills and using it as a renewable feedstock for bioenergy production, GHG emissions can be reduced in two key ways. First, it decreases the volume of waste and the associated methane emissions. Second, it produces eco-friendly and sustainable biofuels that serve as alternatives to fossil fuels, thereby reducing carbon emissions and contributing to a cleaner, more sustainable energy system [29].

6 Economic Feasibility

The economic feasibility of synthesizing biofuels from FVW is highly promising. One of the most significant contributors to the overall cost of biofuel production is the feedstock itself. In many conventional biofuel production processes, feedstock expenses can account for as much as 70–90% of the total production costs [30]. This high cost is a major barrier to making biofuels competitive with fossil fuels. However, using FVW as a raw material offers a cost-effective alternative, as it is often available at little to no cost and helps reduce waste management burdens. The remaining production expenses are primarily associated with the processing and conversion technologies involved. When utilizing FVW, the total cost of biodiesel production is significantly reduced, making it a more affordable and sustainable option compared to the continually rising prices of fossil fuels. Thus, converting FVW into bioenergy not only promotes environmental sustainability but also supports economic viability and energy security [31].

The commercial production of biofuel derived from FVW plays a significant role in supporting economic stability and environmental sustainability. By converting organic waste into clean, renewable energy, this process offers a cost-effective alternative to fossil fuels, helping to reduce reliance on non-renewable resources. As a result, it not only lowers GHG emissions but also contributes to a circular economy by transforming waste into valuable products. Furthermore, the growth of the FVW-based biofuel industry opens up new avenues for investment and economic development. Establishing and maintaining production facilities, transportation networks, and research initiatives related to biofuel technologies lead to the creation of thousands of job opportunities across various sectors. These include roles in waste collection, processing, engineering, and plant operation. Ultimately, investing in FVW-based biofuel not only addresses environmental concerns but also promotes energy security and economic resilience in both local and global contexts [32].

7 Future Outlook and Challenges

The utilization of FVW for bioenergy production presents several notable challenges. One of the primary issues is the significant variability in the feedstock physical and chemical properties. For example, some types of FVW have a high moisture content, making them unsuitable for certain thermochemical processes, while others are dry and rigid, posing handling and processing difficulties. Furthermore, the lipid content of FVW differs widely. Some varieties are rich in oil, whereas others contain almost none, necessitating the use of different conversion technologies to optimize energy yield. In addition to these compositional variations, the availability of FVW is inconsistent and often dependent on seasonal patterns, agricultural cycles, and regional factors. This irregular supply can disrupt continuous bioenergy production. Another critical challenge is the rapid degradation of FVW, as it is prone to spoilage and microbial contamination if not processed quickly. Lastly, the efficient collection of FVW requires substantial labor input and specialized equipment, increasing operational complexity and costs [33].

As a future outlook, promoting the utilization of FVW in the bioenergy industry requires a comprehensive and coordinated approach. First and foremost, advancements in conversion technologies are essential to ensure efficient, cost-effective, and sustainable energy production from FVW. Investment in research and development can lead to optimized processes tailored to the diverse characteristics of the waste. In parallel, governments play a crucial role by establishing supportive policies and regulatory frameworks. Offering financial incentives, such as subsidies, tax breaks, or grants, can encourage industries and stakeholders to adopt FVW-based bioenergy solutions rather than disposing of the waste in landfills. Moreover, strong collaboration is vital. Partnerships between local communities, government agencies, research institutions, and international organizations can foster knowledge exchange, resource sharing, and the development of scalable models. Such multi-level cooperation ensures that the environmental and economic benefits of FVW utilization are maximized, contributing to a more circular economy and reducing the ecological impact of organic waste disposal.

8 Conclusion

Large quantities of fruit and vegetable waste (FVW) are generated daily, posing a significant challenge for effective waste management. Conventional disposal methods such as landfilling and incineration are unsustainable, as they contribute to the pollution of water, soil, and air. In contrast, utilizing FVW for bioenergy production presents a promising solution, transforming waste into valuable biofuels and supporting the circular economy through the waste-to-energy concept. Various conversion technologies were successfully applied to FVW, enabling its transformation into useful bioenergy forms. Thermochemical methods such as pyrolysis, hydrothermal liquefaction, transesterification, and gasification use heat to convert FVW into bio-oil, biodiesel, biochar, and syngas. Biochemical processes including anaerobic digestion and alcoholic fermentation rely on microbial activity to produce biogas and bioethanol. Physical conversion, represented by densification, compresses FVW under specific temperature and pressure into compact, energy-dense units. These biofuels can be used for heating, cooking, electricity generation, or further refined into transportation fuels. Ultimately, this approach reduces greenhouse gas (GHG) emissions from landfills and fossil fuel use, contributing to environmental sustainability. Progressive technological advancements and supportive policies are key to maximizing the potential of FVW as a sustainable bioenergy source.

References

1. Panda J, Amrit R, Mishra A, Chakraborty A, Rustagi S, Nath P, Sarabandi K, Sarma H, Wagh M, Mohanta Y (2025) Sustainable valorization of fruit and vegetable waste for bioactive compounds: Advancing functional food and wellness. Waste Biomass Valorization. https://doi.org/10.1007/s12649-025-02937-6.
2. Batool F, Kurniawan T, Mohyuddin A, Othman M, Aziz F, Al-Hazmi H, Goh H, Anouzla A (2024) Environmental impacts of food waste management technologies: a critical review of life cycle assessment (LCA) studies. Trends Food Sci Technol 143:104287. https://doi.org/10.1016/j.tifs.2023.104287
3. Sarker A, Ahmmed R, Ahsan S, Rana J, Ghosh M, Nandi R (2024) A comprehensive review of food waste valorization for the sustainable management of global food waste. Sustain Food Technol 2:48–69. https://doi.org/10.1039/D3FB00156C
4. Esparza I, Jiménez-Moreno N, Bimbela F, Ancín-Azpilicueta C, Gandía L (2020) Fruit and vegetable waste management: conventional and emerging approaches. J Environ Manag 265:110510. https://doi.org/10.1016/j.jenvman.2020.110510
5. Saini A, Panesar P, Bera M (2019) Valorization of fruits and vegetables waste through green extraction of bioactive compounds and their nanoemulsions-based delivery system. Bioresour Bioprocess 6:26. https://doi.org/10.1186/s40643-019-0261-9.
6. Nayak A, Bhushan B (2019) An overview of the recent trends on the waste valorization techniques for food wastes. J Environ Manag 233:352–370. https://doi.org/10.1016/j.jenvman.2018.12.041
7. Sagar N, Pareek S, Sharma S, Yahia E, Lobo M (2018) Fruit and vegetable waste: bioactive compounds, their extraction, and possible utilization. Compr Rev Food Sci Food Saf 17:512–531. https://doi.org/10.1111/1541-4337.12330
8. Solyom K, Lopez P, Esquivel P, Lucia A (2020) Effect of temperature and moisture contents on dielectric properties at 2.45 GHz of fruit and vegetable processing by-products. RSC Adv 10:16783–16790. https://doi.org/10.1039/C9RA10639A
9. Dey D, Kundu K, Singh D (2024) *Bioenergy production from vegetable and fruit waste.* In: From waste to wealth. Springer Nature Singapore, Singapore, pp 225–245. https://doi.org/10.1007/978-981-99-7552-5_11
10. Zhu Y, Luan Y, Zhao Y, Liu J, Duan Z, Ruan R (2023) Current technologies and uses for fruit and vegetable wastes in a sustainable system: a Review. Foods 12:1949. https://doi.org/10.3390/foods12101949
11. Binhweel F, Hossain M, Ahmad M (2023) Recent trends, potentials, and challenges of biodiesel production from discarded animal fats: a comprehensive review. BioEnergy Res 16:778–800. https://doi.org/10.1007/s12155-022-10527-w
12. Osman A, Mehta N, Elgarahy A, Al-Hinai A, Al-Muhtaseb A, Rooney D (2021) Conversion of biomass to biofuels and life cycle assessment: a review. Environ Chem Lett 19:4075–4118. https://doi.org/10.1007/s10311-021-01273-0
13. Hao J, Xu F, Yang D, Wang B, Qiao Y, Tian Y (2025) Analytical pyrolysis of biomass using pyrolysis-gas chromatography/mass spectrometry. Renew Sustain Energy Rev 208:115090. https://doi.org/10.1016/j.rser.2024.115090
14. Wang H, Srinivasan R, Yu F, Steele P, Li Q, Mitchell B, Samala A (2012) Effect of acid, steam explosion, and size reduction pretreatments on bio-oil production from sweetgum, switchgrass, and corn stover. Appl Biochem Biotechnol 167:285–297. https://doi.org/10.1007/s12010-012-9678-8
15. Xu X, Li Z, Sun Y, Jiang E, Huang L (2018) High-quality fuel from the upgrading of heavy bio-oil by the combination of ultrasonic treatment and mutual solvent. Energy Fuels 32:3477–3487. https://doi.org/10.1021/acs.energyfuels.7b03483
16. Zhu L, Li K, Zhang Y, Zhu X (2017) Upgrading the storage properties of bio-oil by adding a compound additive. Energy Fuels 31:6221–6227. https://doi.org/10.1021/acs.energyfuels.7b00864

17. Zhang L, Liu R, Yin R, Mei Y, Cai J (2014) Optimization of a mixed additive and its effect on physicochemical properties of bio-Oil. Chem Eng Technol 37:1181–1190. https://doi.org/10.1002/ceat.201300786
18. Ocampo E, Beltrán V, Gómez E, Ríos L, Ocampo D (2023) Hydrothermal liquefaction process: review and trends. Hydrothermal Liq Process: Rev Trends 7:100382. https://doi.org/10.1016/j.crgsc.2023.100382
19. Qiu B, Tao X, Wang Y, Zhang D, Chu H (2025) Hydrothermal liquefaction for producing liquid fuels and chemicals from biomass-derived platform compounds: a review. Environ Chem Lett 23:81–115. https://doi.org/10.1007/s10311-024-01791-7
20. Binhweel F, Ahmad M, Zaki S (2022) Utilization of polymeric materials toward sustainable biodiesel industry: a recent review. Polymers 14:3950. https://doi.org/10.3390/polym14193950
21. Singh D, Sharma D, Soni S, Sharma S, Kumari D (2019) Chemical compositions, properties, and standards for different generation biodiesels: a review. Fuel 253:60–71. https://doi.org/10.1016/j.fuel.2019.04.174
22. Hanchate N, Ramani S, Mathpati C, Dalvi V (2021) Biomass gasification using dual fluidized bed gasification systems: a review. J Clean Prod 280:123148. https://doi.org/10.1016/j.jclepro.2020.123148
23. Shahabuddin M, Alam M, Krishna B, Bhaskar T, Perkins A (2020) A review on the production of renewable aviation fuels from the gasification of biomass and residual wastes. Bioresour Technol 312:123596. https://doi.org/10.1016/j.biortech.2020.123596
24. González C, Sandoval L (2020) Sustainability aspects of biomass gasification systems for small power generation. Renew Sustain Energy Rev 134:110180. https://doi.org/10.1016/j.rser.2020.110180
25. Wang J, Liu S, Feng K, Lou Y, Ma J, Xing D (2025) Anaerobic digestion of lignocellulosic biomass: process intensification and artificial intelligence. Renew Sustain Energy Rev 210:115264. https://doi.org/10.1016/j.rser.2024.115264
26. Ndaba B, Rama H, Bingwa N, Roopnarain A (2025) Unravelling the role of nanoparticles during bioethanol production: a review on pretreatment, hydrolysis and fermentation. Fuel 396:135336. https://doi.org/10.1016/j.fuel.2025.135336
27. Sarker T, Nanda S, Meda V, Dalai A (2023) Densification of waste biomass for manufacturing solid biofuel pellets: a review. Environ Chem Lett 21:231–264. https://doi.org/10.1007/s10311-022-01510-0
28. Rashid S, Sultan H, Rashid W, Talpur B, Tulcan R, Khan M, Bohnett E, Korai M, Zhang L (2025) A critical review of opportunities and challenges of solid waste management in an emerging economy- evidence from Pakistan. Environ Develop 55:101182. https://doi.org/10.1016/j.envdev.2025.101182
29. Paddy E, Tchudjo-Tchuente A, Ibrahim F, Fopah-Lele A, Kumar P, Tanyi E (2025) Analysis of waste-to-energy and GHG emissions at the Nkolfoulou sanitary landfill in Yaoundé, Cameroon. Int J Environ Sci Technol. https://doi.org/10.1007/s13762-024-06292-2
30. Bhuiya M, Rasul M, Khan M, Ashwath N (2020) Biodiesel production and characterisation of poppy (*Papaver somniferum L.*) seed oil methyl ester as a source of 2nd generation biodiesel feedstock. Ind Crop Prod. 152:112493. https://doi.org/10.1016/j.indcrop.2020.112493
31. Kiteto M, Vidija B, Mecha C (2025) Production of bioethanol from citrus peel waste: a techno—economic feasibility study. Energy Convers Manag: X 26:100916. https://doi.org/10.1016/j.ecmx.2025.100916
32. Pires V, Cavalcante R, Young A (2023) Process simulation and economic evaluation of biodiesel production from several feedstocks in Brazil. Braz J Chem Eng 42:317–344. https://doi.org/10.1007/s43153-023-00412-0
33. Khan A, López-Maldonado E, Alam S, Khan N, López J, Herrera P, Abutaleb A, Ahmed S, Singh L (2022) Municipal solid waste generation and the current state of waste-to-energy potential: State of art review. Energy Convers Manage 267:115905. https://doi.org/10.1016/j.enconman.2022.115905

Utilizing Eggshell Waste to Develop Composite Materials for Sustainable Water Purification

Rahmawati Munir, Dadan Hamdani, Darnah, Ratna Kusuma, Eva Marliana, Sahara Hamas Intifadhah, and Siti Fatimah

Abstract This chapter discusses the utilization of eggshell waste for developing composite materials aimed at sustainable water purification. The main focus is on the removal of methylene blue (MB), a synthetic dye commonly found in wastewater from the textile industry. Eggshells, as a calcium-rich natural material, are combined with titanium dioxide (TiO_2) to create a composite that facilitates both adsorption and photocatalytic degradation processes. This chapter outlines the synthesis method, characterization techniques, and photocatalytic performance testing of the composite on MB-contaminated samples. This approach not only enhances water purification efficiency but also contributes to environmental sustainability and waste reduction.

Keywords Eggshell waste · Methylene blue (MB) · Textile wastewater · Adsorption · Photocatalysis

1 Organic Waste Problems in the Environment

Agro-food waste such as eggshells has become a global concern due to high production volumes and suboptimal waste management. The decomposition of this waste releases ammonia and hydrogen sulfide gases, producing foul odors and attracting disease vectors such as flies and rodents [1, 2]. Eggshell waste from large-scale food

R. Munir (✉) · D. Hamdani · Darnah · R. Kusuma · E. Marliana · S. H. Intifadhah · S. Fatimah
Theoretical and Material Physics, Physics Department, FMIPA Mulawarman University, Samarinda, Indonesia
e-mail: rahmawati@fmipa.unmul.ac.id

Darnah
Study Program of Statistics, Department of Mathematics, FMIPA Mulawarman University, Samarinda, Indonesia

R. Kusuma
Department of Biology, FMIPA Mulawarman University, Samarinda, Indonesia

E. Marliana
Department of Chemistry, FMIPA Mulawarman University, Samarinda, Indonesia

M. A. Kassim et al. (eds.), *Sustainable Biorefinery*, Green Energy and Technology,
https://doi.org/10.1007/978-981-95-8181-8_8

industries is known to be a major environmental pollutant due to its high volume and partially degradable nature [3]. Eggshells are composed of more than 96% calcium carbonate ($CaCO_3$), with small portions of protein, minerals, and organic matter [4]. If left unmanaged, these components can raise the alkalinity of soils and disturb the pH balance of surrounding water bodies [5].

Recent studies emphasize the relevance of circular economy-based approaches in agro-food waste management Fig. 1 [6]. These approaches focus on converting waste into raw materials for value-added products, thus mitigating waste accumulation and reducing carbon emissions from conventional material production. Eggshell-based composites have been reported to effectively adsorb heavy metals and synthetic dyes in wastewater [7, 8], with adsorption efficiencies ranging between 20 and 80%, depending on process conditions. These materials are also regenerable, supporting long-term use and reducing operational costs.

Furthermore, mechanochemical treatment of eggshells has been developed to increase their surface area and reactivity as adsorbents or photocatalysts [9, 10]. This method breaks down the solid shell structure into ultrafine particles with enhanced surface reactivity, making them suitable for both laboratory-scale and industrial-scale water purification applications.

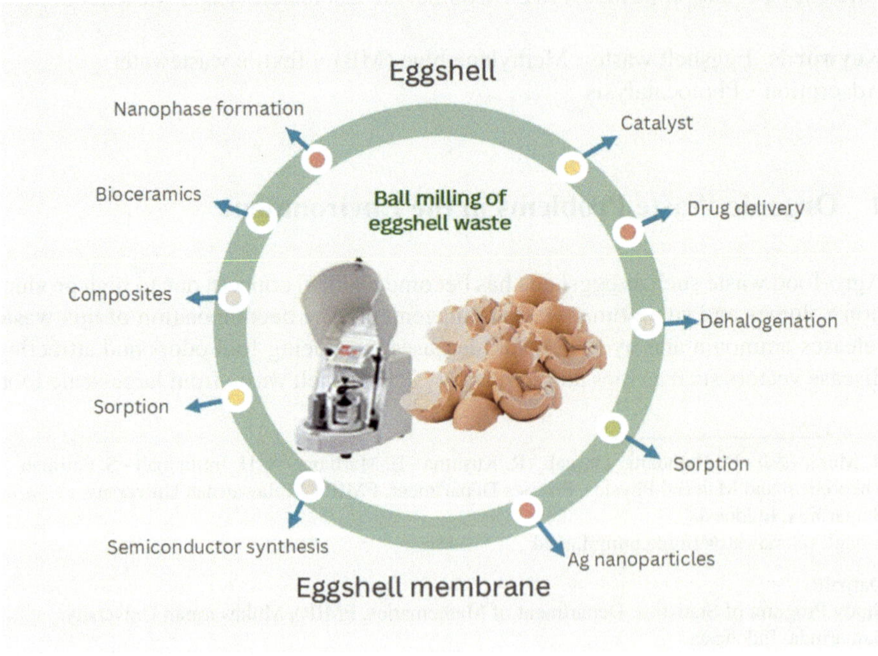

Fig. 1 The effectiveness of mechanochemistry in preparing functional materials

2 Properties and Potential of Eggshell as a Composite Material

Eggshells, primarily composed of calcium carbonate ($CaCO_3$), exhibit unique chemical and structural properties that make them promising candidates for composite material development. Their natural porosity, biocompatibility, and abundant availability—often as agro-food waste—offer a sustainable and low-cost alternative for engineering functional materials [11]. When processed and combined with other active substances such as titanium dioxide (TiO_2) or metal oxides, eggshells can serve as a supportive matrix or active component in photocatalytic, adsorptive, or structural composites [12]. Recent research has demonstrated their applicability in water purification, biomedical scaffolds, and environmentally friendly packaging, emphasizing their versatility in both environmental and industrial applications. The comparison of different lighting technologies should be based on capital cost of the lamp, conversion efficiency, lighting performance, lamp life, various depreciation factors for light output as the lamp ages, as well as labor cost to replace lamps [13].

Eggshells, primarily composed of calcium carbonate ($CaCO_3$), exhibit unique structural and chemical features that make them promising composite matrices. Several studies have demonstrated their use in hydrothermal synthesis of hydroxyapatite–TiO_2 composites with antimicrobial properties [14], as photocatalytic substrates degrading antibiotic contaminants under sunlight-induced carbonate radicals [15], and as supporting matrices for hematite nanoparticle biomedical applications [16]. They further act as osteoconductive fillers within hydrogel scaffolds for bone regeneration and their versatility has been reviewed in the context of sustainable biomaterial applications [17].

3 Development of Composite Materials from Eggshell Waste

The transformation of eggshell waste into value-added composite materials represents a sustainable strategy to address both environmental pollution and waste management challenges. Eggshells, predominantly composed of calcium carbonate ($CaCO_3$) in the form of calcite, exhibit high mechanical strength, large specific surface area, and inherent biocompatibility. These characteristics make them an attractive matrix for the development of composites aimed at water purification applications [18, 19].

In the raw form, eggshells already demonstrate potential as low-cost adsorbents for heavy metal ions, organic dyes, and other aqueous contaminants. Nevertheless, their standalone performance can be further enhanced through structural modification, surface functionalization, and combination with other active components such as metal oxides, polymers, or carbon-based materials. The resulting composites not only improve adsorption efficiency but also introduce new functionalities, such as photocatalytic degradation and antibacterial activity [20, 21].

Several synthesis methods have been explored for the fabrication of eggshell-based composites, including

- Solid-State Dispersion (SSD) [10],
- Sol-Gel Methods [22],
- Hydrothermal Treatments [14],
- Chemical Surface Modification [21].

For instance, composites combining eggshell and TiO_2 nanoparticles have been demonstrated to effectively degrade organic dyes under ultraviolet (UV) or visible light irradiation, owing to the synergistic effect between the photocatalytic activity of TiO_2 and the adsorptive capabilities of the eggshell matrix [22, 23].

Moreover, eggshell composites incorporated with magnetic particles like Fe_3O_4 allow for easy separation and recovery after water treatment, thereby improving reusability and process efficiency. The utilization of eggshell waste thus provides a dual environmental benefit: reducing the burden of agro-industrial waste disposal and supplying an economical, sustainable material platform for water remediation technologies.

4 Synthesis Characterization of Eggshell/TiO_2 Composite

Preparation of eggshell powder is illustrated in Fig. 2. Firstly, washing and membrane removal. Waste eggshells were collected, rinsed under running water, and thoroughly cleaned to remove residual albumen and organic contaminants.

This pre-treatment step is essential in biogenic waste valorization to ensure surface purity prior to drying or further processing. Furthermore, Hot water sterilization is done. Clean eggshells were soaked in hot water at 80 °C for approximately 15 min. Brief immersion at this sub-boiling temperature reduces microbial load effectively without altering the calcite structure. Similar protocols have been reported to sanitize $CaCO_3$-based materials without initiating decomposition [24].

Stepwise drying. After sun-drying to reduce surface moisture, the eggshells were oven-dried at 110 °C for 12 h. Literature recommends drying in the 100–120 °C range for 2–12 h to eliminate moisture while preserving the $CaCO_3$ phase stability. Grinding and sieving, dried samples were pulverized using a household blender, followed by sieving through a 200-mesh screen (~75 μm). This ensures uniform particle size distribution, which enhances specific surface area and improves reproducibility in subsequent composite synthesis.

Synthesis procedures of Eggshells/TiO_2 Composite via Solid State Dispersion (SSD) is illustrated in Fig. 3.

Eggshell powder and commercial TiO_2 (anatase phase) were weighed at a 9:1 mass ratio and mechanically mixed using the Solid-State Dispersion (SSD) method. This low TiO_2 loading preserves the inherent porosity of $CaCO_3$ while introducing

Fig. 2 Eggshell powder preparation through washing, drying, grinding, and sieving

Fig. 3 Synthesis of Eggshells/TiO_2 composite via Solid State Dispersion (SSD) and calcination

sufficient photocatalytic sites. Ethanol dispersion and stirring, the mixed powder was suspended in 95% ethanol (100 mL) and stirred using a magnetic stirrer for 1 h at room temperature. Ethanol reduces surface tension, facilitating homogeneous TiO_2 attachment onto the calcite surface. The suspension was then allowed to settle for 24 h to ensure complete sedimentation [10].

Drying and calcination, the precipitate was dried at 110 °C for 12 h to remove ethanol, followed by calcination at 450 °C for 1 h. Calcination at this temperature enhances particle bonding and TiO_2 anchoring while preventing excessive sintering or phase transformation [4]. Next is final product and significance. The resulting Eggshells/TiO_2 composite is a fine cream-colored powder with well-dispersed TiO_2 nanoparticles anchored on the eggshell matrix. This 9:1 formulation demonstrated >95% methylene blue degradation efficiency while maintaining heavy metal adsorption capacity—validating its selection as the optimal composite for our sustainable water purification study [10].

UV–Vis analysis

Based on the UV-Vis measurement data obtained at a wavelength of 665 nm, the results are presented in Table 1.

Sample 1, which was untreated with any adsorbent or photocatalyst, exhibited the highest absorbance and no color degradation, with a degradation percentage of 0%. This serves as a control to demonstrate that in the absence of an adsorbent or photocatalyst material, methylene blue (MB) remains stable in solution. Sample 2 (TiO_2) showed a significantly lower absorbance value, indicating its strong ability to degrade MB, with a degradation rate of 93.82%. This aligns with the well-known photocatalytic properties of TiO_2 in degrading organic compounds.

In Sample 3 Eggshells showed a higher absorbance than TiO_2, with a degradation percentage of 91.75%. While Eggshells possesses a notable adsorption capacity, the absence of photocatalytic activity like that of TiO_2 results in a slightly lower degradation performance. Sample 4 (composition 9:1) demonstrated the lowest absorbance value, reflecting the highest degradation efficiency at 95.63%. This composition approaches the performance of pure TiO_2, with the added advantage of the adsorptive properties of eggshells which enhances the overall dye removal efficiency.

Sample 5 (composition 8:2) exhibited higher absorbance than the 9:1 composite, with a degradation rate of 93.34%. While still effective, the increased proportion of CT appears to slightly reduce the photocatalytic efficiency of the composite. Sample 6 (composition 7:3) showed even higher absorbance and a degradation rate of 92.16%. With a greater eggshells content, the photocatalytic capability decreases, although adsorption still plays a role in dye removal.

The 9:1 composition is considered the most optimal for methylene blue degradation, closely matching the performance of pure TiO_2 while simultaneously utilizing the adsorptive benefits of eggshell powder. In addition to MB degradation, this study

Table 1 Absorbance, degradation percentage, and pH

Sample	Formula	Absorbansi (nm)	%D	pH
1	MB solution	1442	0	6.72
2	TiO_2	0.089	93.82	7.47
3	Eggshells	0.119	91.75	7.09
4	Eggshells/TiO_2 (9:1)	0.063	95.63	7.91
5	Eggshells /TiO_2 (8:2)	0.069	93.34	7.07
6	Eggshells /TiO_2 (7:3)	0.113	92.16	7.25

also recorded changes in solution pH during the photocatalytic process. It was found that the use of the Eggshells/TiO_2 composite increased the solution pH from 6.72 to 7.91. This indicates that the photocatalytic degradation reaction not only removes pollutants but may also generate byproducts that contribute to an increase in pH. Further analysis of the chemical nature of these byproducts would help in elucidating the underlying degradation mechanisms.

The HRTEM micrograph (Fig. 4a) reveals that raw eggshell (ES) powder is built from loosely agglomerated nano- to sub-micron platelets of calcium carbonate (calcite). The primary crystallites are tens of nanometres across, but they cluster into fluffy, highly porous aggregates roughly 200 nm in dimension, as indicated by the scale bar. Such open, pore-rich morphology is typical for eggshell that has only been ground and low-temperature–calcined; it offers a large external surface and abundant sorption sites that are advantageous for water treatment applications. The selected-area electron diffraction (SAED) pattern beside it displays faint concentric rings, confirming a polycrystalline but relatively low-order calcite phase—consistent with previous reports on eggshell-derived $CaCO_3$ powders prepared below 800 °C [25].

By contrast, the composite Eggshell/TiO_2 (9: 1) micrograph (Fig. 4b) is darker and shows finer, high-contrast domains decorating the eggshell matrix. These darker spots correspond to well-crystallized TiO_2 nanoparticles (NPs) that are anchored on—or partly embedded in—the calcite platelets. The accompanying SAED pattern contains sharper, more intense rings whose measured d-spacings ($\approx$3.5 Å and $\approx$2.4 Å) match the (101) and (004) planes of anatase TiO_2, respectively, superimposed on the broader calcite rings. Such mixed rings verify the intimate heterojunction between anatase TiO_2 and the eggshell support. High-resolution studies of anatase often show 3.52 Å lattice fringes for the (101) planes, identical to those detected here [10]. The nanoscale dispersion of TiO_2 increases interfacial contact area, promotes efficient charge separation, and mitigates NP aggregation—features that directly translate to higher photocatalytic activity.

The porous $CaCO_3$ skeleton derived from waste eggshell not only furnishes a low-cost adsorbent for heavy-metal ions, but also acts as a rugged scaffold that stabilizes photo-active TiO_2. Among the compositions we have screened, a 9:1 Eggshell/TiO_2 ratio strikes the optimal balance: the eggshell fraction maximizes adsorption and light scattering, while the judicious TiO_2 loading supplies enough catalytic sites without inducing particle agglomeration. This synergy explains the superior methylene-blue degradation (>95%) previously reported for the 9:1 composite and underscores the broader circular-economy narrative: upgrading an abundant biowaste into a multifunctional, high-performance material that addresses both dye pollution and heavy-metal remediation in a single, scalable platform.

Morphology and native elemental profile of eggshell powder is shown in Fig. 5a. The SEM micrograph reveals an irregular assembly of angular plate-like particles ranging from sub-micrometer to a few micrometers in size. Inter-particle voids give the powder a highly porous texture that is favorable for adsorptive interactions. EDS quantification shows O $\approx$ 53 at %, C $\approx$ 41 at %, and Ca $\approx$ 6 at %, a ratio close to the $CaCO_3$ stoichiometry reported for raw eggshells in the literature, confirming that the surface is dominated by calcite together with a carbonate-rich organic matrix.

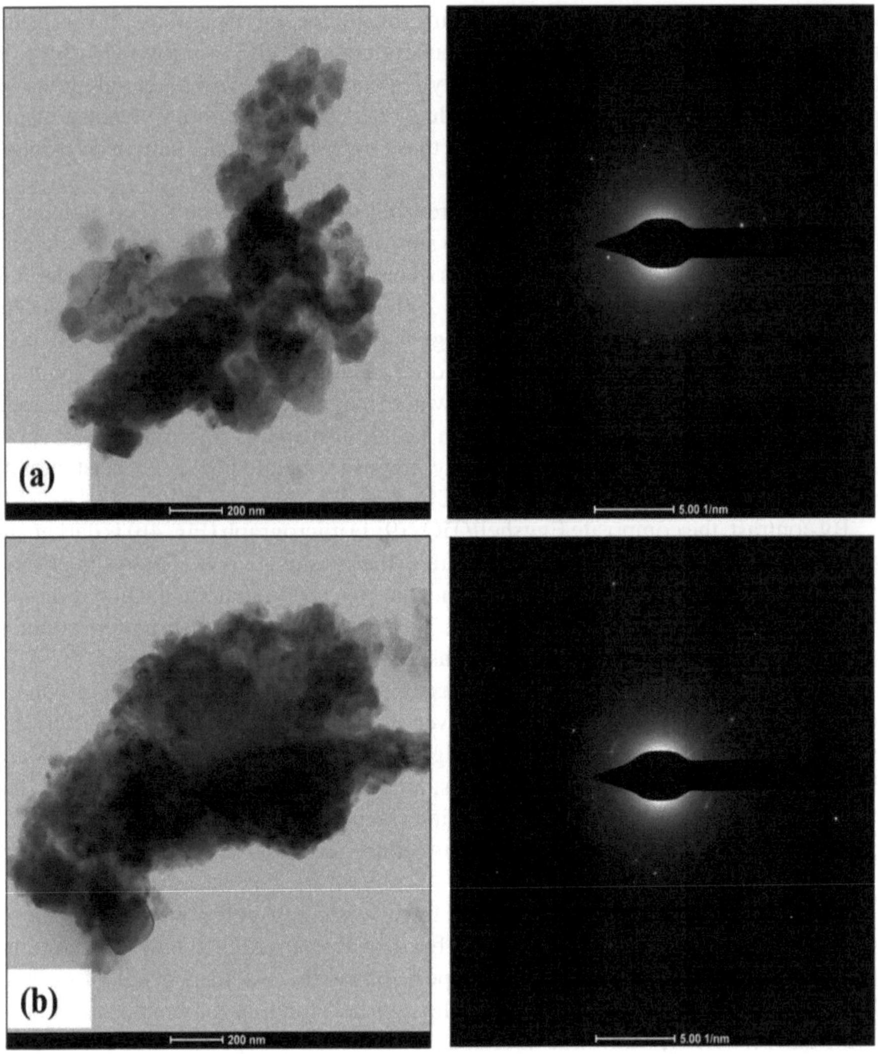

Fig. 4 HRTEM characterization results, **a** HRTEM and SAED of eggshell (ES); **b** HRTEM and SAED of Eggshel/TiO_2 (9:1) composite

Structural changes after TiO_2 incorporation is shown in Fig. 5b. When TiO_2 is introduced at a 9: 1 Eggshells/TiO_2 mass ratio, the SEM image still displays the characteristic calcite fragments, but the surfaces appear rougher and more densely coated by finer particulates—an indication that nano-sized TiO_2 has nucleated on, and partially filled, the eggshell pores. EDS now records a higher Ca signal ($\approx$12 at %) and a slight reduction in carbon, while titanium is detected at trace levels ($<$1 at %). The compositional shift, together with the morphological contrast, points to a conformal TiO_2 layer that enhances interfacial contact without obscuring the $CaCO_3$ framework.

(a)

Element Number	Element Symbol	Element Name	Atomic Conc.	Weight Conc.
8	O	Oxygen	52.82	52.75
6	C	Carbon	40.62	30.46
20	Ca	Calcium	5.79	14.50
22	Ti	Titanium	0.77	2.29

(b)

Element Number	Element Symbol	Element Name	Atomic Conc.	Weight Conc.
8	O	Oxygen	52.68	48.49
6	C	Carbon	35.68	24.65
20	Ca	Calcium	11.65	26.86

Fig. 5 SEM-EDS characterization results, **a** SEM-EDS of eggshell (ES); **b** SEM-EDS of eggshell/TiO_2 (9:1) composite

Such homogeneous dispersion of anatase domains at this low loading has been shown to suppress nanoparticle agglomeration and maximize the catalytic surface area [10].

5 Applications of Eggshells/TiO_2 Composite in Water Purification: Adsorption and Photocatalysis

Composite materials developed from eggshell waste have shown promising potential in water purification, particularly through two primary mechanisms: adsorption and photocatalysis. These approaches target various contaminants such as dyes, heavy metals, and organic pollutants in wastewater. Eggshells, primarily composed of calcium carbonate ($CaCO_3$), possess a porous structure that facilitates adsorption, while the incorporation of photocatalytic agents like titanium dioxide (TiO_2) enhances the degradation of pollutants under light irradiation. The synergy between adsorption and photocatalytic activity makes eggshell-based composites efficient, sustainable, and cost-effective alternatives for environmental remediation [25–28].

5.1 Adsorption Applications

The natural porosity and high surface area of eggshells contribute significantly to their adsorption capabilities. When processed into fine powders or composites, eggshells can effectively adsorb pollutants such as heavy metals (e.g., Pb^{2+}, Cd^{2+}), dyes (e.g., methylene blue), and organic compounds from aqueous solutions. The presence of carbonate and hydroxyl groups on the surface provides active binding sites that interact with various ionic and molecular species in water. In one study, modified eggshell powders demonstrated removal efficiencies exceeding 90% for lead ions, highlighting their applicability in low-cost water treatment systems [28, 29].

In composite form, eggshells can be combined with other functional materials—such as activated carbon, zeolites, or metal oxides—to improve selectivity and adsorption capacity. The integration of TiO_2 or ZnO, for example, not only supports pollutant adsorption but also prepares the material for further photocatalytic action. These composites can be tailored to suit specific environmental conditions and target pollutants, making them adaptable for both industrial wastewater and household greywater treatment [30, 31].

5.2 Photocatalytic Applications

Photocatalysis is another key application of eggshell-based composites, especially when doped or coated with semiconducting materials like TiO_2. Under UV or visible light, TiO_2 generates electron-hole pairs that react with water and oxygen to form reactive oxygen species (ROS), such as hydroxyl radicals (•OH) and superoxide anions ($O_2^{\bullet -}$), which degrade organic pollutants. The role of the eggshell matrix in these composites is crucial—it acts as a support structure, enhancing surface dispersion and preventing nanoparticle agglomeration, thereby improving overall photocatalytic efficiency [32].

Additionally, eggshell-TiO_2 composites have been tested for the degradation of common dye pollutants like methylene blue and rhodamine B. These studies report degradation efficiencies of over 90% under UV light exposure, confirming the potential of these materials in real-world water purification scenarios. Beyond dye degradation, photocatalytic eggshell composites are being explored for pharmaceutical contaminant removal, bacterial disinfection, and even air purification, indicating their broad environmental relevance [33].

6 Conclusion

The work presented in this chapter underscores the promising role of eggshell waste as a functional and sustainable resource in the design of composite materials for water purification. By transforming a commonly discarded biogenic material into a reactive substrate for TiO_2 deposition, we demonstrate how waste valorization can contribute meaningfully to the development of eco-friendly treatment technologies. The eggshell matrix, rich in calcium carbonate and naturally porous, offers an excellent platform for supporting photocatalytically active nanoparticles.

Through a systematic synthesis using the Solid-State Dispersion (SSD) method, the Eggshells/TiO_2 composite with a 9:1 mass ratio was identified as the most effective formulation. This optimal ratio not only preserved the structural integrity and adsorptive capacity of the eggshell but also facilitated efficient photocatalytic degradation of methylene blue (>95%)—a model organic pollutant. The composite exhibited dual-functionality, simultaneously adsorbing metal ions and degrading dye molecules, thereby providing a multifaceted approach to water purification.

Beyond technical performance, the approach presented in this study aligns with broader sustainability goals: minimizing waste, reducing reliance on costly synthetic materials, and promoting circular economy principles. The observed shift in solution pH during photocatalysis further points to underlying chemical processes that merit future exploration. Overall, this research affirms that eggshell-derived composites are not merely waste-based materials, but strategic tools in the advancement of green water treatment solutions.

References

1. Li C, Peirong Y, Zeng Y, Qiu B, Chen M, Cai D, Zhao Z, Chen Z'a, Yaron S, Achmon Y (2025) Impacts of soil biosolarization with eggshell waste on soil health, microbial communities, and gas emissions. Front Sustain Food Syst 9:1543216
2. Ahmad W, Sethupathi S, Munusamy Y, Kanthasamy R (2021) Valorization of raw and calcined eggshell for sulfur dioxide and hydrogen sulfide removal at low temperature. Catalysts 11(295):2021
3. Iftikhar L, Ahmad I, Saleem M, Rasheed A, Waseem A (2024) Exploring the chemistry of waste eggshells and its diverse applications. Waste Manag 189:348–363
4. Ngayakamo B, Onwualu AP (2022) Recent advances in green processing technologies for valorisation of eggshell waste for sustainable construction materials. Heliyon 8(6)
5. Hosny A, Wael N, Hossam MA, Abdelmonem M, El-Sayed SM, Abu-Hussien SH, Abd-Elhalim BT (2025) Eggshell waste as a bioremoval agent for potentially toxic elements/metals and microbial contaminants from raw water of the nile river in Egypt. BMC Res Notes 18(1):171
6. Vigoroso L, Cavallo E, Pampuro N (2024) Recycling of agro-food and urban wastes according to the circular economy and sustainable paradigms. Agronomy 14(7):1465
7. Mensah K, Mahmoud H, Fujii M, Samy M, Shokry H (2024) Dye removal using novel adsorbents synthesized from plastic waste and eggshell: mechanism, isotherms, kinetics, thermodynamics, regeneration, and water matrices. Biomass Convers Biorefinery 14(12):12945–12960
8. Sutherland C, Chittoo BS (2024) Recent developments in the application of Gallus domesticus eggshells for the biosorption of dyes and heavy metals. Adv Environ Eng Res 5(4):1–20

9. Calcite–aragonite transformation in an eggshell: a crucial role of organics and assessment of the impact of milling conditions on its extent using Taguchi design
10. Munir R, Hamdani D, Marliana E, Munir R, Intifadhah SH, Kusuma R (2025) The solid state dispersion method for synthesizing eggshells ES/TiO_2 composite to enhance photocatalytic degradation methylene blue. MethodsX 14:103150
11. Tang, Daniel. KH Valorization of organic waste as biosorbents for wastewater treatment. Water Emerg Contam Nanoplastics 3(4)
12. Priya AK, Muruganandam M, Suresh S (2024) Bio-derived carbon-based materials for sustainable environmental remediation and wastewater treatment. Chemosphere:142731
13. Sharma P, Bano A, Singh SP, Varjani S, Tong YW (2024) Sustainable organic waste management and future directions for environmental protection and techno-economic perspectives. Curr Pollut Rep 10(3):459–477
14. Noviyanti AR, Asyiah EN, Permana MD, Dwiyanti D, Suryana, Eddy DR (2022) Preparation of hydroxyapatite-titanium dioxide composite from eggshell by hydrothermal method: characterization and antibacterial activity. Crystals 12(11):1599
15. Huang Z, Wang J, Yang M-Q, Qian Q, Liu X-P, Xiao L, Xue H (2021) Construction of TiO_2-eggshell for efficient degradation of tetracycline hydrochloride: sunlight induced in-situ formation of carbonate radical. Materials 14(7):1598
16. Ayub M, Bashir M, Majid F, Shahid R, Khan BS, Saeed A, Shaik MR, Kuniyil M, Shaik B, Khan M (2023) Eggshell-mediated hematite nanoparticles: synthesis and their biomedical, mineralization, and biodegradation applications. Crystals 13(12):1699
17. Gouveia T, Eduarda M, Milhans C, Gezek M, Camci-Unal G (2025) Eggshell-based unconventional biomaterials for medical applications. Adv NanoBiomed Res 2400120
18. Gouveia, Maria Eduarda Torres, Charles Milhans, Mert Gezek, Gulden Camci-Unal (2024) Eggshell-based unconventional biomaterials for medical applications
19. Zhang X, He X, Kang Z, Cui M, Yang D-P, Luque R (2019) Waste eggshell-derived dual-functional CuO/ZnO/eggshell nanocomposites (photo) catalytic reduction and bacterial inactivation. ACS Sustain Chem Eng 7(18):15762–15771
20. Alamillo-López VM, Sánchez-Mendieta V, Olea-Mejía OF, González-Pedroza MG, Morales-Luckie RA (2020) Efficient removal of heavy metals from aqueous solutions using a bionanocomposite of eggshell/Ag-Fe. Catalysts 10(7):727
21. Gupta PK, Palanisamy S, Gopal T, Rajamani R, Pandit S, Sinha S, Thakur VK (2021) Synthesis and characterization of novel Fe_3O_4/PVA/eggshell hybrid nanocomposite for photodegradation and antibacterial activity. J Compos Sci 5(10):267
22. Habte L, Shiferaw N, Mulatu D, Thenepalli T, Chilakala R, Ahn JW (2019) Synthesis of nano-calcium oxide from waste eggshell by sol-gel method. Sustainability 11(11):3196
23. Bullo TA, Bayisa YM, Bultum MS (2022) Optimization and biosynthesis of calcined chicken eggshell doped titanium dioxide photocatalyst based nanoparticles for wastewater treatment. SN Appl Sci 4(1):17
24. Liu R, Huang S, Zhang X, Song Y, He G, Wang Z, Lian B (2021) Bio-mineralisation, characterization, and stability of calcium carbonate containing organic matter. RSC Adv 11(24):14415–14425
25. Haq N, Ibrahim MA, Alsarra IA, Alshehri S, Alam P, Shakeel F (2024) Utilization of waste eggshell powder as an excipient for vitamin D3 tablet preparation. ACS Omega 9(3):3980–3987
26. Zein, Annisa D, Mahatmanti FW, Wahyuni S, Widiarti N (2025) Synthesis of TiO_2/hydroxyapatite composite based on chicken egg shells for methylene blue photodegradation. Jurnal Kimia Sains dan Aplikasi 28(1):31–38
27. Ali Ahmed A, Hattab Z, Berredjem Y, Giordana A, Cerrato G, Djellabi R (2024) Heterojunction CuO thin coated CaO drived from eggshell waste for fast removal of organic pollutants from water under visible light. Int J Environ Sci Technol 21(16):9843–9852
28. Zonato d O, Rayane, Estevam BR, Perez ID, dos Santos Ribeiro VA, Boina RF (2022) Eggshell as an adsorbent for removing dyes and metallic ions in aqueous solutions. Cleaner Chem Eng 2:100023

29. Hossain M, Islam R, Rahman MN, Khan MIS, Firoz Ahmed M, Al-Amin, Rabbi MA (2024) A novel approach for the modification of eggshell powder and its application for lead and methylene blue removal. Heliyon 10(16)
30. Ahmad A, Jini D, Aravind M, Parvathiraja C, Ali R, Kiyani MZ, Alothman A (2020) A novel study on synthesis of egg shell based activated carbon for degradation of methylene blue via photocatalysis. Arab J Chem 13(12):8717–8722
31. Salah W, Djeridi W, Houas A, Elsellami L (2024) Synergy between activated carbon and ZnO: a powerful combination for selective adsorption and photocatalytic degradation. Mater Adv 5(4):1667–1675
32. Samsuddin EK, Aida IS, Noraini M, Rus AZM (2020) Preparation of TiO_2 decorated eggshell membrane for photocatalytic dye degradation. In: Materials science forum, vol 1010. Trans Tech Publications Ltd, pp 508–513
33. Ramya S, Vijayakumar S, Vidhya E, Bukhari NA, Hatamleh AA, Nilavukkarasi M, Pham TH (2022) TiO_2 nanoparticles derived from egg shell waste: eco synthesis, characterization, biological and photocatalytic applications. Environ Res 214:113829
34. Aláž M et al (2021) State-of-the-art of eggshell waste in materials science: recent advances in catalysis, pharmaceutical applications, and mechanochemistry. Front Bioeng Biotechnol 8:612567

Transforming Waste into Wealth: The Role of Biorefineries in Advancing a Circular Economy

Fatin Shafiqa Shazili, Qaisara Izzaty Mohd Azni, Siti Nur Halifah Mohd Sani, Wan Zafira Ezza Wan Zakaria, and Noor Aziah Serri

Abstract The increasing generation of organic waste from agricultural, municipal, industrial, and household activities presents both a critical environmental challenge and a valuable opportunity. In the pursuit of a sustainable and circular economy, biorefineries have emerged as transformative systems capable of converting diverse waste streams into high-value products such as biofuels, bioplastics, platform chemicals, fertilizers, and bio-based materials. This chapter explores the pivotal role of biorefineries in advancing circular economy principles by shifting the linear waste-disposal paradigm toward a regenerative model of resource recovery and reuse. The study begins by defining the conceptual framework of the circular economy and examining how biorefineries align with its goals. It then delves into the various thermochemical and biochemical technologies employed in modern biorefineries, including pyrolysis, anaerobic digestion, fermentation, and integrated process systems. Different types of feedstocks, ranging from agricultural residues and food waste to sewage sludge, are evaluated for their suitability and conversion potential. The chapter also highlights the spectrum of value-added products derived from these feedstocks. It assesses their environmental and economic impacts using tools such as life cycle assessment (LCA) and cost–benefit analysis. Further discussion addresses policy support, business models, and community engagement strategies necessary to scale biorefinery solutions. By identifying current challenges and future directions, this research underscores the potential of waste valorisation through biorefineries as a key enabler of sustainable development, resource efficiency, and climate resilience in the global bioeconomy.

Keywords Biorefinery · Circular economy · Life-cycle assessment · Waste

---

F. S. Shazili · Q. I. M. Azni · S. N. H. M. Sani · W. Z. E. W. Zakaria · N. A. Serri (✉)
Bioprocess Technology Division, School of Industrial Technology, Universiti Sains Malaysia, Penang, Malaysia
e-mail: aziah_serri@usm.my

M. A. Kassim et al. (eds.), *Sustainable Biorefinery*, Green Energy and Technology,
https://doi.org/10.1007/978-981-95-8181-8_9

1 Introduction to Circular Economy and Biorefineries

The circular economy has gained significant attention in an era where resources are used intensively but then disposed of after a single use, leading to unsustainable practices that will ultimately harm humans, the environment, and our planet. Therefore, the paradigm shift from linear economy to circular economy is imperative in order to save the planet for future generations' use. Circular economy is a concept of minimizing the products being disposed off to landfills by bringing them back to the production stage [1]. According to Esposito et al., while the linear economy is based on rapid and mass production [2], the circular economy however, is based on four principles:

- Preservation of the natural capital
- Optimization of the available resources
- Risk reduction
- Renewable flow of resources and products

Circular economy is highly interconnected with biorefineries in terms of reducing waste, which directly minimizes the impacts of land use and the emissions of greenhouse gases (GHG) [3]. Biorefinery is a notion of comprehensively utilizing material waste and changing it into a useful resource in a sustainable manner. Biorefineries approach includes various processes such as pyrolysis, fermentation, and anaerobic digestion, where these approaches produce certain goods (chemical, materials, or fuels), depending on the type and feedstock availability [3, 4]. Biorefineries can be categorized into several types, including Conventional Biorefineries (CBR), Green Biorefineries (GBR), Whole Crop Biorefineries (WCBR), Lignocellulosic Feedstock Biorefineries (LCFBR), Marine Biorefineries (MBR), Two-Platform Concept Biorefineries (TPCBR), and Thermo-Chemical Biorefineries (TCBR) [5].

Different biorefineries technologies and processes yield different products and materials depending on the type of feedstock. For example, incineration is typically applied to general waste streams, while pyrolysis and gasification are more effective for organic and inorganic waste. On the other hand, anaerobic digestion is more suitable for organic waste [3]. Recently, the anaerobic digestion process has been widely carried out in managing waste biorefineries due to its simplicity and cost-effectiveness compared to other processes [2]. Figure 1 shows several examples of organic waste sources that are applicable for waste biorefineries [3].

Organic waste stream comprises approximately 42% animal and vegetable waste, 26% of the organic fraction of municipal solid waste, 20% wood waste, and 9% non-hazardous sludge originating from sewage treatment and food processing plants [4]. Moreover, waste biorefineries have been a focus of studies, especially in developing countries such as India, Bangladesh, and Pakistan, to reduce the environmental impact from the impractical waste disposal, especially when the population and urbanization in these countries are rapidly growing. This concept and practice may be a promising solution to improve environments while increasing economic and energy demands [3]. Indeed, biorefineries are an approach that provides many significant

Fig. 1 Examples of organic waste that can be a feedstock for the biorefinery

benefits such as reducing the environmental impacts by minimizing pollution and hazardous products emissions [5]. Moreover, biorefineries are an environmentally friendly method due to their ability to generate green energy as well as lower carbon and water footprint, and producing sustainable products.

2 Biorefinery Technology and Processes

Biorefinery technology leverages a wide array of chemical, biochemical, and thermochemical processes to convert complex biomass feedstocks such as agricultural residues, forestry waste, algae, or food processing by-products into multiple value-added products in a sustainable and efficient manner. The primary objective of a biorefinery is to maximize resource utilization, ensure zero waste generation, and enable the development of a circular bioeconomy.

To achieve this, biorefineries are often designed as multi-output systems, where the various components of biomass (e.g., cellulose, hemicellulose, lignin, proteins, lipids) are fractionated and converted through integrated technologies. According to the technology used in the biomass conversion, biorefineries implement two main types of platforms, which can be used in a single way or combined, which are thermochemical platforms and biochemical platforms [6].

2.1 Thermochemical Process

Thermochemical processes are high-temperature chemical reforming processes with bio-oil, syngas, and biochar as final products. This process includes combustion, gasification, pyrolysis, and hydrothermal liquefaction (HTL) and has its advantages and disadvantages [7]. Figure 2 shows the biomass thermochemical process and the main product produced from the process.

The combustion process refers to the thermal degradation of fuels by reacting with oxygen at high temperature, generating commercially useful intermediate gases, liquids, or solids [7]. During biomass combustion, the moisture in the biomass first

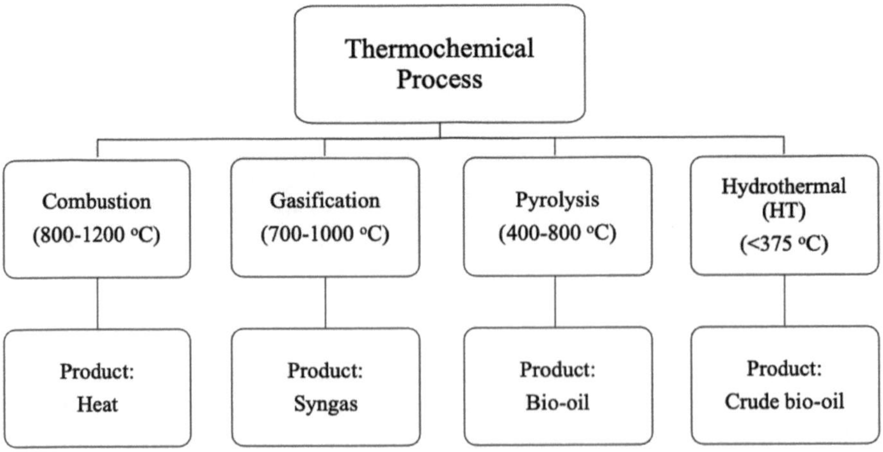

Fig. 2 The main product from the thermochemical process

evaporates. Then, gases released from the biomass burn with oxygen to produce energy. Finally, at higher temperatures, the remaining solid carbon reacts with oxygen to release more energy and complete the combustion process [8]. Although simple and efficient, combustion primarily generates heat and does not yield liquid fuels or chemicals, making it less versatile than other thermochemical options.

Gasification is a thermochemical process that converts carbon-rich biomass materials, such as coal and peat, into a gaseous product primarily composed of synthesis gas (syngas), a mixture of carbon monoxide (CO), hydrogen (H_2), carbon dioxide (CO_2), and small amounts of methane (CH_4) [9]. The process typically operates at high temperatures ranging from 700 to 1000 °C, using a controlled amount of oxygen, air, or steam as the gasifying agent. Gasification begins with the drying of the biomass, followed by pyrolysis, which decomposes the material and releases volatile gases. The volatile gases then undergo further reactions with the gasifying agent to form syngas. Simultaneously, the residual char reacts with CO_2 or steam in reduction reactions, generating additional syngas [10]. Gasification is considered more efficient than combustion and allows better control over emissions.

Pyrolysis is the thermal decomposition of organic matter that occurs in the absence of oxygen, or in conditions where the amount of oxygen is significantly lower than what is required for complete combustion. Under these conditions, the biomass polymer chains break down, producing three main products which is bio-oil, gases, and biochar. Pyrolysis processes are generally classified into slow, fast, and flash pyrolysis, depending on the operating conditions such as temperature, heating rate, and residence time [11]. Fast pyrolysis is optimized to maximize the production of bio-oil and biochar, making it suitable for liquid fuel applications. In contrast, slow pyrolysis favours the generation of biochar and pyrolytic gases, due to the long residence time and low heating rate [7]. The choice of the pyrolysis method depends on the desired end-products and feedstock characteristics, making it a versatile technique within thermochemical conversion processes.

Hydrothermal (HT) processing refers to the thermochemical conversion of biomass in the presence of water as a fluid phase and solid biomass at elevated temperatures, which facilitates the hydrolysis and degradation of lignocellulosic materials into biofuels [10]. HT processes are generally classified into three categories based on the operating temperature and pressure: hydrothermal carbonization (180–250 °C), hydrothermal liquefaction (250–375 °C), and hydrothermal gasification (above 375 °C). In the hydrothermal liquefaction (HTL) process, biomass undergoes a complex series of physical and chemical transformations, leading to the formation of liquefied products. During HTL, small, unstable, and reactive molecules are initially formed, which may undergo repolymerization reactions to produce oily compounds with a broad range of molecular weight distributions [8]. This makes HTL a versatile and promising method for converting wet biomass into high-energy liquid fuels.

2.2 *Biochemical Process*

On the other hand, biochemical conversion technologies are also receiving a lot of interest since they are inexpensive, environmentally friendly, pure, and effective. In addition, biochemical conversion technologies allow for the screening of alternative enzymes or microorganisms, resulting in a wide variety of intermediates that may be used as platforms in the production of sustainable fuels, minerals, and chemicals [6]. Anaerobic digestion and fermentation are common methods in biochemical conversion.

Anaerobic digestion involves a series of four key biochemical stages: hydrolysis, acidogenesis, acetogenesis, and methanogenesis [12]. Each stage is facilitated by distinct groups of microorganisms, including bacteria and archaea, which function together in a complex microbial consortium. The composition and activity of these microbial communities depend on several factors, such as the type of substrate, operating conditions, and environmental parameters [6]. In a single-stage batch reactor, all the waste is introduced at once, allowing the four stages of anaerobic digestion to occur sequentially within the same reactor. After a specified retention period or when biogas production ceases, the digested material, often referred to as compost or digestate, is removed from the reactor. This method is widely used for organic waste treatment and renewable energy generation.

Fermentation is one of the most widely used methods in food processing and biotechnology, with nearly all food commodities such as meat, fish, milk, grains, fruits, and vegetables. Although its primary application was initially aimed at preserving food and extending shelf life, fermentation also offers several additional benefits, including enhanced food safety, nutritional value, and organoleptic qualities (taste, aroma, and texture) [13]. The process is carried out under anaerobic conditions by various microorganisms, including filamentous fungi, yeasts, bacteria, or

Table 1 Summary of four types of fermentation process

Fermentation process	Microorganism used	End product	References
1. Alcoholic	Yeast	Ethanol	[14]
2. Lactic acid	Lactic acid bacteria (LAB)	Lactic acid	[12, 14]
3. Acetic acid	*Acetobacter* species	Acetic acid	[13, 14]
4. Alkaline	*Bacillus* species	Amino acid	[14]

a combination of these. These microorganisms convert fermentable carbohydrates into key end-metabolites such as organic acids, alcohols, and carbon dioxide [14]. There are four main types of fermentation processes commonly employed: alcoholic fermentation, lactic acid fermentation, acetic acid fermentation, and alkaline fermentation, as summarized in Table 1.

Biorefinery technology represents a sustainable approach to converting biomass into a wide range of valuable products, including biofuels, chemicals, and materials [4]. Through both thermochemical and biochemical processes, biorefineries enable the efficient and environmentally friendly utilization of renewable resources, reducing dependency on fossil fuels and minimizing waste [6]. As advancements in processing technologies and feedstock utilization continue to evolve, biorefineries will play a crucial role in supporting a circular bioeconomy and driving the transition toward greener industrial systems.

3 Input Materials and Value-Added Product from Waste

Biorefineries offer integrated, advanced solutions for converting various types of residues and crop waste into biofuels, biochemicals, and other value-added products, supporting a circular bioeconomy and reducing environmental burdens. These facilities utilize a range of technologies such as anaerobic digestion, pyrolysis, and enzymatic hydrolysis to process diverse feedstocks efficiently, maximizing resource recovery and product diversity while minimizing waste and emissions [15]. This section explores the types of feedstocks commonly used in biorefineries and highlights the value-added products that can be derived from these waste materials.

3.1 Agriculture Residues and Crop Waste

Agricultural residues and crop waste are among the most significant by-products generated globally each year due to crop cultivation, harvesting, and processing. These materials are classified into field residues, such as stalks, leaves, stubble, and seed pods, and process residues, including husks, shells, and bagasse. The improper or limited management of these wastes has raised substantial environmental and health concerns, necessitating the development of strategies for their sustainable utilization and valorisation [16].

The primary sources of agricultural waste are diverse, encompassing crop residues, agro-industrial by-products, livestock excreta, and aquaculture waste. Crop residues are abundant, with global production estimated at approximately 2802 million tonnes annually. Notably, rice straw contributes about 731 million tonnes, predominantly in Asia. Wheat straw production ranges from 1 to 3 tons per acre per year, with global output at around 354 million tonnes annually. Corn stover, another major residue, is produced at approximately 4 tons per acre per year, totalling around 128 million tonnes annually. Agro-industrial waste, such as sugarcane bagasse, molasses, and fruit peels, further augments the biomass pool, with sugarcane bagasse alone accounting for nearly 181 million metric tons worldwide [17, 18].

Traditionally, agricultural residues have been utilized as animal feed, soil amendments, fertilizers, and feed for livestock. Rice straw, for instance, is widely used for thatching roofs and as bedding material for livestock. Surface mulching with crop residues helps retain soil moisture, moderate temperature, and suppress weed growth, enhancing soil health and crop productivity. Composting is another common practice, where the organic matter is decomposed under controlled conditions to produce nutrient-rich compost for agricultural use [17].

Agricultural residues are increasingly considered valuable raw materials for various industrial and energy applications. Their transformation into value-added products addresses waste management challenges and supports the transition to a more sustainable and circular bioeconomy. Among the most prominent value-added products are biofuels include bioethanol, biodiesel, and biogas.

Bioethanol: Crop residue is an abundant and valuable biomass resource globally, with its conversion to bioethanol offering significant potential to reduce reliance on non-renewable energy sources. Bioethanol is produced through the fermentation of sugars from starchy, sugary, or cellulosic crops such as sugarcane, corn, and wheat, as well as from plant residues and food-processing by-products. Utilizing these materials in bioethanol production helps address agricultural waste management and supports environmental sustainability [17].

Biodiesel: Biodiesel is another important biofuel derived from agricultural by-products such as waste vegetable oil, rice bran, and non-edible oilseeds. For instance, rice bran production is estimated to be 38.5 million metric tons from rice of 482 million metric tons worldwide, and it contains approximately 17.5% oil [19]. With a free-fatty acid (FFA) content of 8%, this unrefined rice bran oil is processed through chemical reactions to yield a renewable diesel substitute that can be used in conventional engines. The production of biodiesel provides a sustainable energy source and utilizes waste streams that might otherwise contribute to environmental pollution.

Biogas/biomethane: Biogas is a reliable form of bioenergy obtained from anaerobic decomposition or thermochemical conversion of biomass and has CO_2 (20–25%) and CH_4 (50–75%) as the primary constituent gases [20]. It can be produced from various biomass resources, including livestock manure, food-industry waste, household wastes, sewage, energy crops, and crop residues. Several crop residues have proved effective for biogas production, such as rice straws, fallen leaves, wheat straws, and stover [17].

Within biorefineries, these value-added products can be produced in parallel or cascading processes, enabling the extraction of multiple commodities such as biofuels, biochemicals, and biofertilizers from a single feedstock. This multi-product approach maximizes resource use, reduces waste, and supports rural economies by creating new markets and job opportunities.

3.2 Food and Kitchen Waste

Food is a precious commodity whose production is resource-intensive. According to the Food and Agriculture Organization (FAO) of the United Nations, nearly 1.3 billion tons of food products are lost annually along the food supply chain, and this amount is projected to increase exponentially over the next 25 years [21]. Food waste occurs at every stage, from agricultural production through processing to the retail market. Typically, food and kitchen waste are a complex mixture containing carbohydrates such as starch, cellulose, hemicellulose, and lignin, as well as proteins, lipids, organic acids, and smaller inorganic components [22].

Currently, most food waste is recycled as animal feed or compost. However, a significant portion is still incinerated or sent to landfills, where it decomposes anaerobically and emits methane, a greenhouse gas 23 times more potent than carbon dioxide, contributing significantly to climate change. This highlights the urgent need for effective management strategies that follow the principles of reduce, reuse, and recycle [23].

Food waste biorefineries provide innovative solutions by converting these wastes into biofuels, bioplastics, enzymes, and proteins, supporting a circular bioeconomy and minimizing environmental impacts. Modern biorefineries employ microbial fermentation, enzymatic hydrolysis, and anaerobic digestion to process food waste efficiently, enabling the recovery of high-value products and reducing reliance on landfills [15].

Kitchen waste can be transformed into a wide range of value-added products through various biotechnological processes [21]. These products include bioplastics, biopolymers, enzymes, and protein sources like edible insects [8]. Valorising food waste reduces environmental impacts, creates new economic opportunities, and supports a circular economy.

Bioplastics and biopolymers: Bioplastics and biopolymers, particularly polylactic acid (PLA), represent one of the most promising avenues for converting kitchen waste into useful materials. PLA is a biodegradable polymer that can substitute for conventional petrochemical plastics in many applications [24]. Its mechanical and thermal properties depend on the specific types and proportions of lactic acid monomers used in its synthesis. Microbial fermentation is the key technology for producing optically pure lactic acid from food waste, such as orange peel. For example, selected strains of *Lactobacillus delbrueckii* have demonstrated the ability to convert orange peel hydrolysate into D-lactic acid with high yield and optical purity, which is essential for high-quality PLA production. This process not only recycles waste but also generates a valuable material with broad industrial utility [21, 25]. In addition to PLA, other biopolymers can be derived from the carbohydrates, proteins, and lipids present in food waste. The use of microbial consortia can further enhance the degradation of complex waste materials, such as mandarin orange peel, to yield the necessary building blocks for biopolymer synthesis. These advances highlight the versatility of kitchen waste as a feedstock for sustainable material production [21].

Enzymes and Protein Sources: Beyond bioplastics, kitchen waste is a valuable source for the production of enzymes and protein-rich products. For instance, native fungi such as *Xylaria* species have been shown to effectively decompose residues from essential oil extraction, producing enzymes like laccase, cellulase, and xylanase under solid-state fermentation. These enzymes have significant applications in industrial processes and help to valorise otherwise discarded organic waste [26].

Edible insect sources: Protein recovery from food waste is particularly important in addressing both waste management and global protein demand. Spent yeast from brewing processes is a notable example, serving as a raw material for yeast extract production. The nutritional quality of the extract depends on the diversity of the yeast used, highlighting the importance of careful selection for specific applications [21]. Another innovative approach is the use of edible insects, such as black soldier fly larvae, to convert contaminated or biologically unstable food waste into high-quality protein. For example, cereal crops affected by mycotoxin contamination can be fermented and then fed to black soldier fly larvae, which efficiently converts the waste into insect biomass without accumulating toxins. This not only recovers valuable nutrients but also provides a sustainable protein source for animal feed, helping to address protein shortages in the market [23].

Overall, Biorefineries play a pivotal role in transforming food waste into these value-added products by integrating advanced biotechnological processes. This integrated approach not only maximizes resource recovery but also supports sustainable industrial development and addresses global challenges such as plastic pollution and protein shortages.

3.3 Municipal Solid Waste and Sewage Sludge

Municipal Solid Waste (MSW) and sewage sludge present significant management challenges in urban environments. MSW, generated daily by households, businesses, and institutions, consists of various materials such as food waste, paper, plastics, metals, glass, and textiles. Its high organic content makes it particularly suitable for energy recovery through technologies like anaerobic digestion and incineration, which generate energy and reduce landfill dependency [27, 28]. However, managing MSW is complicated by increasing waste volumes due to population growth and urbanization, as well as by inadequate collection and disposal infrastructure, insufficient funding, and inconsistent regulations [29]. The need for public cooperation and adopting sustainable practices compounds these challenges.

Sewage sludge, the by-product of wastewater treatment, contains concentrated pollutants such as heavy metals (including chromium, nickel, copper, zinc, lead, and cadmium), pathogens, and persistent organic compounds [30]. Its complex physical and chemical properties, such as variable moisture content, particle size, and density, make treatment and disposal technically demanding. Improper management can lead to water contamination, public health risks, soil degradation, air pollution from foul odours, and greenhouse gas emissions [31]. Addressing these issues requires robust treatment systems, careful monitoring, and regulated disposal or reuse practices.

Biorefineries offer innovative, integrated solutions to these challenges by transforming MSW and sewage sludge into a spectrum of valuable products such as energy, biofertilizers, and recovered nutrients. These facilities combine multiple conversion technologies like anaerobic digestion, incineration, pyrolysis, and nutrient recovery to maximize resource efficiency and minimize environmental impacts [15]. Notable real-world examples include the URBIOFIN project in Europe, which has demonstrated the techno-economic and environmental viability of converting the organic fraction of MSW into chemical building blocks, bio-based polymers, and additives at a semi-industrial scale [32]. Similarly, wastewater treatment plants in Sweden and Germany have been transformed into biorefineries by integrating processes for biogas production, nutrient extraction, and microalgae cultivation, supporting sustainable urban development and the circular economy [33].

Municipal solid waste (MSW) and sewage sludge are increasingly viewed as valuable resources rather than just waste. Through advanced processing technologies, these materials can be transformed into useful products that support sustainable development and the circular economy. The main categories of value-added products include:

Energy: Municipal solid waste (MSW) and sewage sludge present significant management challenges in urban environments. MSW consists of food waste, paper, plastics, metals, glass, and textiles, with a high organic content suitable for energy recovery through technologies like anaerobic digestion and incineration. Sewage sludge contains concentrated pollutants, including heavy metals and pathogens, making its treatment technically demanding. Improper management can lead to environmental and public health risks, necessitating robust treatment and regulated reuse practices [33, 34].

Biofertilizer: Both MSW and sewage sludge are rich in organic matter and nutrients, making them promising sources for biofertilizer production. Composting mixtures containing sewage sludge, sawdust, and vegetable waste can be treated with microbial inoculants, resulting in nutrient-rich biofertilizers that improve soil health and crop productivity. Composting involves controlled decomposition and fermentation, stabilizing organic matter, and enriching the final product with essential nutrients such as nitrogen, phosphorus, and potassium. Biofertilizers from these sources support sustainable agriculture and help close the nutrient loop, reducing the need for chemical fertilizers and enhancing soil quality [30].

Recovered nutrients: Advanced technologies enable the recovery of valuable nutrients and heavy metals from sewage sludge. Phosphorus, for example, can be recovered as struvite (magnesium ammonium phosphate), which is a valuable raw material for fertilizer production and is both scalable and cost-effective. Biochar, produced through the pyrolysis of sewage sludge, serves as a soil amendment that improves soil quality and can adsorb heavy metals, further reducing environmental risks. Electrochemical and other innovative processes also allow for the extraction of nutrients such as nitrogen and phosphorus, as well as the recovery of heavy metals (e.g., chromium, nickel, copper, zinc, lead, and cadmium) for industrial use, thereby minimizing pollutant release and maximizing resource utilization. These approaches not only mitigate environmental risks but also contribute to the circular bioeconomy and sustainable urban waste management [30].

Through integrated biorefinery processes, MSW and sewage sludge are converted into energy, biofertilizers, and nutrient-rich products, reducing landfill dependency and maximizing resource recovery. Biorefineries enable the simultaneous production of multiple outputs from a single feedstock, supporting a closed-loop system that transforms urban waste into valuable commodities and promotes sustainable urban development.

4 Biorefinery Environmental and Economic Impacts

The biorefinery concept has gained a recognition as a strategic to combat the escalating environmental issue that associated with the waste generation. However, biorefinery approaches needs a reliable systematic methodology to accurately assess their impacts (e.g. greenhouse gases emission, water usage) and validate the sustainability [35]. Such frameworks are found to be imperative to minimize the intensified uncertainties result that arise from the complexity of biorefinery processes at the same time evaluating the viable recycling systems [36].

Life Cycle Assessment (LCA) is notably one of the most widely recognized and accepted tools for evaluating the environmental impacts of products, systems and processes as well as those developed in biorefinery contexts based on the International Organization for Standardisation (ISO) guidelines [35, 37]. Figure 3 show the LCA based on the biorefinery concept. In addition, LCA also has been designed to systematically identify environmental hotspots and impact metrics [36]. LCA embodied four major procedures such as [37]:

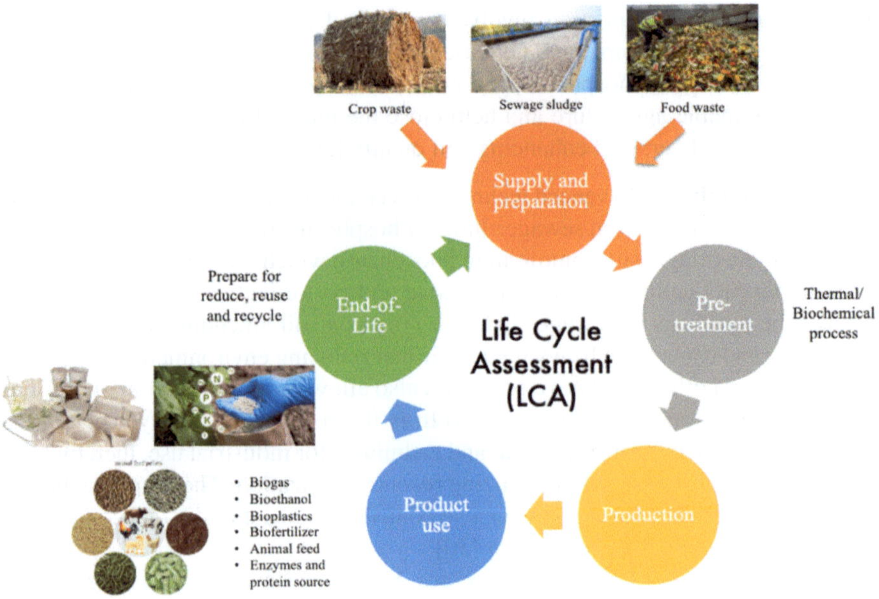

Fig. 3 Life cycle assessment (LCA) based on biorefinery concept

- The goal and scope
- Life cycle inventory
- Life cycle impact assessment
- The results interpretation

However, there are several limitations of the LCA, such as rigid system boundaries, a lack of data, too many differences on statistical methods, which causes uncertainty, and changes in product usability [3]. Despite these challenges, biorefinery approaches are proven to be able to minimize the environmental impact, while offering green energy and products that are sustainable and align with the circular economy principles. Furthermore, the output from the LCA has outlined the crucial impacts of both production processes, such as land occupation, global warming potential, non-renewable energy use and respiratory inorganics [3]. Life Cycle Assessment (LCA) identified four major key environmental impact categories shared by both production systems: land occupation, Global Warming Potential (GWP), non-renewable energy consumption, and emissions of respiratory inorganics, and these findings are consistent with many existing literature [4].

Significantly, biorefinery concepts seem to be promising for a circular bioeconomy to address the current challenges, such as high demand for fuel, depleting resources, and a high amount of waste generated from the development and urbanization [37]. A study by Nizami et al. [3] states that approximately 1.3 billion tons per year of solid waste were generated from the urban cities worldwide, and is predicted to increase up to 2.2 billion tons per year by 2025. The authors also state that municipal solid waste (MSW) comprises a wide range of discarded materials, including electronic waste (e.g., televisions and computers), household and medical waste, as well as waste generated from schools, industries, offices, commercial establishments, and agricultural activities. Therefore, biorefinery approaches present viable solutions for converting MSW into value-added bioproducts such as bioenergy, bioplastics, and biofertilizers, which can be commercialized across industrial, agricultural, and energy sectors [32]. This shows that biorefineries are able to contribute to environmental sustainability by transforming not only renewable biomass but also waste materials into valuable bio-based green products and energy.

Primarily, one of the biorefinery environmental advantages is the potential to reduce the emissions of greenhouse gases (GHG) by replacing fossil-based fuels [36]. This reduction of GHG may be able to mitigate the intense climate change that notably occurred frequent than usual. However, Nizami et al. [3] pointed out that traditional biorefinery approaches were believed to generate more of GHG, such as incineration, that generate high amount of heat and electricity. Hence, incineration was only operated at certain regions with rapid urbanization, for example in the Kingdom of Saudi Arabia. The authors also have outlined several environmental advantages such as minimizing the waste disposed at the landfills, increasing environmental aesthetics, improving the agricultural sectors, and producing green energy [3].

5 Policy in ASEAN Countries

Circular economy is a sustainable model that focuses on resource efficiency, closed-loop systems, and waste reduction to support economic growth and reduce environmental impacts. ASEAN (Association of Southeast Asian Nations) countries consist of ten member countries, including Indonesia, Vietnam, Laos, Brunei, Thailand, Myanmar, Philippines, Cambodia, Singapore, and Malaysia in Southeast Asia, intending to promote economic growth, social progress, and cultural development [38]. Due to rapid urbanization, industrialization, and resource overconsumption, the circular economy becomes a strategic plan to face environmental pressure such as air and water pollution, deforestation, climate change, plastic waste, and inefficient waste management [39]. Effective measures are crucial to address these interlinked issues while promoting a sustainable future in the ASEAN region, because presently, circular economy policies are being adopted globally [39, 40].

The current state and policies related to circular economy in ASEAN region showing Vietnam as the most promising country for circular economy implementation following Indonesia, Malaysia, Philippines and Singapore, which have been actively integrating circular economy principles into their sustainable development strategies, focusing on reducing municipal waste per capita and improving resource efficiency [40, 41]. Furthermore, Brunei, Laos, and Myanmar remain stagnant in circular economy progress, indicating a need for more robust policy frameworks and initiatives [39].

The key policy approaches are Eco-innovation policies where the ASEAN countries are leveraging regulations, incentives and awareness campaigns to promote eco-innovation within the food sector with a goal to reduce food waste and enhance sustainability [40]. The Hanoi 3R declaration (reduce, reuse, recycle) provides a strategic roadmap for establishing circular economy practices in the Asia–Pacific region, including some Southeast Asian countries. Lastly, governments are extensively implementing the Extended Produces Responsibility (EPR) schemes to promote recycling and upcycling programs, encouraging businesses to adopt circular economy practices.

The ASEAN framework for circular economy (2021) is moving towards a resilient, low-carbon and resource-efficient economy, which focuses on harmonization of standards, trade facilitation, innovation, sustainable finance, and efficient resource use. To further understand, as an example, the 12th Malaysia Plan (2021–2025), integrates circular economy, and with ESG and Sustainable Development Goals (SDGs), targeting to reduce greenhouse gas (GHG) intensity, increase recycling waste, and forest cover retention. In addition, the Malaysia Plastic Sustainability Roadmap (2021–2030) aims for zero single-use plastics. Lastly, the Green Technology Master Plan and National Solid Waste Policy focus on construction waste and Extended Producer Responsibility (EPR). Table 2 shows the policy implemented around ASEAN countries presently [38].

Table 2 List of circular economy policies implemented in ASEAN countries

Country	Circular economy policy tools
1. Vietnam	Law Decision 687, CE Hub, National Planning
2. Singapore	Zero waste Masterplan (2019), Green Plan 2030, 10Rs model
3. Malaysia	12th Malaysia Plan (2021–2025), Plastic Sustainability Roadmap (2021–2030), Green Master Plan and National Solid Waste Policy
4. Indonesia	National Action Plan (2020–2024), 9Rs draft, Vision 2045
5. Philippines	Circular economy Act of 2020 (pending), Green Employment Act (2016), Philippine Development Plan (2017–2022), EU support
6. Thailand	Bio-Circular-Green (BCG) economy model, 20-year National Strategy (2017–2036), NESDP (2023–2027)
7. Cambodia	Circular economy strategy and action plan (2020–2034), Clean Air Strategy (2021)
8. Laos	Circular economy strategy for Lao PDR (2017), eco-tourism and bio-economy initiatives
9. Myanmar	National Waste Management Strategy (2018–2030), RecyGlo (startup implementing blockchain-based circular economy waste tracking), Circular Economy Week 2021 for SME's
10. Brunei	Dynamic and Sustainable Economy Blueprint (2020)

In terms of the implementation of circular economy in ASEAN countries, each country can be seen have a different circular economy commitment in which Vietnam and Singapore showing a high commitment to the application of circular economy while Malaysia, Indonesia, Philippines, Thailand, and Cambodia are at a moderate level while Laos, Myanmar and Brunei at the lowest level in the implementation of circular economy. The notable gaps and barriers of circular economy implementation can be attributed to bureaucratic delays, capacity constraints, coordination and enforcement, political instability, weak infrastructure, and policy enforcement and follow-through [39].

To summarize, the ASEAN area recognizes the value of the circular economy, but there are still barriers to overcome, such as behavior change, budget limits, and technology transfer limitations. Thus, all sectors must make commitments, and additional policies should be adopted by government institutions in conjunction with the business community to ensure that the circular economy benefits of all stakeholders.

6 Challenges and Future Directions

As biorefineries emerge as a pivotal solution for sustainable waste management and bio-based production, they face a complex set of challenges that must be addressed to unlock their full potential. While the concept of converting waste into wealth is compelling, practical implementation requires overcoming technical, economic, and societal hurdles. This chapter outlines four primary areas of concern: technical scalability, feedstock logistics, market dynamics, and systemic integration, and offers insights into future opportunities for research, policy, and industrial collaboration.

6.1 Technical Barriers and Scalability

Biorefineries rely on sophisticated biochemical and thermochemical processes to convert various types of biomasses into fuels, chemicals, and materials. However, many of these processes remain limited by low efficiency and high operational costs. Among the most pressing technical challenges are:

Pre-treatment limitations: Lignocellulosic biomass, which constitutes a large fraction of agricultural and forestry waste, requires extensive pre-treatments to break down its tough, fibrous structure. Current pre-treatment methods are energy-intensive and expensive, often involving harsh chemicals or high temperatures [42].

Enzyme performance: The enzymes used in hydrolysis and fermentation processes are sensitive to industrial conditions such as temperature, pH, and contaminant presence. Moreover, enzyme costs remain high, particularly for large-scale applications [43].

Downstream processing: Separation and purification of target products from complex biomass mixtures can be both challenging and costly, especially when yields are low or when co-products interfere with recovery processes [44].

Scalability gap: While many bioconversion technologies have shown promise at the laboratory or pilot scale, translating these results to industrial scale often leads to unforeseen inefficiencies or bottlenecks. Issues such as reactor design, heat and mass transfer limitations, and control systems need to be carefully adapted for scale-up [45].

To overcome these barriers, ongoing research focuses on process intensification, genetically engineered microbes, and consolidated bioprocessing, where multiple stages of biomass conversion are integrated into a single step. Additionally, digital twins and process simulation tools are being explored to optimize design and operations before physical implementation.

6.2 Feedstock Availability and Supply Chain Logistics

The success of a biorefinery is inherently tied to the availability, cost, and consistency of its biomass feedstock. Unlike fossil-based refineries, which deal with uniform raw materials, biorefineries must contend with variable and decentralized biomass sources. Key logistical issues include:

Seasonal and regional variability: Agricultural residues such as corn stover or rice husks are only available during harvest seasons, and their supply may fluctuate due to climate, crop rotation, or market forces. Additionally, different regions produce different types of biomasses, complicating standardization [45].

Transportation and storage: Biomass is bulky and has low energy density, making long-distance transport economically unviable. Storage conditions must also be carefully controlled to prevent microbial degradation or moisture-related spoilage [46].

Competing uses: Biomass resources may also be used for animal feed, soil conditioning, or traditional energy applications, creating competition that affects pricing and availability [47].

Addressing these challenges will require the development of regionally adapted supply chains, biomass densification technologies (e.g., pelletization or torrefaction), and feedstock blending strategies to maintain process consistency. Policymakers can also support biomass aggregation centres and logistics hubs that serve as intermediaries between producers and refineries.

6.3 *Market Acceptance and Consumer Perception*

Even when biorefinery technologies are proven and feedstocks are secured, success depends heavily on market demand and consumer support. Several factors influence the adoption of biobased products:

Cost competitiveness: Many biobased products currently have higher production costs compared to their petroleum-derived counterparts, making them less attractive in markets driven by price competition [48].

Awareness and education: Consumers and industries often lack information about the environmental and performance benefits of biobased alternatives. Misconceptions about quality, durability, or safety can hinder adoption [49].

Policy support: Without strong policy frameworks, including green procurement, subsidies, and labelling standards, biobased products struggle to gain traction in mainstream markets [50].

Efforts to improve market acceptance should focus on public awareness campaigns, eco-labelling, and green certification schemes. Furthermore, life cycle assessments (LCA) and carbon footprint labelling can help demonstrate the environmental benefits of biorefinery products to both consumers and policymakers.

6.4 *Integration with Other Circular Economy Initiatives*

Biorefineries function best when embedded within a broader system of circular economy strategies. This includes collaboration with waste producers, utilities, and other industries to enable resource sharing, energy recovery, and symbiotic operations. However, several integration challenges persist:

Systemic fragmentation: Biorefineries often operate in isolation from municipal waste systems, agricultural networks, or industrial parks, missing opportunities for feedstock access, utility exchange, or by-product valorisation [51].

Lack of standardized metrics: There is no universally accepted method to quantify circularity or compare the sustainability of different processes, making it difficult to align goals or evaluate trade-offs [52].

Regulatory gaps: Policies governing waste classification, product safety, and industrial collaboration are not always harmonized, especially when spanning sectors like agriculture, energy, and manufacturing [53].

Future progress will require cross-sector partnerships, smart industrial symbiosis, and the use of digital platforms to track material flows and coordinate resource sharing. Integrating biorefineries with eco-industrial parks and municipal waste valorisation programs can help create closed-loop systems that maximize resource efficiency and minimize waste.

While biorefineries offer a powerful pathway to convert waste into valuable products, their long-term success hinges on addressing several critical challenges. By focusing on technical innovation, robust logistics, market engagement, and systemic integration, stakeholders can unlock the full potential of biorefineries in a circular economy. The future of sustainable production and waste management depends not only on technology but also on collaborative efforts among researchers, industries, governments, and consumers.

7 Conclusion

The transformation of waste into value-added products through biorefineries represents a powerful and practical pathway toward achieving a circular economy. By harnessing advanced thermochemical and biochemical technologies, biorefineries can convert diverse organic waste streams such as agricultural residues, food waste, municipal solid waste, and industrial by-products into a wide array of sustainable products, including biofuels, bioplastics, platform chemicals, fertilizers, and more. This process not only diverts waste from landfills and reduces environmental pollution but also contributes to resource efficiency, energy security, and economic growth.

The integration of biorefineries into national and regional circular economy frameworks requires strong policy support, innovative business models, and collaboration across sectors. Life cycle assessment and economic evaluations further confirm the environmental and financial viability of such systems, particularly when designed with local feedstock availability and community needs in mind. Despite existing challenges related to technology scalability, feedstock logistics, and market acceptance, continued innovation, public–private partnerships, and regulatory incentives can accelerate the deployment of biorefinery systems.

Ultimately, biorefineries exemplify the core principles of circularity closing loops, minimizing waste, and regenerating value from discarded materials. As the world moves toward more sustainable and resilient economic systems, the strategic development of biorefineries will be essential in turning the concept of "waste" into an opportunity for wealth, well-being, and planetary health.

References

1. Nikolaou IE, Tsagarakis KP (2021) An introduction to circular economy and sustainability: some existing lessons and future directions. Sustain Prod Consum 28:600–609. https://doi.org/10.1016/j.spc.2021.06.017
2. Esposito M, Tse T, Soufani K (2018) Introducing a circular economy: new thinking with new managerial and policy implications. Calif Manage Rev 60(3):5–19. https://doi.org/10.1177/0008125618764691

3. Nizami AS, Rehan M, Waqas M, Naqvi M, Ouda OKM, Shahzad K, Miandad R, Khan MZ, Syamsiro M, Ismail IMI, Pant D (2017) Waste biorefineries: enabling circular economies in developing countries. Biores Technol 241:1101–1117. https://doi.org/10.1016/j.biortech.2017.05.097
4. Alibardi L, Astrup TF, Asunis F, Clarke WP, De Gioannis G, Dessì P, Lens PNL, Lavagnolo MC, Lombardi L, Muntoni A, Pivato A, Polettini A, Pomi R, Rossi A, Spagni A, Spiga D (2020) Organic waste biorefineries: looking towards implementation. Waste Manage 114:274–286. https://doi.org/10.1016/j.wasman.2020.07.010
5. Awasthi MK, Sindhu R, Sirohi R, Kumar V, Ahluwalia V, Binod P, Juneja A, Kumar D, Yan B, Sarsaiya S, Zhang Z, Pandey A, Taherzadeh MJ (2022) Agricultural waste biorefinery development towards circular bioeconomy. Renew Sustain Energy Rev 158:112122. https://doi.org/10.1016/j.rser.2022.112122
6. Manikandan S, Vickram S, Sirohi R, Subbaiya R, Krishnan RY, Karmegam N, Sumathijones C, Rajagopal R, Chang SW, Ravindran B, Awasthi MK (2023) Critical review of biochemical pathways to transformation of waste and biomass into bioenergy. Biores Technol 372:128679. https://doi.org/10.1016/j.biortech.2023.128679
7. Dermibas A (2010) Thermochemical processes. In: Biorefineries. Green energy and technology, pp 135–192. https://doi.org/10.1007/978-1-84882-721-9_6
8. Koçer AT, Özçimen D (2021) Determination of combustion characteristics and kinetic parameters of Ulva lactuca and its biochar. Biomass Conver Biorefin 14:5913–5922. https://doi.org/10.1007/s13399-020-01245-4
9. Gholizadeh M, Castro C, Fabrega SM, Clarens F (2023) A review on thermochemical based biorefinery catalyst development progress. Sustain Energy Fuels 7:4758–4804. https://doi.org/10.1039/d3se00496a
10. Schmid JC, Benedikt F, Fuchs J, Mauerhofe AM, Hofbauer S, Hermann M (2021) Syngas for biorefineries from thermochemical gasification of lignocellulosic fuels and residues—5 years' experience with an advanced dual fluidized bed gasifier design. Biomass Conver Biorefin 11:2405–2442. https://doi.org/10.1007/s13399-019-00486-2
11. Seo MW, Lee SH, Nam H, Lee D, Tokmurzin D, Wang S, Park Y-K (2022) Recent advances of thermochemical conversion processes for biorefinery. Biores Technol 343:126109. https://doi.org/10.1016/j.biortech.2021.126109
12. Meegoda JN, Li B, Patel K, Wang LB (2018) A review of the processes, parameters, and optimization of anaerobic digestion. Environ Res Public Health 15(10):2224. https://doi.org/10.3390/ijerph15102224
13. Allwood JG, Wakeling LT, Bean DC (2021) Fermentation and the microbial community of Japanese koji and miso: a review. Concise Rev Hypoth Food Sci 86(6):2194–2207. https://doi.org/10.1111/1750-3841.15773
14. Kaur P, Ghoshal G, Banerjee UC (2019) Traditional bio-preservation in beverages: fermented beverages. In: Preservatives and preservation approaches in beverages. pp 69–113. https://doi.org/10.1016/B978-0-12-816685-7.00003-3
15. Fadzil NF, Othman SA (2021) The growing biorefinery of agricultural wastes: a short review. J Sustain Nat Resourc 2(2):46–51
16. Jena S, Singh R (2022) Agricultural crop waste materials–a potential reservoir of molecules. Environ Res 206:112284
17. Koul B, Yakoob M, Shah MP (2022) Agricultural waste management strategies for environmental sustainability. Environ Res 206:112285
18. Kumar P, Singh RK (2021) Selection of sustainable solutions for crop residue burning: an environmental issue in northwestern states of India. Environ Dev Sustain 23:3696–3730
19. Hoang AT, Tabatabaei M, Aghbashlo M, Carlucci AP, Ölçer AI, Le AT, Ghassemi A (2021) Rice bran oil-based biodiesel as a promising renewable fuel alternative to petrodiesel: a review. Renew Sustain Energy Rev 135:110204. https://doi.org/10.1016/j.rser.2020.110204
20. Czekała W, Frankowski J (2023) Agricultural plant residues as potential co-substrates for biogas production. Energies. https://doi.org/10.3390/en16114396

21. Mamma D (2020) Food wastes: feedstock for value-added products. Fermentation 6(2):10–13. https://doi.org/10.3390/FERMENTATION6020047
22. Moonsamy TA, Rajauria G, Priyadarshini A, Jansen MAK (2024) Food waste: analysis of the complex and variable composition of a promising feedstock for valorisation. Food Bioprod Process 148:31–42. https://doi.org/10.1016/j.fbp.2024.08.012
23. Polruang S, Talang RPN, Sirivithayapakorn S, Sorn WN (2024) Alternative crop residue management practices to mitigate the environmental and economic impacts of open burning of agricultural residues. Sci Rep 14. https://doi.org/10.1038/s41598-024-65389-3
24. Ahmad A, Banat F, Alsafar H, Hasan S (2022) An overview of biodegradable poly (lactic acid) production from fermentative lactic acid for biomedical and bioplastic applications. Biomass Convers Biorefin 14:1–20. https://doi.org/10.1007/s13399-022-02581-3
25. Shalem A, Yehezkeli O, Fishman A (2024) Enzymatic degradation of polylactic acid (PLA). Appl Microbiol Biotechnol 108(1):413. https://doi.org/10.1007/s00253-024-13212-4
26. Sun P-P, Chen C, Sharma V, Kuo C-H, Tsai M, Dong C, Nargotra P (2022) Agro-industrial food waste as a low-cost substrate for sustainable production of industrial enzymes: a critical review. Catalysts. https://doi.org/10.3390/catal12111373
27. Ashani PN, Shafiei M, Karimi K (2020) Biobutanol production from municipal solid waste: technical and economic analysis. Biores Technol 308:123267. https://doi.org/10.1016/j.biortech.2020.12326
28. Lu J-S, Chang Y, Poon C-S, Lee D-J (2020) Slow pyrolysis of municipal solid waste (MSW): a review. Biores Technol 312:123615. https://doi.org/10.1016/j.biortech.2020.123615
29. Ebrahimian F, Karimi K (2020) Efficient biohydrogen and advanced biofuel coproduction from municipal solid waste through a clean process. Biores Technol 300:122656
30. Jumasheva K, Syrlybekkyzy S, Serikbayeva A, Nurbaeva F, Kolesnikov A (2023) Study on the composition and environmental impact of sewage sludge. https://consensus.app/papers/study-on-the-composition-and-environmental-impact-of-jumasheva-syrlybekkyzy/ea20d047907254e09ff1c844733de667/
31. Okopi S, Li Y, Xu F (2024) Biomass digestion. In: Encyclopedia of sustainable technologies, vol 1–4, 2nd edn. Elsevier, pp V3:236–V3:251. https://doi.org/10.1016/B978-0-323-90386-8.00051-6
32. Vyas S, Shah A, Varjani S, Prajapati P (2022) Municipal solid waste management: dynamics, risk assessment, ecological influence, advancements, constraints and perspectives. Sci Total Environ 152802. https://doi.org/10.1016/j.scitotenv.2021.152802
33. Appleby G, Lima D, Li L (2024) A review of renewable energy technologies in municipal wastewater treatment plants (WWTPs). Energies. https://doi.org/10.3390/en17236084
34. Venkiteshwaran K, McNamara P, Zitomer D, Seyedi S, Mayer B, Raju A, Liu Z (2020) The state of technologies and research for energy recovery from municipal wastewater sludge and biosolids. Curr Opin Environ Sci Health. https://doi.org/10.1016/j.coesh.2019.12.004
35. Duan Y, Tarafdar A, Kumar V, Ganeshan P, Rajendran K, Shekhar Giri B, Gómez-García R, Li H, Zhang Z, Sindhu R, Binod P, Pandey A, Taherzadeh MJ, Sarsaiya S, Jain A, Kumar Awasthi M (2022) Sustainable biorefinery approaches towards circular economy for conversion of biowaste to value added materials and future perspectives. Fuel 325. https://doi.org/10.1016/j.fuel.2022.124846
36. Katakojwala R, Mohan SV (2021) A critical view on the environmental sustainability of biorefinery systems. Curr Opin Green Sustain Chem 27:100392. https://doi.org/10.1016/j.cogsc.2020.100392
37. Ubando AT, Felix CB, Chen WH (2020) Biorefineries in circular bioeconomy: a comprehensive review. Bioresour Technol 299. https://doi.org/10.1016/j.biortech.2019.122585
38. Association of Southeast Asian Nations (ASEAN) (2021) Framework for circular economy for the ASEAN economic community, Jakarta. https://asean.org/wpcontent/uploads/2021/10/Brochure-Circular-Economy-Final.pdf
39. Herrador M, Van ML (2024) Circular economy strategies in the ASEAN region: a comparative study. Sci Total Environ 908:168280. https://doi.org/10.1016/j.scitotenv.2023.168280

40. Hadfield P, Ningrum D, Aditya B, et al (2025) Transformative principles for circular economy transitions in the Global South. npj Urban Sustain 5:34. https://doi.org/10.1038/s42949-025-00225-9
41. Jermsittiparsert K, Pamornmast C, Sriyakul T (2020) Sustainable development and circular economy: functional vs. economic wellbeing in ASEAN. J Sec Sustain Issues 10:414–425. https://doi.org/10.9770/jssi.2020.10.Oct(33)
42. Jain A (2017) UN environment summary report; waste management in ASEAN countries 2017. United Nations Environment Programme, Thailand
43. Kululo WW et al (2025) Advances in various pretreatment strategies of lignocellulosic substrates for bioethanol: a comprehensive review. Discov Appl Sci 7:476
44. Mosier N et al (2005) Features of promising technologies for pretreatment of lignocellulosic biomass. Biores Technol 96(6):673–686
45. Nair SK, et al (2024) Biomass supply chain risk: feedstock availability and variability. In: Handbook of biorefinery research and technology. pp 285–312
46. Munonye WC (2025) Towards circular economy metrics: a systematic review. Circ Econ Sustain 1–43
47. Cookey PE, Cofie O, Koottatep T, Polprasert C (2022) Sanitation biomass recovery and conversion. Integr Funct Sanit Value Chain 125
48. Pandey A, Sharma YC (2024) Advancements in biomass valorization in integrated biorefinery systems. Biofuels, Bioprod Biorefin 18(6):2078–2090
49. Zhang C, Singh RP, Yadav P, Kumar I, Kaushik A, Roychowdhury R, Wang J (2025) Recent advances in biotechnology and bioengineering for efficient microalgal biofuel production. Fuel Process Technol 270:108199
50. Mohammad REA, Abdullahi SSA, Muhammed HA, Musa H, Habibu S, Jagaba AH, Birniwa AH (2024) Recent technical and non-technical biorefinery development barriers and potential solutions for a sustainable environment: a mini review. Case Stud Chem Environ Eng 9:100586
51. Sulis DB, Lavoine N, Sederoff H, Jiang X, Marques BM, Lan K, Wang JP (2025) Advances in lignocellulosic feedstocks for bioenergy and bioproducts. Nat Commun 16(1):1244
52. Kim S, Dale BE, Basso B, Thelen K, Maravelias CT (2023) Supply chain system for a centralized biorefinery system based on switchgrass grown on marginal land in Michigan. Biofuels, Bioprod Biorefin 17(6):1502–1514
53. Bahnaru R, Hlaciuc E (2024) Circular economy measurement tools, challenges, and global standardization. Europ J Account Fin Bus 102–110.e.com/k110

Biomass to Bioenergy: Thermochemical Conversion of Lignocellulosic and Microalgal Feedstock

R. R. Dirgarini Julia Nurlianti Subagyono, Alan L. Chaffee, and Marc Marshall

Abstract Thermochemical conversion of lignocellulosic and microalgal biomass offers a viable alternative to fossil fuels. This chapter discusses three primary pathways: pyrolysis, liquefaction and CO/H_2O reaction, highlighting their mechanisms, operating conditions and catalytic enhancements. Wood and microalgae function as model feedstocks, with their structural and compositional properties affecting conversion efficiency and product quality. Pyrolysis, especially when catalysed, improves oil quality by reducing oxygenated molecules and increasing hydrocarbon content. Liquefaction techniques, both direct and indirect, provide alternate means for generating liquid fuels with potential biofuel uses; nonetheless, the products require enhancement due to concerns over product viscosity and stability. The CO/H_2O reaction pathway is notable for its ability to produce substantial oil yields with limited hydrogen input, particularly from microalgae abundant in proteins and lipids. Methods for upgrading bio-oil, including hydrotreating, hydrocracking and zeolite catalysis, are evaluated to enhance bio-oil compatibility with current engine systems. The chapter underscores the importance of feedstock selection, reaction conditions and catalyst selection in enhancing biofuel production.

Keywords Thermochemical conversions · Microalgae · Wood · Biofuel

R. R. Dirgarini Julia Nurlianti Subagyono (✉)
Physical Chemistry Laboratory, Faculty of Mathematics and Natural Sciences, Mulawarman University, Samarinda, East Kalimantan, Indonesia
e-mail: dirgarini@fmipa.unmul.ac.id

A. L. Chaffee · M. Marshall
School of Chemistry, Monash University, Melbourne, Australia

M. A. Kassim et al. (eds.), *Sustainable Biorefinery*, Green Energy and Technology,
https://doi.org/10.1007/978-981-95-8181-8_10

1 Biomass Source

Research in biofuels has evolved throughout the years, mostly focusing on comparing various kinds of biomass and production methodologies. First-generation biofuels are mostly derived from triglyceride-producing plants, predominantly edible varieties such as palm, peanut, coconut and soy [1, 2]. The utilisation of food—source biomass prompted a global discourse, mostly due to competition for land with food crops and the application of pesticides and fertilisers. The utilisation of palm oil as a biofuel is thought to cause rainforest destruction due to the significant expansion of palm farms [3].

Second-generation biofuels originate from non-consumable biomass, typically lignocellulosic and starch-based plants [2]. Ligno-cellulosic biomass (wood) has gathered greater interest than triglyceride biomass due to its lower cost, greater availability in substantial amounts [4] and lesser significance as a human food source. Recently, third-generation biofuels generated from microalgae have emerged as an alternative [5].

1.1 Wood

Wood primarily consists of lignin, cellulose and hemicellulose. Wood also contains other components, including inorganic minerals, pectins, starch and proteins, albeit in minimal proportions [6, 7]. Hardwood and softwood differ in their lignin and hemicellulose composition; hardwood has a lower lignin content and a higher hemicellulose content compared to softwood [8]. The name 'lignin' was initially coined by Schulze in 1865 [9], denoting an amorphous, very intricate aromatic compound characterised by an uneven arrangement of hydroxy- and methoxy-substituted phenylpropane units with diverse bonding configurations [6, 7]. The precursors of lignin production include p-coumaryl alcohol (p-hydroxyphenyl unit), coniferyl alcohol (guaiacyl unit) and sinapyl alcohol (syringyl unit) [6, 7, 9] (Fig. 1). Coniferyl alcohol serves as a principal precursor of softwood lignin, whereas both coniferyl alcohol and sinapyl alcohol are significant precursors of hardwood lignin; p-coumaryl alcohol is a minor precursor for both softwood and hardwood lignins [6, 7, 9]. Wood lignin primarily has three structural units (p-hydroxyphenyl, guaiacyl and syringyl), which are polymerisation derivatives of their respective lignin precursors [7] (Fig. 1). Softwood lignins predominantly consist of 'guaiacyl lignin', whereas hardwood lignins primarily comprise 'syringyl-guaiacyl lignin' [7]. Proposed configurations of softwood lignin are illustrated in Fig. 3.

Cellulose is a glucan polymer composed of linear chains of 1,4-β-linked anhydroglucose units [6]. Wood-derived cellulose exhibits a mostly highly crystalline structure due to its elevated packing density [6, 7]. Wood cellulose possesses an average degree of polymerisation (DP) of 9000–15,000 glucose units, whereas hemicellulose, also a polysaccharide polymer generated from wood, has a lower average

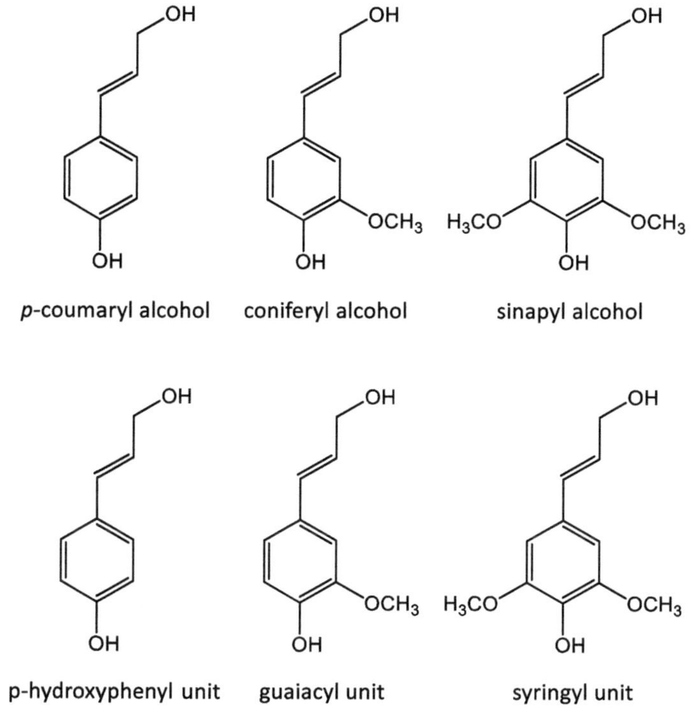

Fig. 1 The precursors of lignin biosynthesis and the three main building blocks of wood lignin

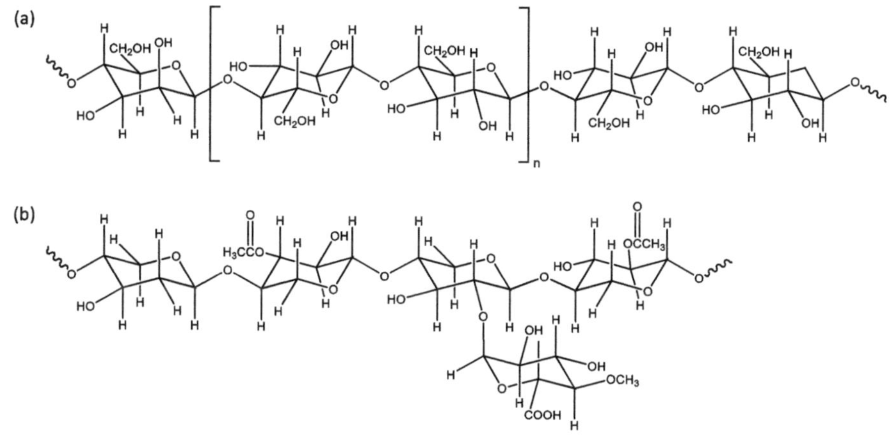

Fig. 2 Partial molecular structures of **a** cellulose and **b** hemicellulose

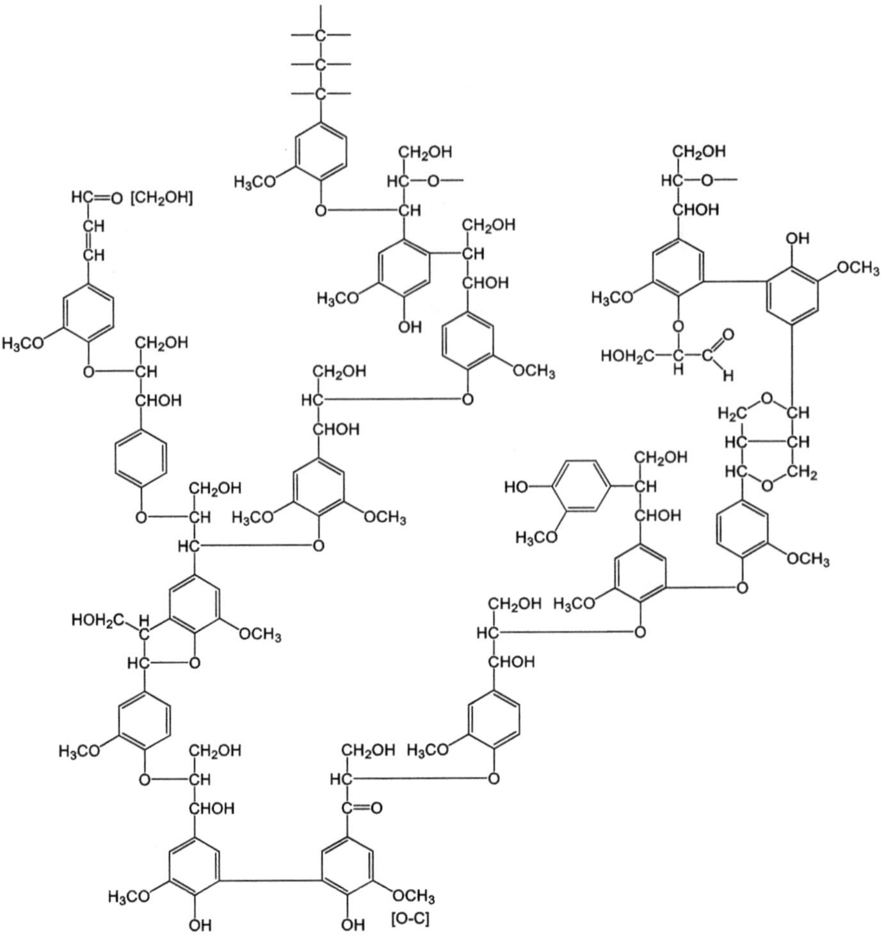

Fig. 3 Structures of softwood lignin

DP of 100–200 glucose units [7] and may exhibit branching [6]. Figure 2 illustrates partial structures of cellulose and hemicellulose (glucuronoxylan).

1.2 *Microalgae*

Microalgae are microscopic photosynthetic organisms primarily made of carbohydrates, lipids, proteins and chlorophyll. Carbohydrates in several algae consist of substantial molecules of starch or starch-like substances, primarily featuring a molecular backbone of α-1,4-linked glucan with α-1,6-linked side chains [10]. Algal lipids

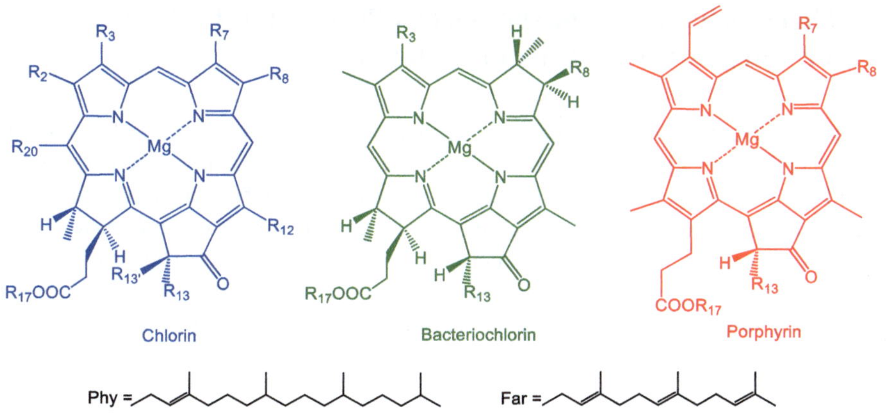

Fig. 4 Structures of common chlorophylls (R = CH_3, C_2H_3, C_2H_5, $COOCH_3$ or other combinations)

and proteins exhibit significant diversity and can be altered by environmental variations, such as differing nutritional medium compositions [10, 11]. An elevation in nitrogen concentration in the medium led to a significant augmentation in total protein and chlorophyll content, alongside a reduction in total lipid for green algae, while blue-green algae exhibited a gradual increase in chlorophyll content [11]. The fundamental structure of chlorophylls is a porphyrin macrocycle composed of four interconnected pyrrole rings linked by carbon bridges [12, 13]. A coordinating metal ion, typically magnesium, is situated at the centre of the four pyrrole rings, with the four nitrogen atoms oriented inward [12] (Fig. 4).

Microalgae present numerous advantages over terrestrial biomass; they exhibit superior photosynthetic efficiency, accelerated growth rates resulting in enhanced productivity, significantly greater oil output, no competition with food production systems, capacity to sequester atmospheric CO_2, ability to thrive on marginal and non-arable lands, contaminated waters, saline or brackish environments, and adaptability to fluctuating climatic conditions [4, 14–16]. Nonetheless, in the growth of microalgae, some challenges must be addressed for large-scale biofuel production. This encompasses the production cost, which presently exceeds that of lignocellulosic biomass, as well as the accessibility of water, CO_2, sunlight and level terrain [4, 17].

2 Thermochemical Conversions as Production Pathways for Alternative Fuels from Biomass

Numerous established techniques for producing alternative fuels exist, including transesterification [18–20] and thermochemical conversions; nevertheless, certain liquid compounds generated by these processes are unsuitable for direct use as fuel and cannot easily be upgraded to usable fuels. Thermochemical conversions represent a distinct group of methods for generating fuel from biomass. The primary subclasses of thermochemical conversion are pyrolysis, liquefaction and gasification, which vary in their products, reaction parameters and processing technologies.

2.1 *Pyrolysis*

Pyrolysis is a thermochemical process occurring in the absence of air and reducing gases, at temperatures ranging from 650 to 800 K, typically under low pressure, converting biomass into liquid (bio-oil), solid (charcoal) and non-condensable gases [21–23]. During pyrolysis, vapour phase products (the light molecular weight decomposed pieces) are formed and then condensed by cooling, resulting in a complex liquid mixture [22, 23]. The composition of pyrolysis products (solid, liquid, gas) depends on the heating rate, contact time and reaction temperature. A substantial yield of solid product can be reached at low temperatures and extended reaction times (slow pyrolysis), but a significant yield of bio-oil can be attained at around 500 °C with short residence times (rapid pyrolysis) [21, 22, 24].

Research has been conducted on fuel production from various algae species (*Tetraselmis chui, Chlorella like, Chlorella vulgaris, Chaetocerous muelleri, Dunaliella tertiolecta, Synechococcus*) utilising pyrolysis at 500 °C [25]. Research indicated that biofuels derived from several algae species exhibited differences in total calorific value, average molecular weight of the liquid fraction and the proportions of liquid, gas and char. Research on the pyrolysis of *Chlorella* in a fixed-bed reactor indicated that elevated temperatures led to a reduction in solid residues and an increase in liquid conversion [26]. The oil yield was reduced in the presence of a Na_2CO_3 catalyst. The catalytic bio-oil exhibited a superior heating value, increased hydrocarbon and aromatic content and reduced acidity compared to the non-catalytic bio-oil.

Subagyono et al. [27] examined the pyrolysis of *Macaranga gigantea* wood, demonstrating that elevated temperatures and times improved the yield of liquid products abundant in phenolics and lignin derivatives, as indicated by TG-FTIR, GC–MS and NMR studies. In another work, Subagyono et al. [28] utilised Py-GC/MS to examine bio-oils obtained from CO/H_2O-treated microalgae, grape marc and wood chips, revealing complex conversions of lipids, proteins and carbohydrates into nitrogenous and oxygenated chemicals. Together, the findings illustrate that feedstock composition and pyrolysis parameters substantially affect product distribution and chemical complexity, providing essential insights for enhancing thermochemical biomass conversion.

The catalytic pyrolysis of *Botryococcus braunii* with Ni/SBA-15 synthesised via an ultrasonic-assisted sol-gel technique has demonstrated an increase in the production of aromatic compounds, including furans and nitrogen-containing aromatics, while concurrently diminishing the fatty acid content in the resultant bio-oil [29]. A related study demonstrated that using Ni/Al-SBA-15 and Ni/Ga-SBA-15 catalysts in the pyrolysis of *Spirulina platensis* resulted in a notable decrease in activation energy and pyrolysis temperature, facilitating the breakdown of proteins, lipids and carbohydrates into smaller compounds, while simultaneously enhancing the yield of alkyl nitriles [30]. These findings highlight the significance of catalyst composition and microalgal species in influencing pyrolysis behaviour and enhancing the yield and quality of bio-oil derived from microalgal biomass.

The liquid products of pyrolysis must be improved for use as transport fuel. Bio-oils generated during pyrolysis are unsuitable for direct application due to their elevated levels of oxygenated chemicals [22]. Bio-oils generated from slow and flash pyrolysis exhibit high viscosity, limited water tolerance, acidity and chemical instability [22, 24, 31]. The application of catalysts in pyrolysis can enhance oil yields, reduce the concentration of oxygenated chemicals and facilitate the production of high-octane aromatic gasoline products [31].

2.2 Liquefaction

A different approach to biofuel production is the liquefaction process, which operates at moderate temperatures (250–450 °C) and often high pressures (up to 150 bar), generally for longer times than pyrolysis. This process often involves catalysts and reducing gases, such as hydrogen or carbon monoxide (H_2, CO), to improve conversion efficiency [32–36]. Liquefaction and pyrolysis convert biomass into liquid products, but their methods are distinct. In liquefaction, liquids are predominantly generated via the repolymerization of unstable low-molecular-weight fragments, rather than through the condensation of vapours or light liquid fractions, as observed in pyrolysis [23]. However, a constraint of liquefaction is its projected greater capital expenditure than that of pyrolysis [4].

Liquefaction techniques are classified into direct and indirect approaches. Indirect liquefaction initially transforms biomass into synthesis gas (a mixture of carbon monoxide and hydrogen), which is then processed using Fischer–Tropsch (FT) synthesis into liquid hydrocarbons [37]. FT-derived bio-oils are defined by the high content of straight-chain hydrocarbons and the lack of aromatic, sulphur and nitrogen-containing components, facilitating clean combustion with little particle emissions [38, 39]. Although FT synthesis is considered to be one of the most energy-efficient ways to make biofuels [38, 40], issues such as obtaining enough feedstocks and improving how the process technology operates more effectively remain to be solved.

Moreover, FT biofuels generally possess densities that fall below jet fuel specifications, and their absence of aromatic compounds may lead to elastomer shrinkage, potentially causing fuel system leakage [38]. These difficulties may be overcome by combining FT fuels with conventional jet fuels.

In contrast, direct liquefaction produces hydrocarbons through the thermal decomposition of biomass, which includes interactions with reducing gases or hydrogen-donor solvents [37]. The reducing agents commonly used are H_2 and $CO—H_2O$ mixtures. Prior research on algae-based direct liquefaction has investigated the effects of algal composition, temperature and catalyst types on oil output. For example, in the $Fe(CO)_5-S$ catalysed liquefaction of *Spirulina*, oil production was mostly due to thermal polypeptide breakdown and reaction-induced hydrolysis [41]. In contrast, hydrothermal liquefaction (HTL) of *Nannochloropsis* sp. generated oil from various macromolecular sources (lipids, proteins and carbohydrates), with yields above those anticipated from lipid content alone [15]. Research indicated that bio-oils derived from different algal strains had roughly 55–80% lipids, 11–18% proteins and 6–15% carbohydrates [42]. Although high-lipid algae often produce greater quantities of bio-oil, this pattern is influenced by the specific responses of different macromolecules to processing conditions. For instance, carbohydrates required Na_2CO_3 as a catalyst for effective liquefaction, whereas proteins and lipids were more efficiently transformed without it [42].

Temperature is essential in the liquefaction of algae. Most studies function were at below 400 °C, with optimal yields achieved at or slightly below water's critical point (374 °C, 22.1 MPa) when employing wet algae or water as a solvent [15, 16, 43, 44]. Maximum yields for *Spirulina platensis* and *Nannochloropsis* sp. liquefied under N_2 were both recorded at 350 °C [16]. Temperatures beyond this threshold frequently resulted in diminished oil output due to the generation of gases or more complex molecules [44]. Moreover, longer reaction periods have been associated with high gas fractions due to the conversion of lighter hydrocarbons. In water-assisted algal liquefaction [16], bio-oil is produced below the critical point by isomerisation, depolymerisation and condensation, with the possibility of carbonisation as well. Beyond the critical point, gas generation increases due to decarboxylation, cracking and reforming processes.

The use of catalysts in liquefaction is optional and can sometimes lead to decreased oil yields relative to non-catalysed methods [15]. However, when the ultimate objective is superior fuel quality, the advantages of catalytic liquefaction, such as enhanced volatility or combustion properties, may warrant the utilisation of catalysts, despite the potential reduction in conversion efficiency. For instance, the liquefaction of *Botryococcus braunii* using Na_2CO_3 did not yield any catalytic improvement, presumably due to its intrinsic high hydrocarbon content [43]. Alternative catalysts (e.g. Pd/C, Pt/C, Ru/C, Ni/SiO_2-Al_2O_3, $CoMo/\gamma$-Al_2O_3, zeolites) had negligible or detrimental impacts on oil output during the hydrothermal liquefaction of *Nannochloropsis* sp. in the presence of hydrogen [15]. Notably, in the presence of helium rather than hydrogen, certain catalysts exhibited improved conversion, indicating that hydrogen might contribute to catalyst deactivation or be restricted by insufficient metal loading and diffusion limitations [15]. Although the catalytic gain

Table 1 Effect of catalyst on hydrothermal liquefaction of *Nannochloropsis* sp. (350 °C, 60 min) [15]

Catalyst	Effect
Pt, Pd, Ru	Increased hydrogen to carbon ratio, reduced viscosity of bio-oil
Pt, Ni, CoMo	Reduced oxygen per carbon ratio (effective as a deoxygenation catalyst)
Ru, CoMo	Showed desulphurization activity
Ni	Produced sulphur-free bio-oil
Pt	Showed activity in fatty acids to alkane conversion
Ni, Ru	Increased H_2 and CH_4 yields
Zeolite	Increased N_2 yield

in yield was minimal, catalysts such as Pd, Pt, Ru and CoMo generated oils with markedly reduced viscosity, an essential characteristic for use in transportation. The effects of catalysts are summarised in Table 1.

Bio-oils derived from liquefaction generally require upgrading due to their high viscosity and limited storage stability. Over time, these oils may experience volatile loss, oxidative polymerisation or reactions induced by residual inorganics from biomass, all of which enhance viscosity and reduce fuel quality [4, 16].

2.3 CO/H_2O Reaction

The carbon monoxide and water (CO/H_2O) system is one of the common methods in direct liquefaction, generating a combination of liquid and solid products with a gas phase rich in CO_2 and H_2. This technique has been widely employed in the liquefaction of coal, especially for low-rank types, where it has demonstrated a substantial increase in the production of liquid fuels [45–47]. The efficacy of CO/H_2O has been shown for lignocellulosic biomass, particularly with the incorporation of alkaline catalysts, which enhances both oil yield and quality [48]. A key element affecting the reactivity of specific feedstocks in CO/H_2O systems is the presence of free hydroxyl (–OH) groups. Low-rank coals, possessing such functional groups in humic compounds, exhibit more reactivity than bituminous coals under these conditions [49].

Despite the extensive use of carbon monoxide in numerous industrial applications [50–53], its utilisation in laboratory and pilot-scale liquefaction necessitates cautious management owing to its significant toxicity. Carbon monoxide exhibits a great affinity for haemoglobin, resulting in diminished oxygen transport in the bloodstream, a condition referred to as hypoxia. In extreme instances, this may lead to lethal consequences [54, 55]. Consequently, appropriate safety protocols are necessary when managing carbon monoxide in reaction systems.

The mechanism of liquefaction with CO/H_2O, originally proposed by Ross [56], comprises a series of steps starting with the interaction of carbon monoxide and

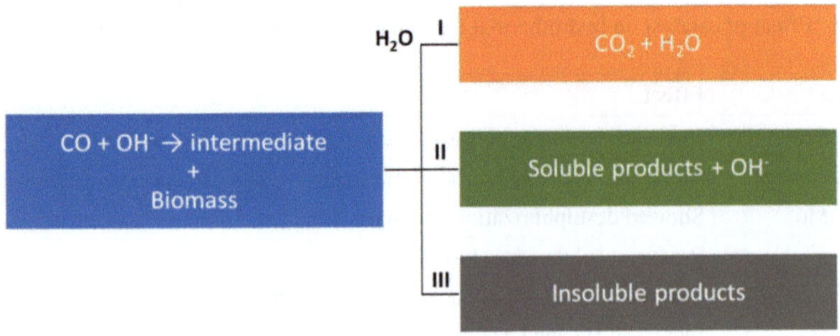

Fig. 5 A reaction sequence for the generation of the intermediate and its subsequent reaction with biomass for liquefaction with CO/H_2O

hydroxide ions, resulting in the generation of reactive intermediates, potentially as formate ions [57]. The intermediates may follow three competing pathways: (I) reacting with water to yield hydrogen and carbon dioxide via the water—gas shift reaction; (II) directly interacting with biomass to generate soluble, oil-like products; or (III) facilitating thermolytic polymerisation of biomass into insoluble, solid residues (Fig. 5). Catalysts can affect this mechanism in two specific ways—either by promoting the formation of reactive intermediates or by preferentially enhancing one of the three subsequent reaction pathways.

The CO/H_2O liquefaction process has several advantages compared to traditional hydrogen-based liquefaction methods [37, 58]. It notably eliminates the necessity for pre-drying of moist feedstocks, facilitates operation at reduced reaction temperatures, and generally yields a high conversion rate to liquid and gaseous products with minimum solid byproducts. The method can operate efficiently without the need for hydrogen-donor solvents, while simultaneously producing hydrogen in situ. A further advantage is the relative simplicity of isolating the resultant oil phase from the aqueous layer.

Similar to other liquefaction techniques, the operating temperature greatly impacts oil yield in CO/H_2O systems. Conversion typically escalates with increasing temperature. For coal-derived feedstocks, optimal conversion typically occurs near the critical point of water [37, 46, 47]. In addition to temperature, parameters such as residence time, water-to-feedstock ratio and CO pressure all influence the degree of conversion. Increased CO pressures can improve yields in certain instances, although this result depends on the characteristics and composition of the biomass being processed [46].

Basic alkali metal catalysts are commonly used in CO/H_2O liquefaction to improve oil conversion by facilitating efficient CO-based reactivity. Sodium aluminate has shown superior efficacy relative to other alkaline catalysts, including sodium silicate, sodium hydroxide, calcium aluminate and magnesium acetate, resulting in higher total oil conversions [37, 45]. The incorporation of sodium aluminate alongside transition metals such as Fe, Co, Ni, Cu and Ni-Mo has been shown to yield a slight enhancement in conversion efficiency [47].

Recent studies on biomass feedstocks have elucidated the responses of various materials to CO/H_2O liquefaction. Subagyono et al. assessed blue gum and fossil wood to represent lignocellulosic feedstocks [36]. Blue gum, with abundant cellulose, exhibited no notable yield enhancement under CO/H_2O compared to processes in N_2 or H_2, indicating that cellulose is essentially resistant to the advantages of this environment. Conversely, fossil wood, characterised by lignin derivatives, demonstrated increased oil production in the presence of CO/H_2O, especially when sodium aluminate served as a catalyst. This is due to the high phenolic content of fossil wood, which exhibits greater reactivity towards CO-derived intermediates.

A further study on grape marc, comprising a blend of carbohydrates, proteins and fatty acids, revealed that the use of CO/H_2O at temperatures between 320 and 425 °C enhanced product quality, albeit not necessarily the oil production [35]. The use of sodium carbonate or sodium aluminate reduced the oxygen concentration of the produced oil, indicating improved decarboxylation and deoxygenation, while increasing the hydrocarbon proportion in the resulting product.

Algal biomass has demonstrated significant potential in reaction with CO/H_2O. In a comparative analysis of algae, grape marc and Pinus radiata wood waste, algae consistently produced superior yields of dichloromethane-soluble materials and bio-oils abundant in aliphatic hydrocarbons [33]. These oils are preferable for aviation fuel uses in comparison to the more aromatic-rich products derived from wood or grape marc. This enhanced performance is likely due to the large concentration of proteins and lipids in algae, which are more susceptible to conversion under CO/H_2O.

Another study of the thermochemical conversions of marine algae species (*Chlorococcum* sp. and a polystrain mixture) under CO/H_2O revealed superior oil yields and enhanced product quality, indicated by reduced nitrogen content and higher hydrogen-to-carbon (H/C) ratios—relative to reactions conducted under H_2 or N_2 [32]. The non-catalytic CO/H_2O process produced significant conversion rates (>90%), indicating that catalysts may primarily serve to enhance product quality rather than increase overall yield. In this respect, iron-based catalysts like $Fe(CO)_5{-}S$, notwithstanding their application, demonstrated minimal supplementary advantage in oil conversion when utilised with *Spirulina* at temperatures ranging from 300 to 425 °C [41].

Together, these studies illustrate that the efficacy of CO/H_2O liquefaction varies considerably based on feedstock composition. Among biomass varieties, algae demonstrate the highest responsiveness to CO/H_2O, yielding greater outputs and enhanced fuel quality, so positioning them as particularly viable options for sustainable liquid fuel production.

3 Bio-oil Upgrading

Bio-oils produced via thermochemical conversions typically show limitations, including high viscosity and poor stability, making them unsuitable for direct application in engines. Upgrading methods are necessary to make these oils compatible with current engine technologies. Numerous approaches have been suggested for this objective, including hydrodeoxygenation (HDO), hydrocracking, zeolite-catalysed upgrading, solvent addition, emulsification and steam reforming [59–65]. Hydrotreating and hydrocracking are hydrogenation techniques that remove heteroatoms (including sulphur, nitrogen and oxygen) and saturate carbon-carbon bonds to improve fuel properties. Hydrotreating generally operates at pressures ranging from 70 to 200 bar, using heterogeneous catalysts mainly composed of transition metals such as Ni, Co and Mo, which are supported on high-acidity materials [22, 23, 40, 66]. In hydrotreating and hydrocracking, the common catalysts used are heterogeneous and generally contain transition metals such as cobalt, nickel or molybdenum, supported on substrates with elevated surface acidity [4]. These catalysts promote the elimination of oxygen from bio-oil by reactions with hydrogen, yielding water and saturated hydrocarbons. Despite their efficacy, these catalysts tend to be costly and require high-pressure hydrogen, which may hinder their application on a large scale. Hydrocracking involves bond cleavage under more extreme conditions than hydrotreating [67].

On the other hand, zeolite-catalysed upgrading has attracted attention for its ability to operate at atmospheric pressure without requiring hydrogen [4, 68]. Zeolites such as H-ZSM-5 are widely used to transform bio-oil into gasoline-range hydrocarbons; nonetheless, this procedure often results in higher coke yields and reduced hydrocarbon yields compared to hydrogen-based methods [65, 68]. To enhance performance, zeolites can be modified by coating them with mesoporous materials like MCM-41 or SBA-15, which improve conversion rates and enhance oil yield [64]. Zeolite-based upgrading generally leads to reduced hydrocarbon yields and increased coke formation [4], however, its operation at atmospheric pressure and absence of hydrogen make it a more economical option compared to hydrodeoxygenation.

The catalytic upgrading of biocrude derived from *Chlorococcum* sp. with NiMo supported on Al-SBA-15 with different Si/Al ratios has been studied [69]. It was found that support acidity and the presence of NiMo greatly affected product quantity and quality. Thermal hydrotreating alone reduced nitrogen content by approximately 50%; nevertheless, Al-SBA-15 supports without modification with NiMo induced coking and decreased yields. The addition of NiMo, particularly with more acidic supports, enhanced oil yield, especially within the kerosene range (C_{12}–C_{17}), and decreased coke formation. NiMo/Al-SBA-15 altered the product composition by increasing the concentrations of branched and linear alkanes while reducing oxygenated and heavy chemicals. Comparisons of different upgrading techniques are given in Table 2.

These findings indicate that while significant progress has been achieved in thermochemical conversion and bio-oil upgrading, further studies are necessary to

Table 2 Comparisons of some upgrading methods [4, 67, 68]

	Hydrotreating	Hydrocracking	Zeolite upgrading
Atmosphere	H_2	H_2	No hydrogen is required
Pressure	Low-high pressure	100–2000 psi	Atmospheric pressure
Temperature	300–500 °C	>350 °C	350–500 °C
Catalyst	CoMo, NiMo	Ni/Al_2O_3-TiO_2	H-ZSM-5
Pros	Cheap route, commercially available	High yield of light products	Low operating cost
Cons	High coking, Poor quality fuels	Complicated equipment, high cost and catalyst deactivation	Require the removal of some compounds that can cause coking

maximise their potential. Transitioning from laboratory results to industrial applications remains a major challenge, particularly in terms of cost, reliability and compatibility with existing fuel infrastructure. Additional efforts are required to bridge the gap between research and real-world implementation of thermochemical conversion technologies. Ongoing work should focus on improving system efficiency while ensuring scalability and adaptability to practical industrial environments. Enhancing catalytic performance, reducing hydrogen consumption and improving bio-oil stability will significantly impact outcomes. Finally, transforming scientific findings into practical, cost-effective and sustainable energy solutions represents the next significant advancement in reducing the demand on fossil fuels.

4 Conclusions

Thermochemical conversion methods provide efficient ways for converting lignocellulosic and microalgal biomass into sustainable liquid fuels. The processes of pyrolysis and liquefaction are influenced by feedstock composition, temperature, pressure and catalyst presence, all of which influence product yield and fuel quality. Microalgae demonstrate enhanced conversion potential in CO/H_2O systems, whereas lignocellulosic biomass is more effectively processed through catalytic pyrolysis and liquefaction under carefully chosen conditions. Despite the potential of these approaches, the resulting bio-oils generally require upgrading due to high oxygen content, viscosity and stability concerns. Among the evaluated upgrading strategies, hydrotreating with NiMo-based catalysts shows potential for upgrading microalgal bio-oils, whilst zeolite upgrading offers a more economical alternative. Optimising feedstock selection and processing conditions is important when developing ready-to-use bio-oil and producing economically feasible biofuels for the transportation and aviation industries.

References

1. Ma F, Hanna MA (1999) Biodiesel production: a review. Bioresour Technol 70:1–15
2. Naik SN, Goud VV, Rout PK, Dalai AK (2010) Production of first and second generation biofuels: a comprehensive review. Renew Sustain Energ Rev 14:578–597
3. Goeltenboth F, Timotius KH (1996) Impact of rainforest destruction—the Siberut Island case, Sumatra Indonesia. In: Edwards DS, Booth WE, Choy SC (eds) Tropical rainforest research—current issues. Springer, Netherlands, pp 425–433
4. Huber GW, Iborra S, Corma A (2006) Synthesis of transportation fuels from biomass: chemistry, catalysts, and engineering. Chem Rev 106:4044–4098
5. Firoz A, Abhijit D., Roesfiansjah R., Saleh M., Hazim M., Abdul B. (2013) Biofuel from algae, advances in biofuel production, apple academic press. pp 107–119
6. Pettersen RC (1984) The chemical composition of wood, the chemistry of solid wood, American Chemical Society, pp 57–126
7. Roger MR, Roger P, Mandla AT (2012) Cell wall chemistry, handbook of wood chemistry and wood Composites, second edition. CRC Press, pp 33–72
8. Barton GM (1984) Definition of biomass samples involving wood, bark and foliage. Biomass 4:311–314
9. Adler E (1977) Lignin chemistry—past, present and future. Wood Sci Tech 11:169–218
10. Andersen RA (2013) The microalgal cell, handbook of microalgal culture. Wiley, pp 1–20
11. Piorreck M, Baasch K-H, Pohl P (1984) Biomass production, total protein, chlorophylls, lipids and fatty acids of freshwater green and blue-green algae under different nitrogen regimes. Phytochemistry 23:207–216
12. Scheer H (2013) Chlorophylls and carotenoids. In: Lane WJLD (ed) Encyclopedia of biological chemistry. Academic, Waltham, pp 498–505
13. Humphrey AM (1980) Chlorophyll. Food Chem 5:57–67
14. Chisti Y (2007) Biodiesel from microalgae. Biotechnol Adv 25:294–306
15. Duan P, Savage PE (2011) Hydrothermal liquefaction of a microalga with heterogeneous catalysts. Ind Eng Chem Res 50:52–61
16. Jena U, Das KC, Kastner JR (2011) Effect of operating conditions of thermochemical liquefaction on biocrude production from Spirulina platensis. Bioresour Technol 102:6221–6229
17. Wijffels RH (2008) Potential of sponges and microalgae for marine biotechnology. Trends Biotechnol 26:26–31
18. Johnson MB, Wen Z (2009) Production of biodiesel fuel from the microalga Schizochytrium limacinum by direct transesterification of algal biomass. Energ Fuel 23:5179–5183
19. Levine RB, Pinnarat T, Savage PE (2010) Biodiesel production from wet algal biomass through in situ lipid hydrolysis and supercritical transesterification. Energ Fuel 24:5235–5243
20. Vijayaraghavan K, Hemanathan K (2009) Biodiesel production from freshwater algae. Energ Fuel 23:5448–5453
21. Alonso DM, Bond JQ, Dumesic JA (2010) Catalytic conversion of biomass to biofuels. Green Chem 12:1493–1513
22. Bridgwater AV (1994) Catalysis in thermal biomass conversion. Appl Catal A Gen 116:5–47
23. Demirbas A (2001) Biomass resource facilities and biomass conversion processing for fuels and chemicals. Energ Convers Manag 42:1357–1378
24. Huber GW, Iborra S, Corma A (2006) Synthesis of transportation fuels from biomass: chemistry, catalysts, and engineering. Chem Rev 106:4044–4098
25. Grierson S, Strezov V, Ellem G, McGregor R, Herbertson J (2009) Thermal characterisation of microalgae under slow pyrolysis conditions. J Anal Appl Pyrol 85:118–123
26. Babich IV, van der Hulst M, Lefferts L, Moulijn JA, O'Connor P, Seshan K (2011) Catalytic pyrolysis of microalgae to high-quality liquid bio-fuels. Biomass Bioenergy 35:3199–3207
27. Subagyono RRDJN, Miten PD, Sinaga RJ, Wijayanti A, Qi Y, Marshall M, Sanjaya AS, Chaffee AL (2022) Pyrolysis of fast growing wood *Macaranga gigantea*: product characterisation and kinetic study. Fuel 315:123182

28. Subagyono DJN, Qi Y, Jackson WR, Chaffee AL (2016) Pyrolysis-GC/MS analysis of biomass and the bio-oils produced from CO/H_2O reactions. J Anal Appl Pyrol 120:154–164
29. Subagyono RRDJN, Putri SA, Manawan M, Mollah M, Nugroho RA, Gunawan R (2023) Catalytic pyrolysis of the green microalgae Botryococcus braunii over Ni/SBA-15 prepared by the ultrasonic-assisted sol-gel method. ACS Omega 8:8582–8595
30. Subagyono RRDJN, Madani NM, Prechisilia CZLCB, Sahirah NS, Utami DU, Qosim AM, Kassim MA, Gunawan R, Allo VL (2025) Pyrolysis of microalgae over Ni/Al-SBA-15 and Ni/Ga-SBA-15 catalysts prepared using a low-acidity solvent and ultrasonic-assisted sol-gel method. J Anal Appl Pyrolysis 186:106935
31. Tran NH, Bartlett JR, Kannangara GSK, Milev AS, Volk H, Wilson MA (2010) Catalytic upgrading of biorefinery oil from micro-algae. Fuel 89:265–274
32. Subagyono DJN, Marshall M, Jackson WR, Chow M, Chaffee AL (2014) Reactions with CO/H_2O of two marine algae and comparison with reactions under H_2 and N_2. Energ Fuel 28:3143–3156
33. Subagyono DJN, Marshall M, Fei Y, Jackson WR, Chaffee AL (2015) Thermo-chemical reactions of algae, grape marc and wood chips using a semi-continuous/flow-through system. Fuel 158:927–936
34. Subagyono DJN, Marshall M, Jackson WR, Chaffee AL (2015) Pressurized thermal and hydrothermal decomposition of algae, wood chip residue, and grape marc: a comparative study. Biomass Bioenergy 76:141–157
35. Subagyono DJN, Marshall M, Jackson WR, Chaffee AL (2015) Improvement in liquid fuel product quality from reactions of grape marc with CO/H_2O. Fuel 159:234–240
36. Subagyono DJN, Marshall M, Jackson WR, Fei Y, Chaffee AL (2016) Thermochemical reactions of blue gum and fossil wood with CO/H_2O: some mechanistic comments. Energ Fuel 30:1039–1049
37. Stray GJ (1988) Reactions of coal and model compounds with carbon monoxide and water PhD thesis. Department of Chemistry, Monash University, Melbourne, p 375
38. Hemighaus G, Boval T, Organ CBR, Lind J, Brouette R, Thompson T, Lynch J, Jones J (2006) Alternative jet fuels. Houston, Chevron, pp 1–13
39. Demirbas A (2009) Biofuels Biohydrogen, Springer London, pp 61–84
40. Blakey S, Rye L, Wilson CW (2011) Aviation gas turbine alternative fuels: a review. Proc Combust Inst 33:2863–2885
41. Matsui T-O, Nishihara A, Ueda C, Ohtsuki M, Ikenaga N-O, Suzuki T (1997) Liquefaction of micro-algae with iron catalyst. Fuel 76:1043–1048
42. Biller P, Ross AB (2011) Potential yields and properties of oil from the hydrothermal liquefaction of microalgae with different biochemical content. Bioresour Technol 102:215–225
43. Dote Y, Sawayama S, Inoue S, Minowa T, Yokoyama S-Y (1994) Recovery of liquid fuel from hydrocarbon-rich microalgae by thermochemical liquefaction. Fuel 73:1855–1857
44. Brown TM, Duan P, Savage PE (2010) Hydrothermal liquefaction and gasification of Nannochloropsis sp. Energ Fuel 24:3639–3646
45. Chan JST, Jackson WR, Marshall M (1994) Reactions of brown coals with $CO\text{-}H_2O$ in the presence of alkaline catalysts in a well-stirred autoclave with hot-charge facility. Fuel 73:1628–1631
46. Hughes CP, Sridhar T, Chuan LS, Redlich PJ, Jackson WR, Larkins FP (1993) Reactions of brown coals with CO/H_2O in the presence of sodium aluminate. Fuel 72:205–210
47. Hulston CKJ, Redlich PJ, Roy Jackson W, Larkins FP, Marshall M, Burgess CE (1997) Effects of added transition metals and two-stage heating on reactions of sodium-aluminate-treated coals in water with CO, H_2 and $CO\text{-}H_2$ mixtures. Fuel 76:247–256
48. El-Saied H (1977) Liquefaction of lignochemicellulosic waste by processing with carbon monoxide and water. J Appl Chem Biotechnol 27:443–452
49. Clemow LM, Favas G, Jackson WR, Marshall M, Patti AF, Redlich PJ (1999) Humic acids and methylated humic acids as models for reactions of brown coal with CO/H_2O and with H_2. Fuel 78:567–572

50. Young A, Lesmana D, Dai D-J, Wu H-S (2013) Short review: mitigation of current environmental concerns from methanol synthesis. Bull Chem React Eng Catal 8:1–13
51. Zennaro R, Ricci M, Querci C, Camelli L, Monforte ADA (2013) Syngas: the basis of fischer-tropsch. In: Maitlis PM, De Klerk A (eds) Greener Fischer-Tropsch processes for fuels and feedstocks. Wiley-VCH Verlag GmbH & Co. KGaA, Weinheim, pp 19–51
52. Beller M (2012) Carbonylation reactions. In: Beller M, Renken A, van Santen R (eds) Catalysis. Wiley-VCH Verlag GmbH & Co. KGaA, Weinheim, pp 233–249
53. Tassini R, La Sorella G, Montin D, Paganelii S, Baldi F, Piccolo O (2012) Sustainable syntheses with CO. Chimica e l'Industria 94:157–161
54. Milan JH (2000) Effect of carbon monoxide on work and exercise capacity in humans. In: Carbon monoxide toxicity. CRC Press, pp 101–134
55. Neil BH (2000) Carbon monoxide poisoning and its management in the United States. In: Carbon monoxide toxicity. CRC Press, pp 193–205
56. Ross DS (1984) Coal conversion in carbon monoxide-water system. In: Gorbaty ML, Larsen JW, Wender I (Eds) In: Coal science, Academic Press, Inc, Orlando, Florida, pp 301–335
57. Horvath IT, Siskin M (1991) Direct evidence for formate ion formation during the reaction of coals with carbon monoxide and water. Energ Fuel 5:932–933
58. Molton PM, Miller RK, Russell JA, Donovan JM (1981) Effects of reaction conditions on the aqueous thermochemical conversion of biomass to oil, biomass as a Nonfossil fuel source. American Chemical Society, pp 137–162
59. de la Puente G, Gil A, Pis JJ, Grange P (1999) Effects of support surface chemistry in Hydrodeoxygenation reactions over CoMo/activated carbon Sulfided Catalysts. Langmuir 15:5800–5806
60. Lin Y-C, Li C-L, Wan H-P, Lee H-T, Liu C-F (2011) Catalytic hydrodeoxygenation of guaiacol on Rh-based and sulfided CoMo and NiMo catalysts. Energ Fuel 25:890–896
61. French RJ, Stunkel J, Baldwin RM (2011) Mild hydrotreating of bio-oil: effect of reaction severity and fate of oxygenated species. Energ Fuel 25:3266–3274
62. Honkela ML, Viljava T-R, Gutierrez A, Krause AOI, (2010) Chapter 11 hydrotreating for bio-oil upgrading, thermochemical conversion of biomass to liquid fuels and chemicals. The Royal Society of Chemistry pp 288–306
63. Adjaye JD, Katikaneni SPR, Bakhshi NN (1996) Catalytic conversion of a biofuel to hydrocarbons: effect of mixtures of HZSM-5 and silica-alumina catalysts on product distribution. Fuel Process Tech 48:115–143
64. Ooi Y-S, Zakaria R, Mohamed AR, Bhatia S (2005) Catalytic conversion of fatty acids mixture to liquid fuel and chemicals over composite microporous/mesoporous catalysts. Energ Fuel 19:736–743
65. Sharma RK, Bakhshi NN (1993) Catalytic upgrading of pyrolysis oil. Energ Fuel 7:306–314
66. Elliott DC (2007) Historical developments in hydroprocessing bio-oils. Energ Fuel 21:1792–1815
67. Xiu S, Shahbazi A (2012) Bio-oil production and upgrading research: a review. Renew Sustain Energ Rev 16:4406–4414
68. Elliott DC, Beckman D, Bridgwater AV, Diebold JP, Gevert SB, Solantausta Y (1991) Developments in direct thermochemical liquefaction of biomass: 1983–1990. Energ Fuel 5:399–410
69. Subagyono RRDJN, Marshall M, Jackson WR, Auxilio AR, Fei Y, Chaffee AL (2020) Upgrading microalgal biocrude using NiMo/Al-SBA-15 as a catalyst. Energy Fuel 34:4618–4631

Bioenergy Production from Biomass via Biochemical Technology

Mohd Asyraf Kassim, Tan Kean Meng, and Mardiana Idayu Ahmad

Abstract This chapter aims to discuss on biochemical conversion technologies—including pretreatment, hydrolysis, and fermentation—for producing liquid biofuels such as bioethanol, biobutanol, and biodiesel, as well as gaseous fuels like biohydrogen, biogas, and biomethane. Detailed discussion is provided on the role of microorganisms and enzymes, types of fermentation systems, and process optimization factors influencing yield and efficiency. Case studies highlight the influence of feedstock composition, microbial strains, and operating conditions on biofuel productivity. Despite significant progress, technical and economic challenges remain, particularly in pretreatment, enzyme cost, fermentation of mixed sugars, and large-scale integration. The chapter underscores the importance of advanced bioprocessing strategies and waste valorization approaches to enhance biofuel production, reduce greenhouse gas emissions, and contribute to sustainable energy transitions.

Keywords Biofuel · Microorganisms · Biosugar · Fermentation · Hydrolysis

1 Introduction

Energy is essential for daily life and industrial operations, powering everything from household appliances to global manufacturing and transportation. However, most energy still comes from fossil fuels like coal and petroleum, which contribute significantly to CO_2 emissions and resource depletion. In 2024, global energy demand grew by 2.2%—faster than the previous decade's average—driven by industrial needs, transport electrification, cooling demand, and the rapid growth of data centers and AI. Sustainable energy management is critical to reduce costs, improve competitiveness, and address climate change.

High global energy demand impacts the economy, industry, and environment in interconnected ways. Economically, it drives up energy prices, raising production and living costs, straining public finances, and exposing nations—especially developing

M. A. Kassim (✉) · T. K. Meng · M. I. Ahmad
School of Industrial Technology, Universiti Sains Malaysia, Penang, Malaysia
e-mail: asyrafkassim@usm.my

M. A. Kassim et al. (eds.), *Sustainable Biorefinery*, Green Energy and Technology,
https://doi.org/10.1007/978-981-95-8181-8_11

ones—to inflation and market volatility. Industrially, energy-intensive sectors face rising operational costs, reduced competitiveness, and risks from overused or unstable infrastructure, prompting investments in efficiency and technology upgrades. Environmentally, reliance on fossil fuels intensifies greenhouse gas emissions, climate change, habitat loss, and pollution, threatening ecosystems and public health. Transitioning to renewable and alternative sources such as solar, wind, hydro, geothermal, bioenergy, and nuclear—supported by energy storage and smart grids—offers a path to lower emissions, improved energy security, and greater long-term sustainability.

1.1 Biomass Composition

In biofuel production from biomass via biochemical composition, the presence of chemical composition in the biomass plays a major role in final product production. Biomass refers to organic material that comes from living or recently living organisms, primarily plants and animals, which can be used as a source of energy. It includes a wide range of materials such as wood, agricultural residues (e.g., rice husks, corn stalks), forest residues, animal manure, algae, and even organic municipal waste. Biomass stores chemical energy from the sun through the process of photosynthesis, and this energy can be released through various conversion processes like combustion, fermentation, anaerobic digestion, or gasification to produce heat, electricity, or biofuels. As a renewable resource, biomass plays a crucial role in sustainable energy systems, especially when managed responsibly, as it helps reduce greenhouse gas emissions, recycle waste, and support rural economies. Table 1 shows the chemical composition in different types of biomass.

Table 1 Chemical composition in different types of biomass

Biomass Type	Cellulose (%)	Hemicellulose (%)	Lignin (%)	Ash (%)
EFB	35–45	25–35	20–25	3–7
PKS	10–15	10–20	40–55	1–3
Rice husk	35–40	15–20	20–25	15–20
Rice straw	30–40	20–25	15–20	10–15
Sugarcane bagasse	40–50	20–30	20–25	1–3
Corn residues	35–45	20–25	15–20	3–5
Sago waste	30–40	20–30	15–20	2–5
Rubberwood/acacia	40–50	20–25	20–30	1–2

2 Bioconversion Process

Biochemical conversion is a sustainable approach for producing liquid biofuels from biomass through the use of microorganisms and enzymes to break down complex organic matter into fermentable products. Biomass sources—including agricultural residues, energy crops, forestry waste, and organic municipal waste—contain carbohydrates, proteins, and lipids that can be transformed into fuels such as bioethanol, biobutanol, and biodiesel. The process typically involves pretreatment to release sugars, enzymatic hydrolysis to convert polysaccharides into simple sugars, and microbial fermentation to produce target fuels. Compared to thermochemical methods, biochemical conversion operates under milder conditions, offers higher product selectivity, and can integrate with waste valorization strategies. This technology not only supports the diversification of the energy mix but also contributes to reducing greenhouse gas emissions, enhancing energy security, and promoting circular economy principles through the utilization of renewable resources. Figure 1 illustrates a typical process on liquid biofuel production from biomass via biochemical approaches.

2.1 *Pretreatment*

Biomass pretreatment is a crucial initial step in the conversion of raw biomass into biofuels and biochemicals. Its primary aim is to break down the complex and rigid structure of lignocellulosic biomass to improve the accessibility of cellulose and hemicellulose for further processing. This involves increasing the surface area, reducing cellulose crystallinity, and removing or altering lignin, which acts as a

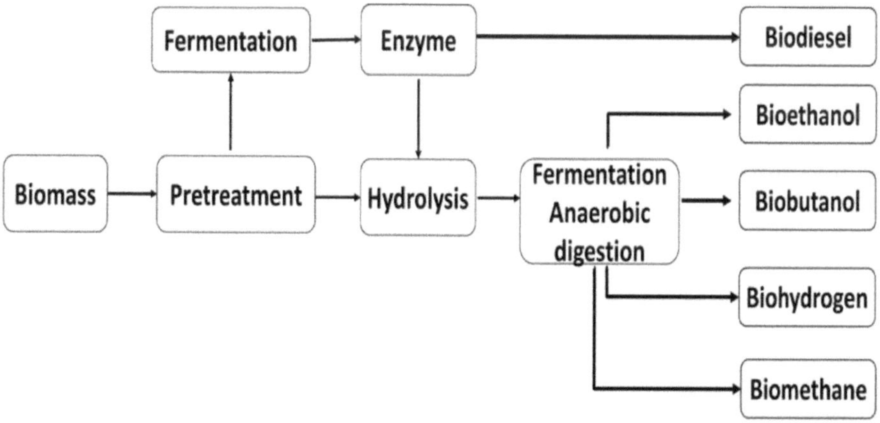

Fig. 1 Typical process on liquid biofuel production from biomass via biochemical approaches

barrier to enzymatic and microbial action. Effective pretreatment enhances the efficiency of downstream processes such as enzymatic hydrolysis and fermentation while minimizing the production of inhibitory compounds. Overall, pretreatment significantly improves the yield and efficiency of biomass conversion technologies.

Biomass pretreatment is an essential step in converting raw biomass into valuable products like biofuels, bioplastics, or bioactive compounds. The primary goal is to break down the complex structure of biomass—especially cellulose, hemicellulose, and lignin—making it more accessible for enzymatic or microbial processing. There are three main categories of pretreatment: physical, chemical, physico-chemical, and biological (Fig. 2). Physical methods such as milling or grinding reduce particle size and increase surface area, while chemical approaches use acids, alkalis, or solvents to break chemical bonds and solubilize specific components. Physico-chemical techniques like steam explosion or ammonia fiber expansion combine heat, pressure, and chemicals to disrupt biomass structure more effectively.

Biological pretreatment, on the other hand, uses fungi, bacteria, or enzymes to selectively degrade lignin or hemicellulose in a more environmentally friendly way, although it is typically slower. The choice of pretreatment depends on the type of biomass—woody materials may require harsher methods like alkaline or steam treatments due to high lignin content, while softer agricultural residues or algae can be treated with milder processes. Effective pretreatment improves yield and efficiency in downstream processes such as fermentation or extraction, but must also consider cost, energy input, and environmental impact to be viable at the industrial scale. Different biomass pretreatment route prior for valorisation via biochemical conversion pathway is shown in Fig. 2.

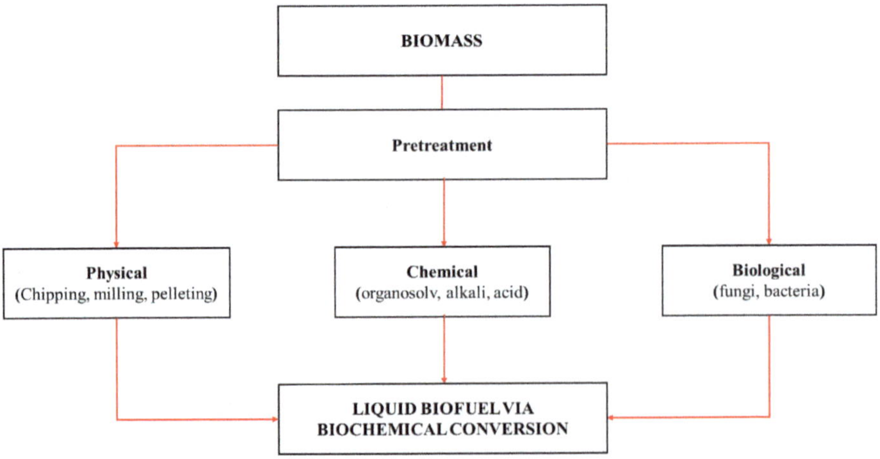

Fig. 2 Different biomass pretreatment route prior for valorisation via biochemical conversion pathway

2.2 Hydrolysis

Another important process for biofuel production via the biochemical conversion process is hydrolysis. In this process, all the carbon-based materials including starch, polysaccharides, cellulose and hemicellulose in lignocellulose materials will be converted into a simple monomer sugar. This sugar will act as the carbon source for microorganisms metabolisms to produce the liquid biofuel.

Biomass hydrolysis is the process of breaking down complex carbohydrates in biomass into simpler, fermentable sugars, typically using enzymes or acids. This step follows pretreatment and is crucial for converting cellulose, hemicellulose, and starch into monosaccharides like glucose and xylose. Enzymatic hydrolysis is preferred in many bioconversion processes due to its specificity, mild conditions, and reduced formation of inhibitory byproducts. The efficiency of hydrolysis largely depends on the activity of specific enzymes tailored to the major polysaccharide components in the biomass.

The key enzymes involved are cellulase, hemicellulase, and amylase. Cellulases (a complex of endoglucanase, exoglucanase, and β-glucosidase) degrade cellulose into glucose. Hemicellulases break down hemicellulose, a heterogeneous polymer, into sugars like xylose, arabinose, and mannose. Meanwhile, amylases hydrolyze starch (common in algal or agricultural biomass) into glucose or maltose units. The synergy between these enzymes ensures the complete saccharification of biomass, enabling efficient downstream fermentation for biofuel or biochemical production. Table 2 summarises sugar release from different biomasses.

Table 2 Sugar release from different biomasses

Biomass	Conditions	Sugar yield (%)	References
EFB	60 °C for 180 min Accelerase XY Accelerase XC Hemicellulase	51.50	[1]
Rice straw	46.45 FPU/g exoglucanase 12% substrate at 45C for 4 days	36.2	[2]
Water hyacinth	49.56 U/g Cellulase 280 U/g xylanase 50 °C for 60 h	56.72	[3]
Palm trunk	50 °C for 24 h 50 FPU/g Celluclast 1.5 L 100 CBU/ml Novozyme 188	73.5	[4]
Sugarcane bagasse	Pretreatment using 0.05 M FeC13 Cellulase 90 FPU/mL Time: 72 h	93	[5]

2.3 Fermentation

Fermentation for producing the liquid biofuel is a biological process that converts fermentable sugars derived from biomass into biofuels—primarily bioethanol and biobutanol—using microorganisms such as yeast or bacteria. After the biomass undergoes pretreatment and hydrolysis to release simple sugars like glucose and xylose, these sugars serve as substrates for microbial fermentation under controlled anaerobic conditions.

The most common form is ethanol fermentation, where *Saccharomyces cerevisiae* (brewer's yeast) converts sugars into ethanol and carbon dioxide. For more complex or mixed sugars, genetically engineered microbes or naturally occurring strains like *Zymomonas mobilis* or *Clostridium* spp. are used. Another example is acetone–butanol–ethanol (ABE) fermentation, performed by *Clostridium* species, which produces biobutanol with higher energy content and better blending properties than ethanol. Fermentation is a key step in the biorefinery process, offering a renewable and relatively low-energy pathway.

There are several types of fermentation systems used in biofuel production, particularly when converting lignocellulosic biomass into bioethanol or other liquid fuels. These systems mainly differ in how the hydrolysis (enzymatic breakdown of biomass) and fermentation (conversion of sugars to biofuels) steps are organized:

Separate Hydrolysis Fermentation

Separate Hydrolysis and Fermentation (SHF) is a two-step process where enzymatic hydrolysis of pretreated biomass is carried out first to release fermentable sugars, followed by a separate fermentation step to convert these sugars into biofuels like ethanol. This system allows independent optimization of both hydrolysis and fermentation, since enzymes and microorganisms often have different optimal conditions for temperature and pH. However, SHF may suffer from issues such as sugar accumulation that can inhibit enzymatic activity, longer processing time, and higher operational costs due to the use of multiple reactors.

Simultaneous Hydrolysis Fermentation

Simultaneous Saccharification and Fermentation (SSF) integrates hydrolysis and fermentation into a single process, where cellulolytic enzymes break down the biomass while microorganisms simultaneously ferment the resulting sugars. This approach minimizes sugar inhibition on enzymes, shortens process time, and reduces equipment costs. However, a key challenge with SSF is the compromise between the optimal operating conditions for enzymes (typically higher temperatures) and fermenting microbes (usually lower temperatures), which may affect the overall efficiency.

Consolidation Hydrolysis and Fermentation

Consolidated Bioprocessing (CBP) is an advanced and integrated approach that combines enzyme production, hydrolysis, and fermentation in a single step using specially engineered microorganisms or microbial consortia. CBP offers the greatest potential for cost reduction and process simplification by eliminating the need for added enzymes. Despite its advantages, CBP is still largely in the research and development stage, as it requires highly efficient microorganisms capable of performing all necessary functions simultaneously while maintaining high yields and productivity.

3 Type of Biofuel

3.1 Bioethanol

Bioethanol is an alcohol-based biofuel produced by the fermentation of biomass using microorganisms under anaerobic condition. This liquid biofuel can be produced via valorization of sugars or starches derived from biomass such as sugarcane bagasse, corn residues, wheat, or lignocellulosic materials that consisted of cellulose as well as hemicellulose as the main chemical component. Bioethanol is chemically identical to ethanol (C_2H_5OH) but this liquid biofuel is classified as "bio" due to it produced from biological sources using the fermentation process rather than fossil fuels. This biofuel has a specific energy between 21 and 27 MJ/L according to the raw biomass used during fermentation [6, 7]. To date, bioethanol is commonly used as a fuel or fuel additive, especially in gasoline blends like E10 (10% ethanol) or E85 (85% ethanol), to reduce greenhouse gas emissions and improve engine combustion efficiency [8].

The production of bioethanol involves several steps: pretreatment to chiefly break down the biomass structure and provide more access for the enzyme to hydrolyse the complex chemical structure in the biomass, hydrolysis process which involves the release of fermentable sugars from the biomass structure, fermentation process that involves the conversion of sugars into ethanol using microbes like Saccharomyces cerevisiae, and distillation mainly to purify the ethanol [9]. Table 3 shows the example of bioethanol production using different biomass feedstock and microorganisms as biocatalysts. According to the finding, it can be concluded that different bioethanol production can be significantly affected by the type of biomass feedstock, process condition and types of microorganism used during the fermentation process.

The main factors affecting the production of bioethanol are the type of feedstock used, the efficiency of the fermentation process, and environmental conditions such as temperature and pH. Feedstocks such as corn, cassava, or agricultural waste contain different sugar or starch contents, which affect the final yield of bioethanol. Feedstocks such as starch are easier to convert than lignocellulosic materials because of their less complex chemical structure. Some biomasses that have high lignin in lignocellulosic material is the main obstacle to obtaining sugar from the biomass [17]. In addition, microorganisms such as yeast need to function optimally in the fermentation process, which is highly dependent on temperature and acidity (pH) control [18]. The most common microbe used in the fermentation process to produce bioethanol is the yeast *Saccharomyces cerevisiae*. This yeast is very efficient in converting sugars

Table 3 Bioethanol production from different biomass

Biomass	Pretreatment	Condition	EtOH (%)	References
Cornstalk	H_2SO_4 Thermal treatment	30 °C for 96 h	32.53 g/L	[10]
Bamboo waste	Dilute alkaline	28 °C without agitation for 48 h	3.08% or 71.34%	[11]
Microalgae	Acid hydrolysis	200 g/L 30 °C for 4 days	43.9 gL	[12]
Chlorella sp.	2% NaOH at 120 °C for 120 min	30 °C at 100 rpm for 96 h	84%	[13]
Empty fruit bunches	Alkaline treatment	30 °C and 100 rpm for 72 h	0.51 g/g	[14]
Banana peels	Sulphuric acid	30 °C and 100 rpm for 72 h	80 ppm	[15]
Sugarcane baggase	Steam-assisted Sequential salt-alkali pretreatment	Enzyme loading 100 U/g 39 °C	0.49 g/g or 4.88 g/L	[16]

such as glucose, fructose, and sucrose into ethanol and carbon dioxide through the alcoholic fermentation process [17]. In addition, *S. cerevisiae* is also resistant to high ethanol concentrations and can survive in a variety of environmental conditions, making it the main choice in the bioethanol industry. In some cases, other microbes such as *Zymomonas mobilis* or genetically modified yeast are also used to increase the efficiency of bioethanol production, especially when using raw materials containing cellulose or hemicellulose [19]. Other microbes have also been reported to be able to produce bioethanol from the biomass, including *Kluyveromyces marxianus*, which is heat-resistant and capable of metabolizing various types of sugars [20]. Furthermore, the use of modern technologies such as special enzymes to accelerate the conversion of starch to sugar also increases the efficiency and yield of bioethanol production [21]. Therefore, the combination of quality raw materials, controlled process conditions, and advanced technology is the main key in ensuring high and effective bioethanol production.

3.2 *Biobutanol*

Biobutanol is also a liquid biofuel in the alcohol group that can be produced through fermentation. This alcohol-based biofuel is chemically similar to butanol ($C_4H_{10}O$) and more similar to gasoline than ethanol, which is why it is considered a more suitable replacement for fossil fuels [22]. Biobutanol has several key advantages as a biofuel compared to ethanol. It has a higher energy density and is almost equivalent to gasoline, making it more efficient and usable in existing engines without major modifications [23]. In addition, biobutanol absorbs less water (is not hygroscopic), is more stable during storage, and is easy to transport through existing systems [24]. From a sustainability point of view, it can be produced from agricultural waste or other biomass sources, thus reducing dependence on fossil fuels and supporting the reduction of carbon emissions [25].

Biobutanol can also be produced through fermentation using specific microbes such as *Clostridium* sp. In the production of biobutanol, the most widely used bacteria are from the genus Clostridium, specifically *Clostridium acetobutylicum* and *Clostridium beijerinckii* due to their ability to produce butanol with high yields through the ABE (Acetone-Butanol-Ethanol) fermentation process [26]. Both of these species are capable of fermenting various carbon sources such as sugars, starches, and lignocellulosic waste, and have better tolerance to solvent stress than other species [22]. In addition, there are also other species such as *C. saccharobutylicum* and *C. saccharoperbutylacetonicum*, but their use is more limited [27]. Overall, *Clostridium* sp. is the main microorganism that supports the development of biobutanol as a more sustainable alternative biofuel. Biobutanol production is usually accompanied by other chemicals such as acetone and ethanol in a process called ABE fermentation (Acetone-Butanol-Ethanol) [23].

Biobutanol production from starch fermentation by *Clostridium* sp. begins with the breakdown of starch by enzymes such as α-amylase and glucoamylase into simple sugars, mainly glucose [28]. This glucose then undergoes glycolysis to produce pyruvate along with energy (ATP) and NADH [24]. Pyruvate is converted to organic acids such as acetic acid and butyric acid in the acidogenesis phase, causing the pH of the medium to decrease [22] When acidic conditions occur, the bacteria switch to the solventogenesis phase, where the acids are converted back into solvents, namely butanol, acetone, and ethanol [23]. The end product of this fermentation is biobutanol as the main product, with small amounts of acetone and ethanol as byproducts [25].

Several studies of biobutanol production from the biomass have been reported by several researchers. Biobutanol can be produced from various feedstocks, including sugar crops, starch-based materials, lignocellulosic biomass, or even algae (Table 4). In general, the biobutanol production process is the same as the bioethanol production process.

As with bioethanol production, several important parameters affect the production of biobutanol from the biomass. In the production of biobutanol from the biomass, several important parameters need to be controlled to ensure optimal yield. Among them is the type of substrate used, as the content of easily fermentable sugars such as glucose, starch, or pretreated lignocellulosic waste will affect the fermentation efficiency [27, 28]. pH plays an important role as it determines the phase transition from acidogenesis to solventogenesis [22]. Temperature also needs to be maintained in the optimal range for the growth of *Clostridium* sp., usually around 30–37 °C [24]. In addition, the concentration of the substrate and toxic products (such as butanol itself which inhibits bacterial growth at high concentrations) need to be monitored [23]. Other parameters such as fermentation time, anaerobic conditions, and the presence of additional nutrients are also critical in maximizing the production of biobutanol from the biomass [25].

Table 4 Biobutanol production from various biomass using different biocatalyst

Biomass	Microbes	Condition	Butanol	References
Cassava waste	*Clostridium* sp. AS3	Acid hydrolysis pH 6.3 Temp; 33 °C Agitation: 85 rpm	ABE: 10.5 g L1	[29]
EFB	*Clostridium acetobutylicum* DSM 1713	Acid hydrolysis 37 °C 168 h	1.262 ± 0.218 g/L	[30]
POME	*Clostridium saccharoperbutylacetonicum* N1-4	5.8, 30 1 °C, 100 rpm and 15%(v/v),	0.9 g/L	[31]
Pomegranate peel	*Clostridium acetobutylicum* s (MTCC no. 11274)	Acid treatment	12 g/L	[32]
Rice straw	*Clostridium acetobutylicum* and *Saccharomyces cerevisiae*	Pretreatment: Alkaline treatment 37 °C and 150 rpm	ABE: 9.5 g/L 6.5 g/L	[33]
Banana peel	*Saccharomyces cerevisiae* and *Pichia* sp.	H_2SO_4 72 h, pH-5.5, Temperature−32 °C	7.4 g/L	[34]

3.3 Biohydrogen

Biohydrogen is a renewable energy in the form of hydrogen gas produced through biological reaction processes. Biohydrogen is considered a new energy source because it is renewable, environmentally friendly and has high potential as an alternative to fossil fuels [35]. It is from biological sources such as microalgae, bacteria, and organic waste through a more sustainable biotechnology process, and only produces water without carbon dioxide emissions [36]. With its high specific energy value and its potential for use in fuel cell technology, biohydrogen can not only reduce pollution, but also support the transition to a low-carbon economy and improve global energy security [37].

Biohydrogen can be produced through several main biological methods, including dark fermentation, photo fermentation and biological photolysis. Dark fermentation uses anaerobic bacteria to convert organic matter or agricultural waste into hydrogen without the need for light [38]. In the dark fermentation process, several types of microbes are used to produce biohydrogen, including *Clostridium* sp., *Enterobacter* sp., *Klebsiella* sp., and *Bacillus* sp. [39]. The mechanism occurs when these microorganisms break down organic substrates such as glucose, starch, or agricultural waste under anaerobic conditions. Through the process of glycolysis, glucose is converted into pyruvic acid, which is then metabolized by enzymes such as pyruvate: ferredoxin oxidoreductase (PFOR). This enzyme produces electrons that are transferred to ferredoxin, which is then transferred again by the hydrogenase enzyme to convert H^+ ions to hydrogen gas (H_2) [39, 40]. The main by-products are usually organic acids such as acetic acid, butyrate, or lactate depending on the dominant metabolic pathway. In short, the mechanism of dark fermentation relies on the activity of hydrogenase enzymes in anaerobic microorganisms to convert organic matter to hydrogen in addition to producing organic acids.

Photofermentation involves photosynthetic bacteria that use light energy in conjunction with organic substrates to produce hydrogen. In photofermentation, the microbes commonly used are purple non-sulfur photosynthetic bacteria (PNSB) such as Rhodobacter sphaeroides, Rhodopseudomonas palustris, and Rhodospirillum rubrum [41, 42]. The mechanism relies on sunlight as an energy source to activate their photosynthetic system. Under anaerobic conditions and in the presence of organic substrates such as organic acids (e.g. acetic, lactate, or malate), electrons from the substrate are transferred through the photosynthetic electron transport chain. The main enzyme involved is nitrogenase, which normally functions in nitrogen fixation, but when nitrogen is limited, nitrogenase reduces H^+ ions to hydrogen (H_2) using energy from light [43]. In short, photofermentation produces biohydrogen through the activity of nitrogenase driven by light energy, with the organic substrate as the main electron donor.

In addition, biological photolysis is carried out by green algae or cyanobacteria that break down water molecules with the help of sunlight to release hydrogen.

In biological photolysis, the microbes used are micro green algae (e.g. *Chlamydomonas reinhardtii*) and cyanobacteria (e.g. *Anabaena* sp., *Nostoc* sp.). The mechanism occurs when these organisms carry out photosynthesis, which uses sunlight to break down water molecules (H_2O) in a photochemical reaction in photosystem II [44].. This process produces oxygen, electrons, and protons (H^+). The released electrons are then passed through the photosynthetic electron transport chain and transferred to the hydrogenase or nitrogenase enzyme. This enzyme converts the proton (H^+) into hydrogen molecules (H_2) [45]. However, the main challenge in this mechanism is that the hydrogenase and nitrogenase enzymes are very sensitive to oxygen, so continuous hydrogen production requires microaerobic conditions or oxygen control strategies [46]. In short, biological photolysis produces biohydrogen when light is used to split water, and the released electrons are utilized by enzymes to convert protons into hydrogen gas. Table 5 shows biohydrogen production via different production route.

The production of biohydrogen currently faces several major challenges, including low yield compared to conventional methods, high production costs due to specialized bioreactors and strict operating conditions, and the oxygen sensitivity of key enzymes such as hydrogenase and nitrogenase. In addition, the biological systems involved often suffer from instability, while the use of food-related substrates like glucose competes with other industries, making the process less sustainable. Scaling up from laboratory to industrial levels also remains difficult, as efficiency and cost-effectiveness are still limited. These barriers highlight the need for further research in

Table 5 Biohydrogen production via different production route

Method	Microorganisms used	Mechanism of biohydrogen production
Dark fermentation	Anaerobic bacteria such as *Clostridium sp., Enterobacter sp., Klebsiella sp., Bacillus sp.*	Organic substrates (e.g., glucose, agricultural waste) are metabolized through glycolysis → pyruvate → electrons transferred to **ferredoxin** → **hydrogenase** enzyme converts H^+ into H_2. By-products: organic acids
Photo-fermentation	Purple non-sulfur photosynthetic bacteria (PNSB) such as *Rhodobacter sphaeroides, Rhodopseudomonas palustris, Rhodospirillum rubrum*	Uses **light** as the energy source; organic acids as electron donors. Electrons pass through the photosynthetic electron transport chain → **nitrogenase** (under nitrogen-limited conditions) reduces H^+ into H_2
Biophotolysis	Green microalgae (*Chlamydomonas reinhardtii*) and cyanobacteria (*Anabaena sp., Nostoc sp.*)	Sunlight splits water molecules (H_2O) in **photosystem II** → produces electrons, protons (H^+), and oxygen. Electrons are transferred via electron transport chain → **hydrogenase/nitrogenase** enzymes convert H^+ into H_2

biotechnology, reactor engineering, and the use of low-cost waste-based substrates to make biohydrogen a viable clean energy option.

3.4 Biogas and Biomethane

Biogas and biomethane are gaseous bioenergy that can be produced through the conversion of biomass through biological activity. In general, biogas is a mixture of gases produced through the anaerobic (without oxygen) decomposition of organic matter by microorganisms. It consists mainly of methane (CH_4, 50–70%) and carbon dioxide (CO_2, 30–50%), with small amounts of other gases such as hydrogen sulfide (H_2S), hydrogen (H_2), and water vapor [47]. Meanwhile, biomethane is a renewable form of methane (CH_4) produced from the anaerobic digestion of biomass and is essentially purified biogas, with a methane content of over 90%, making it comparable in quality to natural gas [48]. This bioenergy material can be produced through the conversion of biomass as a substrate material. Past studies have shown that raw materials (substrates) include agricultural waste, animal waste, food waste, sewage, and other biomass that is easily decomposed and can be used as a substrate for biogas production [49]. The methane gas in biogas has a high energy value and can be used as fuel for cooking, generating electricity, or heating buildings, while the solid and liquid waste after the process is known as digestate, which can be used as the organic fertilizer [50]. Because it is derived from natural waste and reduces greenhouse gas emissions, biogas is considered one of the most environmentally friendly renewable energy sources [51]. Comparison between biogas and biomethane generated from the biochemical reaction is shown in Table 6.

Biogas is produced through the anaerobic digestion of organic matter by various groups of microorganisms. This process involves a complex reaction consisting of four main stages, namely hydrolysis, acidogenesis, acetogenesis, and methanogenesis. In hydrolysis, several types of microorganisms such as *Clostridium* and

Table 6 Comparison between biogas and biomethane generated from biochemical reaction

Aspect	Biogas	Biomethane
Definition	A gas mixture produced from the anaerobic digestion of organic matter	Upgraded biogas with a higher methane content
Composition	50–70% methane (CH_4), 30–50% carbon dioxide (CO_2), plus H_2S, NH_3, water vapor	> 90% methane (CH_4), closely resembling fossil natural gas
Energy quality	Lower calorific value due to the presence of CO_2 and other gases	Higher calorific value, comparable to natural gas
Main uses	Direct use for electricity generation, heating, or cooking	Injection into natural gas grids, vehicle fuel (Renewable Natural Gas, RNG)
Process	Produced directly through anaerobic digestion of organic matter	Purified via **upgrading/scrubbing** to remove CO_2, H_2S, and moisture

Bacteroides break down complex organic matter (carbohydrates, proteins, lipids) into simpler molecules. Next, is the acidogenesis reaction where acidogenic bacterial microorganisms such as *Enterobacter* and *Lactobacillus* convert the hydrolysis products into volatile fatty acids (VFA), alcohol, hydrogen (H_2) and carbon dioxide (CO_2). At this stage, acetogenic bacteria such as *Syntrophomonas* and *Syntrophobacter* convert the VFA into acetic acid, H_2 and CO_2 which are important substrates for methanogenic archaebacteria, the main group in methane production. At the methanogenesis stage, methanogens such as *Methanosaeta* and *Methanosarcina* use the acetoclastic pathway to convert acetic acid to CH_4 and CO_2, while *Methanobacterium* and *Methanospirillum* carry out the hydrogenotrophic pathway by converting H_2 and CO_2 to CH_4. This layered process ensures that organic waste is converted into a mixture consisting of 50–70% methane, 30–50% carbon dioxide, as well as small amounts of other gases such as H_2S, NH_3, and water vapor, also known as biogas [48, 52, 53]. Figure 3 illustrates typical biogas production by microorganisms activity in anaerobic digester.

Biogas production from biomass is influenced by several key factors that determine the effectiveness and stability of the anaerobic digestion process. These include temperature, where the mesophilic (30–40 °C) and thermophilic (50–60 °C) ranges affect the rate of microbial reaction; a suitable pH (around 6.8–7.5) is required for stable methanogenic activity; and a balanced carbon to nitrogen (C/N) ratio (20–30:1) to avoid nutrient deficiency or toxic ammonia production. In addition, the type and composition of biomass play a role, as the high lignocellulose content requires pretreatment before it can be easily decomposed. The concentration of solids, organic loading, and hydraulic residence time (HRT) also affect the amount of gas produced. Additional factors such as the presence of toxic substances (e.g. heavy metals, antibiotics, or detergents) can interfere with microbial activity, while agitation helps ensure substrate homogenization and avoid sedimentation. All of these factors need to be optimized to ensure maximum and consistent biogas production [54–56]. Table 7 shows biogas and biomethane produced from different types of biomass.

To obtain biomethane, the raw biogas must undergo upgrading processes such as water scrubbing, pressure swing adsorption, or membrane separation to remove CO_2, hydrogen sulfide, moisture, and other impurities. The resulting biomethane can be used as a fuel for heating, electricity generation, or transportation, and it can also be injected into natural gas pipelines. Biomethane production not only contributes to renewable energy generation, but also supports waste management and greenhouse gas reduction, making it a key player in circular bioeconomy and sustainable energy strategies.

Biomethane is produced from biogas that has undergone a purification or "upgrading" process to increase the methane content to over 90%. This purification involves the removal of by-products such as carbon dioxide (CO_2), hydrogen sulfide (H_2S), ammonia (NH_3), and water vapor, using techniques such as water scrubbing, chemical scrubbing, membrane separation, or pressure swing adsorption (PSA) [70, 71]. The result is biomethane gas that has a quality equivalent to fossil natural gas in terms of energy content and can be used in a variety of applications. Biomethane is more suitable for being piped into gas pipelines, used as a vehicle fuel (renewable

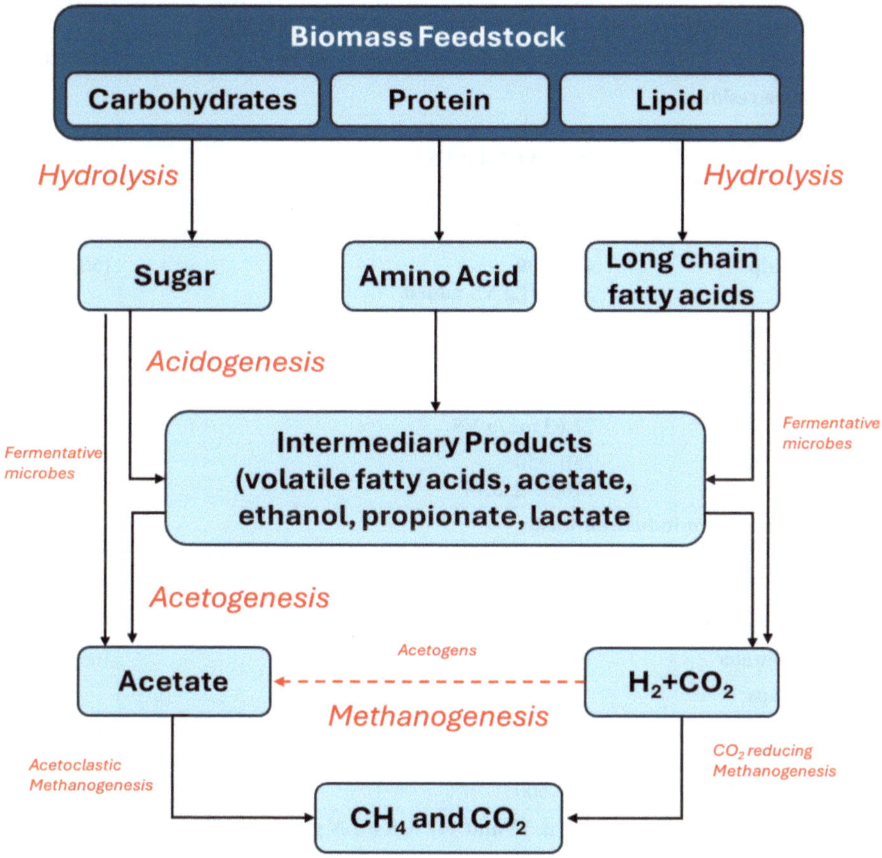

Fig. 3 Typical biogas production by microorganisms activity in anaerobic digester

natural gas, RNG), or as a clean energy source in the industry. In this way, biomethane is considered an "upgraded" version of biogas that is more efficient, versatile, and environmentally friendly [72, 73].

The main challenges in biogas and biomethane production from biomass are related to the characteristics of the feedstock, biological processes, technology, and cost. In terms of feedstock, biomass such as agricultural or food waste often contains high lignocellulose or toxic substances that are difficult to decompose, thus affecting microbial efficiency. In terms of process, the stability of anaerobic digestion can be disrupted by imbalances in pH, temperature, or carbon to nitrogen ratio, which reduce methanogen activity. In terms of technology, biogas to biomethane treatment and upgrading systems such as membranes, pressure swing adsorption (PSA), or chemical scrubbing still require high investment costs and large energy consumption. In addition, challenges related to infrastructure, gas quality standards, and market acceptance also make it difficult to integrate biomethane into existing energy systems.

Table 7 Biogas and biomethane produced from different types of biomass

Biomass types	Methane yield (m^3 CH_4/ton VS*)	References
Agriculture residues		
Wheat straw	200–300 **296 ml CH4/g VS**	[57]
Rice straw	180–290	[58]
Corn stover	250–320	[59]
Cassava pulp	**400–500** **306 m/Lg VS added**	[60]
Animal manure		
Cow manure	150–250	[61]
Pig manure	300–400 **27.63 mL/g VS**	[62]
Poultry manure	250–350 **270.53 ± 3.74**	[63]
Industrial and agro-industrial waste		
Food waste	400–700 **320 ml/g VS**	[64]
Slaughterhouse waste	500–900	[65]
Dairy wastewater	100–300	[66]
Energy crops		
Napier grass	250–450 **169**	[67]
Sorghum	350–500	[68]
Microalgae (e.g., *Chlorella*)	**750 ± 15 and 462 ± 9 mLN g − 1 volatile solids (VS) day**	[69]

Overall, these technical, economic, and logistical issues are major barriers to scaling up biogas and biomethane production on a large scale.

4 Future Prospect and Challenges

Producing liquid biofuels from biomass via biochemical conversion (such as fermentation) faces several key challenges that affect its economic and technical viability. One major challenge is the complex structure of lignocellulosic biomass, which requires costly and energy-intensive pretreatment and enzymatic hydrolysis steps to release fermentable sugars. The presence of lignin and other inhibitors can reduce enzyme efficiency and microbial performance, leading to lower fuel yields.

Another significant challenge is the high production cost, especially for enzymes and downstream processing like purification or distillation. Enzymes used in hydrolysis are expensive, and their performance can be hindered by non-ideal process conditions or by-products. Additionally, the fermentation of mixed sugars, especially from hemicellulose, is inefficient because most conventional microbes preferentially consume glucose over other sugars like xylose or arabinose, limiting overall conversion efficiency.

A third challenge is the scale-up and integration of processes. Combining pretreatment, hydrolysis, and fermentation steps into a continuous and cost-effective system—such as SSF or CBP—requires careful balancing of biological, chemical, and engineering factors. Variability in biomass quality and composition further complicates process control and consistency at the industrial scale.

5 Conclusion

Biochemical conversion of biomass provides a promising pathway toward renewable and sustainable bioenergy production. Technologies such as pretreatment, hydrolysis, and microbial fermentation enable the transformation of diverse biomass feedstocks into valuable fuels including ethanol, butanol, hydrogen, biogas, and biomethane. These processes contribute not only to reducing fossil fuel reliance but also to promoting circular economy principles through waste utilization. However, realizing the full potential of biomass-based bioenergy requires overcoming several barriers, including high processing costs, limited enzyme and microbial efficiency, and the complexities of scaling up integrated systems. Future advancements in biotechnology, microbial engineering, and process optimization—combined with supportive policies and infrastructure—will be critical to achieving cost-effective, high-yield biofuel production. Ultimately, the integration of bioenergy into the global energy mix offers significant opportunities for enhancing energy security, mitigating climate change, and fostering sustainable development.

References

1. Mardawati E, Werner A, Bley T, Kresnowati MTAP, Setiadi T (2014) The enzymatic hydrolysis of oil palm empty fruit bunches to xylose. J Japan Inst Energy 93(10):973–978. https://doi.org/10.3775/jie.93.973
2. Ong LGA, Chan CH, Chew AL (2012) Enzymatic hydrolysis of rice straw: process optimization. J Med Bioeng 1(1):14–16. https://doi.org/10.12720/jomb.1.1.14-16
3. Ganguly A, Das S, Bhattacharya A, Dey A, Chatterjee PK (2013) Enzymatic hydrolysis of water hyacinth biomass for the production of ethanol: Optimization of driving parameters. Indian J Exp Biol 51(7):556–566
4. Zulkefli S, Abdulmalek E, Rahman MB (2017) Pretreatment of oil palm trunk in deep eutectic solvent and optimization of enzymatic hydrolysis of pretreated oil palm trunk. Renew Energy 107:36–41. https://doi.org/10.1016/j.renene.2017.01.037
5. Zhang H, Fan M, Li X, Zhang A, Xie J (2018) Enhancing enzymatic hydrolysis of sugarcane bagasse by ferric chloride catalyzed organosolv pretreatment and Tween 80. Biores Technol 258:295–301. https://doi.org/10.1016/j.biortech.2018.03.004

6. Dos Santos J, et al (2024) Energy cycle assessment of bioethanol production from sugarcane bagasse by life cycle approach using the fermentation conversion process. Biomass Convers Biorefin https://doi.org/10.1007/s13399-023-04288-5
7. Ferrari MD, Guigou M, Lareo C (2013) Energy consumption evaluation of fuel bioethanol production from sweet potato. Biores Technol 136:377–384. https://doi.org/10.1016/j.biortech.2013.03.04
8. Ye Y, Hu J, Zhang Z, Zhong W, Zhao Z, Zhang J (2023) Effect of different ratios of gasoline-ethanol blend fuels on combustion enhancement and emission reduction in electronic fuel injection engine. Polymers 15(19):3932. https://doi.org/10.3390/polym15193932
9. Kassim MA, Tan KM, Kamaludin R, Hussain AH, Bukhari NA (2022) Bioprocessing of sustainable renewable biomass for bioethanol production. In: Value-chain of biofuels. Elsevier, p. 195–234. https://doi.org/10.1016/B978-0-12-824388-6.00004
10. Jehadin F, Rabeya T, Asad MA, Ayodele OO, Adekunle AE, Islam MS (2021) Efficient conversion of cornstalk to bioethanol using dilute H_2SO_4 pretreatment. Int J Energy Environ Eng 12(2):203–211. https://doi.org/10.1007/s40095-020-00366-w
11. Kuttiraja M, Sindhu R, Varghese PE, Sandhya SV, Binod P, Vani S, Pandey A, Sukumaran RK (2013) Bioethanol production from bamboo (Dendrocalamus sp.) process waste. Biomass Bioenerg 59:142–150. https://doi.org/10.1016/j.biombioe.2013.10.01
12. Condor BE, de Luna MDG, Chang Y-H, Chen J-H, Leong YK, Chen P-T, Chen C-Y, Lee D-J, Chang J-S (2022) Bioethanol production from microalgae biomass at high-solids loadings. Biores Technol 363:128002. https://doi.org/10.1016/j.biortech.2022.128002
13. Kassim MA, Tan KM, Serri NA (2018) Enzymatic hydrolysis of dilute alkaline pretreated Chlorella sp. biomass for biosugar production and fed-batch bioethanol fermentation. Asia-Pacific J Sci Technol 24(1):APST-24. https://doi.org/10.14456/apst.2019.4
14. Kassim MA, Loh SK, Abu Bakar N, Abdul Aziz A, Mat SR (2011) Biothanol production from enzymatically saccharified empty fruit bunches hydrolysate using Saccharomyces cerevisiae. Res J Environ Sci 5:573–586. https://doi.org/10.3923/rjes.2011.573.586
15. Yousif YM, Abdulhay HS (2017) Bioethanol production from banana peels using different pretreatments. Asian J Biol Life Sci 6(3):408–412
16. Jugwanth Y, Sewsynker-Sukai Y, Gueguim Kana EB (2019) Valorization of sugarcane bagasse for bioethanol production through simultaneous Saccharification and fermentation: optimization and kinetic studies. Fuel 262:116552. https://doi.org/10.1016/j.fuel.2019.116552
17. Chen X, Zhang Y, Wang Z et al (2024) Engineered Zymomonas mobilis expressing α-amylase for direct starch fermentation. Microb Cell Fact 23(1):112. https://doi.org/10.1186/s12934-024-02539-2
18. Kang A, Kang W, Ma Y et al (2024) Bioethanol production from lignocellulosic biomass using genetically modified microorganisms: a review. Front Bioeng Biotechnol 12:1045021. https://doi.org/10.3389/fbioe.2024.1045021
19. Kumar A, Sharma S, Singh D (2023) Optimization of fermentation parameters for ethanol production using Saccharomyces cerevisiae. Fermentation 9(4):268. https://doi.org/10.3390/fermentation9040268
20. Sun S, Li X, Xu J (2023) Kluyveromyces marxianus: a promising yeast for high-temperature bioethanol fermentation. Fermentation 9(9):840. https://doi.org/10.3390/fermentation9090840
21. Tsioulpas A, Xenopoulos E, Chatzifragkou A (2023) Review of second-generation bioethanol production from residual biomass. Food Technol Biotechnol 61(2):153–168. https://doi.org/10.17113/ftb.61.02.23.7636
22. Dürre P (2007) Biobutanol: an attractive biofuel. Biotechnol J 2(12):1525–1534. https://doi.org/10.1002/biot.200700168
23. Green EM (2011) Fermentative production of butanol—the industrial perspective. Curr Opin Biotechnol 22(3):337–343. https://doi.org/10.1016/j.copbio.2011.02.004
24. Lee SY, Park JH, Jang SH, Nielsen LK, Kim J, Jung KS (2008) Fermentative butanol production by Clostridia. Biotechnol Bioeng 101(2):209–228. https://doi.org/10.1002/bit.22003
25. Nigam PS, Singh A (2011) Production of liquid biofuels from renewable resources. Prog Energy Combust Sci 37(1):52–68. https://doi.org/10.1016/j.pecs.2010.01.003

26. Jones DT, Woods DR (1986) Acetone-butanol fermentation revisited. Microbiol Rev 50(4):484–524. https://doi.org/10.1128/mr.50.4.484-524.1986
27. Qureshi N, Blaschek HP (2001) Recent advances in ABE fermentation: hyper-butanol producing Clostridium beijerinckii BA101. J Ind Microbiol Biotechnol 27(5):287–291. https://doi.org/10.1038/sj.jim.7000199
28. Jang YS, Malaviya A, Cho C, Lee J, Lee SY (2012) Butanol production from renewable biomass by Clostridia. Biores Technol 123:653–663. https://doi.org/10.1016/j.biortech.2012.07.104
29. Johnravindar D, Elangovan N, Olaganathan N, Muthaiyan A, Fei Q, et al (2019) Biobutanol production from cassava waste residue using Clostridium sp. AS3 in batch culture fermentation. Biofuels 12(6):1259–1266. https://doi.org/10.1080/17597269.2019.1608671
30. Noomtim P, Cheirsilp B (2011) Production of butanol from palm empty fruit bunches hydrolyzate by Clostridium acetobutylicum. Energy Procedia 9:140–146. https://doi.org/10.1016/j.egypro.2011.09.022
31. Al-Shorgani NKN, Shukor H, Abdeshahian P, Hamid AA, Anuar N, Rahman NA, Kalil MS, Yusoff WMW (2015) Process optimization of butanol production by Clostridium saccharoperbutylacetonicum N1–4 (ATCC 13564) using palm oil mill effluent in acetone-butanol-ethanol fermentation. Biocatal Agric Biotechnol 4(2):244–249. https://doi.org/10.1016/j.bcab.2015.01.004
32. Motghare KA, Shende DZ, Wasewar KL (2025) Butanol fermentation using pomegranate peels and separation using ionic liquid. J Indian Chem Soc 102(6):101736. https://doi.org/10.1016/j.jics.2025.101736
33. Cortés-Rojas F, Schiffer D, Vidas Reed-Teaza A, Romero N et al (2024) Solvent production from rice straw by a co-culture of Clostridium acetobutylicum and Saccharomyces cerevisiae: effect of pH control. Biomass Conver Biorefin 14:5561–5573. https://doi.org/10.1007/s13399-022-02750-4
34. Mishra RR, Samantaray B, Chandra Behera B, Pradhan BR, Mohapatra S (2020) Process optimization for conversion of waste banana peels to biobutanol by a yeast co-culture fermentation system. Renew Energy 162:478–488. https://doi.org/10.1016/j.renene.2020.08.045
35. Kanwal F, Torriero AAJ (2022) Biohydrogen—a green fuel for sustainable energy solutions. Energies 15(20):7783. https://doi.org/10.3390/en15207783
36. Chader S, Hacene H, Agathos SN (2021) Biohydrogen from microalgae: production and applications. Appl Sci 11(4):1616. https://doi.org/10.3390/app11041616
37. Kumar G et al (2022) Production of biohydrogen from microalgae biomass after wastewater treatment and air purification from CO_2. Processes 11(10):2978. https://doi.org/10.3390/pr11102978
38. Hallenbeck PC, Benemann JR (2002) Biological hydrogen production; fundamentals and limiting processes. Int J Hydrogen Energy 27(11–12):1185–1193. https://doi.org/10.1016/S0360-3199(02)00131-3
39. Azwar MY, Hussain MA, Abdul-Wahab AK (2014) Development of biohydrogen production by photobiological, fermentation and electrochemical processes: a review. Renew Sustain Energy Rev 31:158–173. https://doi.org/10.1016/j.rser.2013.11.022
40. Ghimire A, Frunzo L, Pirozzi F, Trably E, Escudie R, Lens PN, Esposito G (2015) A review on dark fermentative biohydrogen production from organic biomass: process parameters and use of by-products. Appl Energy 144:73–95. https://doi.org/10.1016/j.apenergy.2015.01.045
41. Basak N, Das D (2007) Photofermentative hydrogen production using purple non-sulfur bacteria: a review. Int J Hydrogen Energy 32(16):228–246
42. Ghosh D, Hallenbeck PC, Dutta A (2017) Photofermentative hydrogen production. In: Hydrogen production technologies. Springer
43. Koku H, Eroglu I, Gunduz U, Yucel M, Turker L (2002) Aspects of the metabolism of hydrogen production by Rhodobacter sphaeroides. Int J Hydrogen Energy 27(11–12):1315–1329
44. Melis A, Happe T (2001) Hydrogen production. Green algae as a source of energy. Plant Physiol 127(3):740–748
45. Ghirardi ML, Zhang L, Lee JW, Flynn T, Seibert M, Greenbaum E, Melis A (2000) Microalgae: a green source of renewable H_2. Trends Biotechnol 18(12):506–511

46. Khan MI, Shin JH, Kim JD (2018) The promising future of microalgae: current status, challenges, and optimization of a sustainable and renewable industry for biofuels, feed, and other products. Microb Cell Fact 17(1):36
47. Surendra KC, Takara D, Hashimoto AG, Khanal SK (2014) Biogas as a sustainable energy source for developing countries: opportunities and challenges. Renew Sustain Energy Rev 31:846–859
48. Angelidaki I, Treu L, Tsapekos P, Luo G, Campanaro S, Wenzel H, Kougias PG (2018) Biogas upgrading and utilization: current status and perspectives. Biotechnol Adv 36(2):452–466
49. Zhao J, Wei L, Kumar A, Xu C (2021) Biogas production from lignocellulosic biomass: process optimization and economic analysis. Renew Energy 168:146–156
50. Mata-Alvarez J, Dosta J, Romero-Güiza MS, Fonoll X, Peces M, Astals S (2014) A critical review on anaerobic co-digestion achievements between 2010 and 2013. Renew Sustain Energy Rev 36:412–427
51. Scarlat N, Dallemand JF, Fahl F (2018) Biogas: developments and perspectives in Europe. Renew Energy 129:457–472
52. Appels L, Lauwers J, Degrève J, Helsen L, Lievens B, Willems K, Van Impe J (2011) Anaerobic digestion in global bio-energy production: potential and research challenges. Renew Sustain Energy Rev 15(9):4295–4301
53. Khalid A, Arshad M, Anjum M, Mahmood T, Dawson L (2011) The anaerobic digestion of solid organic waste. Waste Manage 31(8):1737–1744
54. Appels L, Baeyens J, Degrève J, Dewil R (2008) Principles and potential of the anaerobic digestion of waste-activated sludge. Prog Energy Combust Sci 34(6):755–781
55. Achinas S, Euverink GJW (2019) Theoretical analysis of biogas potential prediction from agricultural waste. Resources 8(2):84
56. Sibiya NT, Shabangu S, Faloye F (2023) Factors influencing biogas production: a review. Clean Eng Technol 12:100632
57. Peng X, Nges IA, Liu J (2016) Improving methane production from wheat straw by digestate liquor recirculation in continuous stirred tank processes. Renew Energy 85:12–18. https://doi.org/10.1016/j.renene.2015.06.023
58. Li Z, Wachemo AC, Yuan H, Korai RM, Li X (2020) Improving methane content and yield from rice straw by adding extra hydrogen into a two-stage anaerobic digestion system. Int J Hydrogen Energy 45:3739–3749. https://doi.org/10.1016/j.ijhydene.2019.07.235
59. Li Y, Hou Z, Shi Q, Cheng Y, Zhu W (2020) Methane production from different parts of corn stover via a simple co-culture of an anaerobic fungus and methanogen. Front Bioeng Biotechnol. 8:314. https://doi.org/10.3389/fbioe.2020.00314
60. Panichnumsin P, Nopharatana A, Ahring BK, Chaiprasert P (2010) Production of methane by co-digestion of cassava pulp with various concentrations of pig manure. Biomass Bioenergy 34(8):1117–1124. https://doi.org/10.1016/j.biombioe.2010.02.018
61. Obileke K, Makaka G, Nwokolo N (2022) Efficient methane production from anaerobic digestion of cow dung: an optimization approach. Challenges 13(2):53. https://doi.org/10.3390/challe13020053
62. Wang Y, Zhang J, Li Y, Jia S, Li G (2020) Methane production from the co-digestion of pig manure and corn stover with the addition of cucumber residue: role of the total solids content and feedstock-to-inoculum ratio. Biores Technol 306:123172. https://doi.org/10.1016/j.biortech.2020.123172
63. Paranhos AG, de O, Adarme OFH, Barreto GF, Silva SdQ, Aquino SF (2020) Methane production by co-digestion of poultry manure and lignocellulosic biomass: kinetic and energy assessment. Bioresour Technol 300:122588. https://doi.org/10.1016/j.biortech.2019.122588
64. Yang Q, Liu H, Liu L, Yan Z, Chui C, Yang N, Wang C, Shen G, Chen Q (2025) Enhancing methane production in anaerobic digestion of food waste using co-pyrolysis biochar derived from digestate and rice straw. Molecules 30(8):1766. https://doi.org/10.3390/molecules30081766
65. Khitous M, Saber M, Tirichine N, Berouaken A (2025) Methane production from slaughterhouse animal fat waste using a pilot-scale thermophilic anaerobic digester. Biofuels 1–13. https://doi.org/10.1080/17597269.2025.2528315

66. Zeb I, Ma J, Mehboob F, Kafle GK, Amin BAZ, Nazir R, Ndegwa P, Frear C (2019) Kinetic and microbial analysis of methane production from dairy wastewater anaerobic digester under ammonia and salinity stresses. J Clean Prod 219:797–808. https://doi.org/10.1016/j.jclepro.2019.01.295
67. Prapinagsorn W, Sittijunda S, Reungsang A (2017) Co-digestion of Napier grass and its silage with cow dung for methane production. Energies 10(10):1654. https://doi.org/10.3390/en10101654
68. Richards BK, Cummings RJ, Jewell WJ (1991) High rate low solids methane fermentation of sorghum, corn and cellulose. Biomass Bioenerg 1:249–261. https://doi.org/10.1016/0961-9534(91)90036
69. Shan J, Abreu CM, Capareda SC (2017) Highly efficient methane generation from untreated microalgae biomass. Biotechnol Biofuels 10:86. https://doi.org/10.1186/s13068-017-0871-4
70. Chen XY, Vinh-Thang H, Avalos Ramirez A, Rodrigue D, Kaliaguine S (2015) Membrane gas separation technologies for biogas upgrading. RSC Adv 5:24399–24448. https://doi.org/10.1039/C5RA00666
71. Swinbourne A, et al (2024) A Comprehensive review on biomethane production from biogas separation and its techno-economic assessments. ChemSusChem. Chemistry Europe
72. Brunetti A, Barbieri G (2021) Membrane engineering for biogas valorization. Front Chem Eng 3:775788. https://doi.org/10.3389/fceng.2021.775788
73. BioEnergy Consult (n.d) Biogas upgradation methods. BioEnergy Consult. https://www.bioenergyconsult.com/biogas-upgradation/

Improvement of Liquid Biofuel from Biomass Using Genetic Modified Microbes

Mohd Asyraf Kassim, Nurul Najihah Sham Faizal, Nur Aliah Balqis Roslan, Mardiana Idayu Ahmad, and Norli Ismail

Abstract This chapter discusses recent genetic and metabolic engineering advances to produce liquid biofuel from renewable biomass. It provides a comprehensive overview of the techniques employed, including metabolic pathway engineering, strain improvement and synthetic biology approaches. In this chapter, discussion on metabolic pathway that involves gene expression, gene know-out and also introduction to a novel pathway to enhance biofuel production. In addition, the potential of microorganisms' strain improvement as well as synthetic biology to design new strain catalyst to produce liquid biofuel such as bioethanol, biobutanol and biodiesel from biomass are evaluated. The chapter highlights several investigations, demonstrating the potential of these technologies to produce sustainable and renewable drop-in biofuels. The future of genetic and metabolic engineering for liquid biofuel production appears promising to be an alternative to partially replace energy from fossil fuel. This chapter also discusses the challenge that potentially arises from the development of the liquid biofuel from genetically modified microorganisms. Overall, this chapter serves as a comprehensive insight into production of liquid biofuel, challenges, and future directions through genetic and metabolic engineering.

Keywords Biofuel · Microorganisms · Genetic manipulation · Bioethanol · Lipid

1 Introduction

Liquid biofuel is a liquid biofuel produced from the conversion process of biological resources such as plants, microorganisms, microalgae and organic waste [1]. This liquid biofuel is a fuel that is not based on petroleum or natural gas sources. Now, liquid biofuel is used as a replacement for some of the fossil fuels that can be used in internal combustion engines that are obtained from renewable sources.

Usually, the main sources of liquid biofuel consist of several main sources such as starchy plants, fatty plants, microorganisms and organic waste from agricultural

M. A. Kassim (✉) · N. N. S. Faizal · N. A. B. Roslan · M. I. Ahmad · N. Ismail
School of Industrial Technology, Universiti Sains Malaysia, Penang, Malaysia
e-mail: asyrafkassim@usm.my

M. A. Kassim et al. (eds.), *Sustainable Biorefinery*, Green Energy and Technology,
https://doi.org/10.1007/978-981-95-8181-8_12

and domestic activities. The production of this liquid biofuel can be categorized into several groups based on the feedstock used to produce the liquid biofuel. For example, bioethanol and biobutanol are liquid biofuels produced from the conversion of sugar or starch from feedstock through a fermentation process by microorganisms. Meanwhile, liquid biofuels such as biodiesel and jet-fuel are liquid biofuels produced from the conversion reaction of fats to fuel in the presence of chemical catalysts and enzymes.

Liquid biofuels such as bioethanol, biobutanol and biodiesel have various environmental, energy and economic advantages. Bioethanol and biobutanol produced through the fermentation of starchy or sugary materials can reduce dependence on petrol because they can be blended or replaced with conventional fuels in existing engines without major modifications. Biobutanol has an energy density that is almost the same as petrol and is more resistant to water pollution than bioethanol, making it a more efficient and stable fuel. Biodiesel derived from plant oils or animal fats is a renewable and fully biodegradable fuel, which helps reduce emissions of greenhouse gases, sulphur and other pollutants. Overall, the use of liquid biofuels not only supports the development of green energy and the mitigation of climate change, but can also stimulate the local economy through the use of local biomass resources and the creation of job opportunities in the bioindustrial sector.

The main challenges in the production of liquid biofuels such as bioethanol, biobutanol and biodiesel are closely related to production efficiency, high costs and limited raw materials. Among the main challenges are the low conversion efficiency of biomass materials, for example lignocellulose which is difficult to break down into fermentable sugars [2, 3]; low biofuel production rates due to inefficient natural metabolic pathways [4; and the resistance of producing microorganisms to environmental stresses such as toxicity, pH and temperature [5]. In addition, the production of bioethanol and biobutanol often competes with food sources because raw materials such as corn and sugarcane are used for both.

However, advances in genetic engineering technology now offer a variety of solutions. Through genetic engineering, researchers can modify the genes and metabolic pathways of microorganisms or plants to increase biofuel production. For example, genes that control the production of cellulase and hemicellulase enzymes can be inserted into bacteria or yeast to enable them to break down lignocellulosic materials directly—making the process faster and cheaper [6]. Microorganisms can also be modified to withstand toxic stress or produce more lipids for biodiesel [7]. In addition, technologies such as CRISPR/Cas9 and synthetic biology allow the design of new microorganisms with optimized biosynthetic pathways, increasing bioethanol and biobutanol yields by using a wider range of substrates including agricultural waste and non-food materials [8, 9]. Overall, genetic engineering offers innovative solutions to make liquid biofuel production more efficient, cost-effective, and sustainable, thus strengthening the role of biofuels as a future green energy source. Figure 1 shows an overview of the liquid biofuel from genetically modified microorganisms

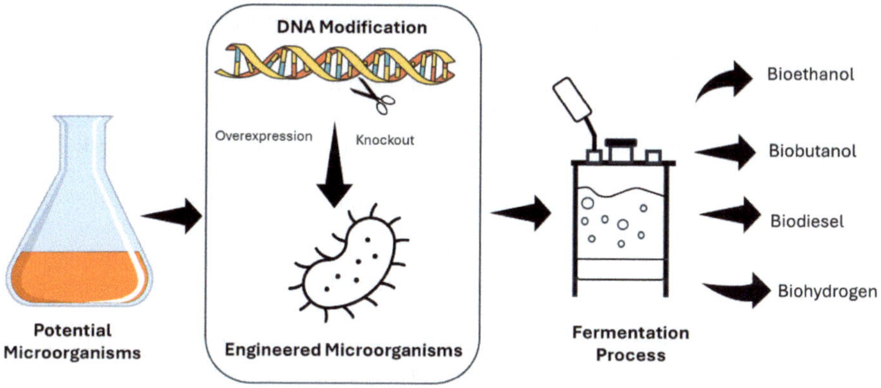

Fig. 1 Overview of liquid biofuel from genetically modified microorganisms

1.1 Genetic Engineering Strategies for Biofuel Feedstocks

Improving the production of Liquid Biofuel from biomass is found among the methods of molecular level manipulation. This method aims to create and optimize the biochemical pathways of the process of converting carbon or substrates such as sugars, glycerol, CO2 and so on into the desired liquid fuel. This involves (a) adding new pathways (heterologous), (b) reducing competing pathways, (c) optimizing enzymes and cofactors, and (d) increasing cell resistance to toxic products. Table 1 shows the previous studies that have been conducted on the utilization of genetically modified microbes on production of various types of liquid biofuel. Figure 2 shows genetic modification of microorganisms for liquid biofuel production.

According to this analysis, it shows that modification is done either by adding new pathways, enhancing existing ones, or eliminating competing pathways—to optimize the efficiency, yield, and/or tolerance for producing various biofuels. Strategies primarily involve manipulating metabolic pathways: either overexpressing key enzymes (like alcohol dehydrogenase for ethanol) to enhance yield, or deleting/knocking out genes responsible for competing or undesirable by-products (like glycerol in yeast or acids in Clostridium). Specific examples include modifying yeast (S. cerevisiae) and Zymomonas mobilis for ethanol tolerance and yield, engineering E. coli and C. acetobutylicum for efficient butanol/acetone production, and altering algae (Chlamydomonas) and cyanobacteria (Synechocystis) to boost lipid content for biodiesel or enable direct biofuel synthesis from the biomass. The overarching goal across all these studies is to increase the efficiency, productivity, and yield of the target biofuel.

Table 1 Different genetic modification strategies for liquid biofuel production using different microbes

Microbe	Biofuel type	Genetic engineering Strategy	Improvement	References
Saccahromyces cerevisiae	Ethanol	Overexpression of pyruvate decarboxylase and alcohol dehydrogenase; deletion of glycerol and acetate pathways	Increased ethanol yield; higher tolerance to inhibitors	[10]
Zymomonas mobilis	Ethanol	Overexpression of stress-response genes; membrane modification; deletion of competing pathways	Enhanced ethanol tolerance and productivity	[11]
Escherichia coli	Butanol	Introduction of heterologous butanol pathway from *Clostridium* spp.; knockout of competing lactate/acetate pathways	Efficient butanol production from sugars	[12]
Clostridium acetobutylicum	Butanol/Acetone	Overexpression of solventogenic genes; deletion of acid production pathways	Improved butanol yield and reduced acid by-products	[13]
Chlamydomonas reinhardtii	Biodiesel (lipids)	Overexpression of acetyl-CoA carboxylase and diacylglycerol acyltransferase; knockout of starch biosynthesis genes	Higher lipid content for biodiesel	[14]
Nannochloropsis sp	Biodiesel (lipids)	CRISPR-Cas9 mediated gene editing to enhance TAG accumulation; overexpression of lipid biosynthesis enzymes	Increased lipid/oil content; improved stress tolerance	[15]

(continued)

Table 1 (continued)

Microbe	Biofuel type	Genetic engineering Strategy	Improvement	References
Corynebacterium glutamicum	Isobutanol	Introduction of heterologous ketoacid pathway; overexpression of valine biosynthesis genes	Improved isobutanol production from glucose	[16]
Synechocystis sp.	Bioethanol/ Isobutanol	Engineering carbon flux to ethanol/ isobutanol pathways; deletion of glycogen storage genes	Direct biofuel production from CO_2 via photosynthesis	[17]

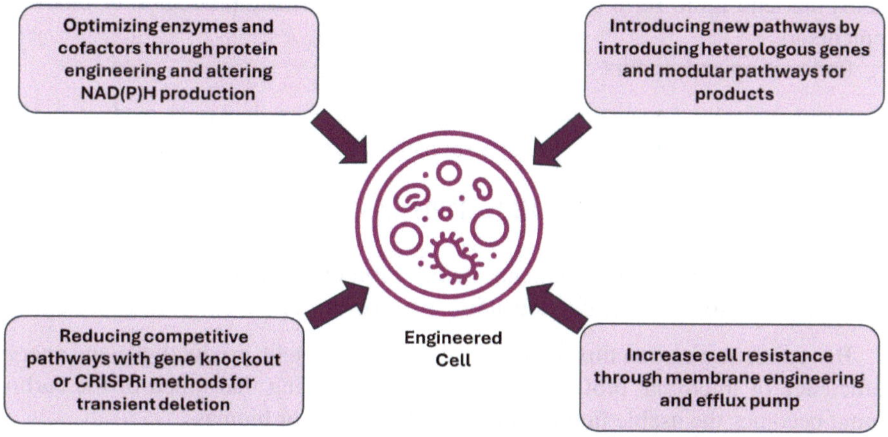

Fig. 2 Genetic Modification of Microorganisms for Liquid Biofuel Production

1.2 Metabolic Pathway Engineering for Biofuel Yield Optimization

Metabolic pathway engineering optimizes biofuel production by genetically modifying microorganisms to efficiently convert renewable carbon sources into fuels. This involves manipulating metabolic pathways to maximize fuel yield by introducing new pathways, overexpressing key enzymes, or deleting competing pathways (Engineering microbes for sustainable biofuel production and extraction of lipids—Current research and future perspectives). Key strategies include redirecting carbon flux, enhancing substrate conversion of substrate from the biomass into targeted products, improving robustness and tolerance, and using tools like CRISPR-Cas9 for precision engineering. Challenges include low yield and titer in large-scale production, but advances in synthetic biology, AI-driven optimization, and consolidated bioprocessing are addressing these issues for more cost-effective biofuel production.

One of the genetic engineering strategy optimization for liquid bioenergy production is focus on precise metabolic pathway manipulation to maximize conversion of substrates into biofuels. Generally, this approach involves redirecting carbon flux toward the desired biofuel precursors by strengthening key metabolic nodes and blocking routes that lead to by-products, ensuring efficient substrate utilization. This technique refers to the deliberate modification of a microorganism's metabolic network so that carbon from the substrate such as glucose, glycerol, or CO_2 flows preferentially into the desired biofuel synthesis pathway rather than into competing pathways or cell maintenance processes.

In simpler terms, it means rerouting the carbon "traffic" inside the cell to increase the proportion of main important metabolites such as acetyl-CoA, pyruvate, or fatty acids that end up to form biofuel molecules such as ethanol, butanol, or biodiesel precursors.

Redirecting carbon flux can be achieved by:

- Overexpressing key enzymes that drive carbon toward the target product.
- Knocking out or repressing genes responsible for undesired by-product formation (e.g., organic acids, CO_2).
- Balancing cofactors like NADH/NADPH to ensure sufficient reducing power for biosynthetic reactions.
- Introducing regulatory elements (e.g., promoters, riboswitches) to control metabolic flux dynamically depending on growth stage.

By redirecting carbon flux, engineers can improve yield, productivity, and energy efficiency of microbial biofuel production, and ensuring that much of the carbon input becomes the usable liquid fuel rather than waste or biomass.

At the same time, redirecting carbon flux also can be done by introducing novel pathways from heterologous organisms that enables the host microbe to gain new metabolic capabilities to synthesize new intermediates or final fuels such as higher alcohols or fatty acid derivatives that it could not naturally synthesize, In biofuel production, this strategy involves transferring genes or entire biosynthetic pathways from a natural biofuel producer into a more genetically tractable or industrially robust host such as *Escherichia coli* or *Saccharomyces cerevisiae*. These heterologous pathways can introduce new enzymatic reactions that convert central metabolites such as pyruvate or acetyl-CoA into biofuel molecules such as n-butanol, isobutanol, fatty acid ethyl esters, or alkanes.

For example, incorporating the Clostridial ABE (acetone–butanol–ethanol) pathway into E. coli enables butanol biosynthesis [18]. Another study by Lin et al. [19] while introducing fatty acyl–reductase and aldehyde-deformylating oxygenase genes from cyanobacteria allows production of long-chain alkanes resembling diesel. These approaches expand the metabolic repertoire of the host by integrating non-native biochemical routes. Together with regulatory optimization and flux balancing, it will enable efficient and sustainable liquid biofuel production from renewable feedstocks.

In order to further enhance productivity, manipulation at molecular level by enhancing gene expression and enzyme function via protein engineering could increase catalytic efficiency as well as reaction stability. This improvement is achieved through several complementary approaches including rationally modifying enzyme amino acid sequences based on structural and mechanistic knowledge to enhance catalytic efficiency [20]. In addition, improvement on thermal stability, or substrate affinity, are other approaches that allow the enzyme to function optimally under industrial fermentation conditions. Directed evolution, on the other hand, mimics natural selection by generating large libraries of enzyme variants through random mutagenesis and screening for improved performance, enabling the discovery of enzymes with superior activity that can lead to microbial tolerance to fermentation by-product such as solvents and fermentation temperature. Meanwhile, expression tuning adjusts the level, timing, or localization of enzyme synthesis using different promoters, ribosome-binding sites, or regulatory elements to balance metabolic flux and prevent pathway bottlenecks. Together, these strategies optimize enzyme kinetics, stability, and cofactor compatibility (e.g., NADH/NADPH usage), resulting in more efficient conversion of substrates into liquid biofuels such as ethanol, butanol, or biodiesel precursors.

Additionally, deleting competing pathways via targeted gene knockouts or CRISPR interference removes metabolic drains, channeling cellular resources exclusively toward the targeted biofuel pathway. Deleting competing pathways in microbes through targeted gene knockouts or CRISPR interference effectively removes metabolic drains, ensuring that cellular precursors, energy, and cofactors that are related in typical metabolic pathway are redirected toward the desired biofuel pathway. Many central metabolites, such as acetyl-CoA or pyruvate, are naturally shared among multiple biosynthetic routes, producing by-products that reduce biofuel yield. By silencing or eliminating enzymes in these competing pathways, more substrate flux is channeled into the engineered biofuel-producing pathway, enhancing product formation without additional inputs. This strategy not only increases carbon efficiency but also minimizes accumulation of inhibitory by-products, and when combined with overexpression of biofuel pathway enzymes, it can substantially improve overall production, effectively focusing the cell's metabolic resources exclusively on the target biofuel.

The use of latest technology such as CRISPR-Cas9 in metabolic engineering allows precise and efficient modification of microbial, plant, and microalgal systems to improve biofuel production [21–23]. For instance, in microbes such as *Saccharomyces cerevisiae* and *Zymomonas mobilis*, adapting CRISPR-based editing can delete competing pathways and optimize fuel-producing routes, enhancing ethanol yield and increasing tolerance to inhibitory compounds, toxins, temperature fluctuations, and other stresses encountered during industrial biofuel processes [22, 24]. Similarly, in plants and microalgae, CRISPR can be applied to increase biomass, lipid productivity and oil content, while reducing lignin in crops like barley to

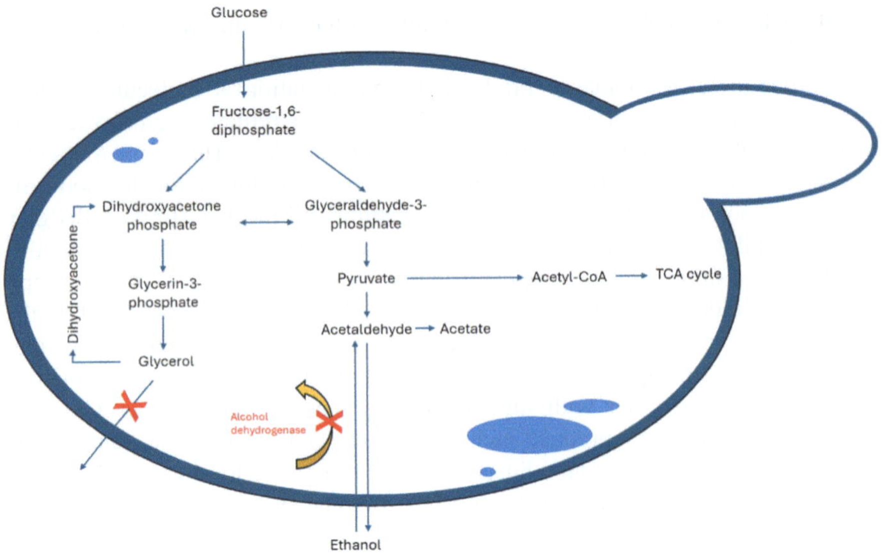

Fig. 3 Genetically modified yeast using CRISOR-Cas9 for bioethanol production

improve sugar accessibility for fermentation. By enabling the parallel and high-throughput optimization of multiple genes, this approach not only channels metabolic resources toward the production of specific biofuel precursors or final products but also improves overall process efficiency, yielding more robust and economically viable biofuel strains. Figure 3 illustrates genetically modified yeast using CRISOR-Cas9 for bioethanol production.

Together, these strategies create an optimized metabolic network that increases yield, tolerance, and efficiency in microbial liquid bioenergy production systems.

1.3 Strain Development for Stress Tolerance and Productivity

Low production of biofuel via fermentation process also could be attributed by the presence of inhibitors and other stress condition that could affect the metabolic pathway of the microbes. Fermentation under stress condition could reduce enzyme activity and productivity of the desired products. Genetic engineering enables the development of microbial and algal strains with enhanced stress tolerance, which is critical for efficient liquid biofuel production. In industrial fermentation, microbes often encounter inhibitory compounds, high biofuel concentrations, temperature fluctuations, osmotic stress, and nutrient limitations that can reduce growth and productivity [25]. Through genetic engineering, strains can be modified to overexpress stress-response proteins such as heat shock proteins, chaperones, or efflux pumps, strengthening their resilience [26]. Additionally, modifying cell membrane composition or introducing detoxification pathways allows microbes to survive and function under conditions that would normally inhibit biofuel synthesis, ensuring more consistent and reliable production [27].

At the same time, genetic engineering can improve productivity by optimizing metabolic pathways for biofuel synthesis. Key strategies include overexpressing enzymes in ethanol, butanol, or lipid biosynthetic pathways, deleting competing pathways that divert precursors, and balancing cofactors like NAD(P)H to maintain efficient flux toward the desired product [28, 29]. Combining these targeted modifications with adaptive laboratory evolution (ALE) can further select for strains that are both highly productive and tolerant to industrial stresses [30–32]. For example, engineered Saccharomyces cerevisiae or Zymomonas mobilis strains exhibit higher ethanol yields under inhibitory conditions, while genetically enhanced microalgae produce greater lipid content for biodiesel. Together, these approaches create robust, high-performance strains that maximize liquid biofuel output under industrially relevant conditions.

Genetic engineering for strain development combines enhancing stress tolerance and optimizing metabolic flux to produce strains that survive harsh industrial conditions while maximizing liquid biofuel yield, creating more robust and economically viable production systems.

1.4 Synthetic Biology and Biofuel Pathway Design

Synthetic biology enables biofuel production by designing microorganisms to act as efficient "cell factories" that convert various feedstocks into biofuels [33]. It uses genetic engineering tools like CRISPR-Cas systems to precisely edit the genomes of microbes such as bacteria, yeast, and algae. This allows for the optimization of metabolic pathways to produce advanced biofuels like butanol and jet fuel analogs, improve feedstock processing, and create sustainable alternatives to fossil fuels.

Synthetic biology contributes significantly to biofuel production by creating "designer microbes" capable of producing advanced, energy-dense biofuels like butanol and isoprenoid-based fuels that are chemically compatible with existing fossil fuel infrastructure [34]. This field enhances microbial performance in organisms like yeast and E. coli by optimizing their ability to convert feedstocks into biofuels, increasing product yields, and improving their tolerance to harsh industrial conditions [35]. Furthermore, synthetic biology enables microorganisms to utilize a wider, more sustainable range of feedstocks, such as non-food crops, lignocellulosic biomass, and algae. For microalgae specifically, it is a crucial tool used to engineer metabolic pathways to simultaneously increase lipid accumulation for biodiesel production and promote overall biomass growth, addressing a major hurdle for the commercial viability of algal biofuels [36, 37]. Figure 4 denotes the production of liquid biofuel from microorganisms via synthetic biology approach.

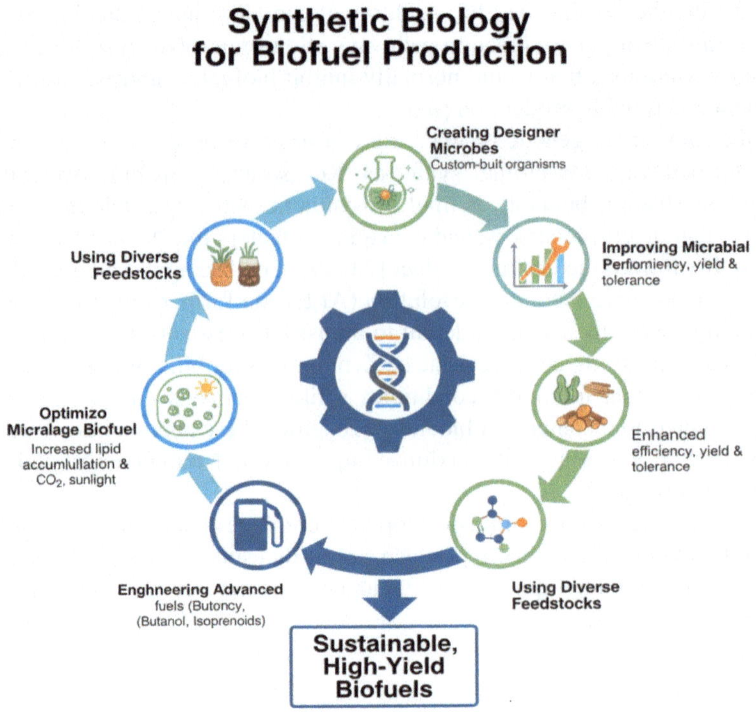

Fig. 4 Production of liquid biofuel from microorganisms via synthetic biology approach

Genetic Engineering is defined as the process of modifying an organism's existing genes by adding, deleting, or altering specific traits primarily to improve characteristics like biofuel yield or stress tolerance. Its scope generally involves enhancing or redirecting native metabolic pathways in microbes, plants, or algae. The complexity of these modifications is typically incremental, relying on natural cellular machinery, such as knocking out competing pathways in S. cerevisiae to increase ethanol yield, or overexpressing native enzymes to increase lipid content in algae. The main goal is to improve or fine-tune natural metabolism for better efficiency.

Synthetic Biology, in contrast, goes beyond simple gene modifications, focusing on designing and constructing entirely new biological systems, pathways, or organisms from scratch, often utilizing standardized genetic "parts.“ The scope centers on creating novel metabolic pathways or synthetic circuits, which may not exist in nature, to optimize and control the production of biofuels or their precursors. This field integrates multiple engineered modules, sensors, and regulatory circuits to create highly optimized and programmable systems. Examples include designing a completely synthetic pathway in E. coli to convert sugars directly into advanced fuels like butanol, or engineering microalgae with synthetic, high-yield lipid-production pathways. The ultimate goal is to build new, predictable, and controllable metabolic

Table 2 Comparison of genetic engineering and synthetic biology

Aspect	Genetic engineering	Synthetic biology
Definition	Involves modifying an organism's existing genes—adding, deleting, or altering specific genes—to improve traits like biofuel yield or stress tolerance	Goes beyond simple gene modifications, designing and constructing entirely new biological systems, pathways, or organisms from scratch, often using standardized genetic "parts."
Scope	Usually focuses on enhancing or redirecting existing metabolic pathways in microbes, plants, or algae	Focuses on creating novel metabolic pathways or synthetic circuits, which may not exist in nature, to optimize production of biofuels or precursors
Goal	Improve or fine-tune natural metabolism for better efficiency	Build new, predictable, and controllable metabolic systems that maximize biofuel production beyond natural limits
Complexity	Modifications are typically incremental and rely on natural cellular machinery	Integrates multiple engineered modules, sensors, and regulatory circuits to create fully optimized and programmable systems
Example in microbes	Knocking out competing pathways in *S. cerevisiae* to increase ethanol yield	Designing a synthetic pathway in *E. coli* to convert sugars directly into butanol, including enzymes sourced from multiple organisms
	Increasing lipid content in algae by overexpressing native enzymes	Engineering microalgae with completely synthetic lipid-production pathways for high-yield biodiesel

systems that maximize biofuel production beyond natural limits. Table 2 summarises the comparison of genetic engineering and synthetic biology.

In short: genetic engineering is like upgrading an existing car engine to make it more efficient, while synthetic biology is like designing a completely new engine from scratch optimized specifically for high performance.

2 Challenges, Bioethics, and Regulatory Considerations

2.1 *Biosafety Concerns with Genetically Modified Organisms (GMOs)*

Biosafety concerns in the use of genetically modified organisms (GMOs) for liquid biofuel production primarily revolve around the potential ecological, environmental, and health risks associated with the release or handling of engineered microbes and plants [38]. Firstly, there is a risk of accidental release of genetically modified microorganisms (such as engineered E. coli, yeast, or cyanobacteria) into the environment during cultivation or processing. These organisms may potentially transfer modified genes to native species through horizontal gene transfer, potentially disrupting local microbial ecosystems. For example, engineered microbes with enhanced metabolic efficiency could outcompete natural strains, altering nutrient cycles or microbial balance in soil or water.

Secondly, the use of GMO-based energy crops such as modified sugarcane or switchgrass with altered lignin content may raise concerns about cross-pollination with wild relatives, leading to unintended genetic flow [39]. This could impact biodiversity and ecosystem stability.

Another concern involves containment and waste management in industrial biofuel facilities. Improper handling of fermentation residues containing live GMOs could lead to contamination of soil or water systems. Therefore, strict biosafety protocols, closed bioreactor systems, and sterilization steps are essential to prevent environmental exposure. To mitigate these risks, researchers and industries follow biosafety level (BSL) guidelines, use biological containment systems (such as auxotrophic strains that cannot survive outside lab conditions), and perform risk assessments under regulatory frameworks like the Cartagena Protocol on Biosafety. Furthermore, advancements in synthetic biology "kill-switches" and biocontainment circuits ensure that engineered microbes can self-destruct or lose function outside controlled environments.

In summary, while GMO-based biofuel production offers tremendous efficiency and sustainability advantages, maintaining robust biosafety management, genetic containment, and regulatory compliance is crucial to ensure environmental and public safety.

2.2 Public Perception and Policy Frameworks

Public perception and policy frameworks present significant challenges in the development and commercialization of liquid biofuels, especially those derived from genetically modified organisms (GMOs) or non-traditional biomass sources [40]. Public skepticism often arises from concerns about food security, environmental impact, and genetic modification, which can influence consumer acceptance and political support [41]. For example, first-generation biofuels made from food crops such as corn or sugarcane are often criticized for competing with food supply and contributing to land-use change, leading to deforestation and biodiversity loss. Similarly, biofuels produced using GMOs may face resistance due to perceived risks of genetic contamination or long-term ecological effects, even when scientific evidence suggests the risks are minimal.

From a policy standpoint, inconsistent or unclear regulatory frameworks across countries can hinder large-scale adoption and investment [42]. Differences in biofuel standards, subsidies, and sustainability certification create uncertainty for producers and investors [43]. Moreover, many policies still favor fossil fuels through tax incentives or subsidies, making biofuels less competitive economically. The absence of strong carbon pricing mechanisms or incentives for low-emission technologies further slows biofuel integration into the energy market.

To overcome these challenges, governments and industries must work together to improve transparency, communicate scientific evidence, and engage stakeholders in policy design [44]. Implementing clear sustainability standards, ensuring traceability of biofuel feedstocks, and promoting public education on the environmental benefits of advanced (non-food, GMO-based) biofuels can enhance public trust. In short, aligning public perception, regulatory clarity, and supportive policy instruments is crucial for accelerating the transition to sustainable liquid biofuel production.

2.3 Scalability and Techno-Economic Feasibility

Scalability and techno-economic feasibility are among the major challenges in liquid biofuel production [45]. While laboratory- and pilot-scale studies often demonstrate promising yields, translating these results into large-scale, cost-effective, and sustainable production systems remains difficult.

One key limitation is that biofuel production costs are still higher than fossil fuels, mainly due to the high cost of feedstock, energy-intensive processing, and downstream purification. For example, producing bioethanol from lignocellulosic biomass requires costly pretreatment and enzyme hydrolysis steps, while biodiesel from microalgae involves expensive cultivation, harvesting, and lipid extraction processes [46–48]. These economic constraints make large-scale biofuel plants less competitive without government subsidies or carbon incentives.

Scalability also faces engineering and bioprocess challenges in liquid biofuel production from either microorganisms or algae [49, 50]. Generally, production of liquid biofuel from these microorganisms is performed efficiently under controlled laboratory conditions often show reduced productivity or stability in industrial-scale bioreactors due to fluctuations in temperature, light, oxygen, or nutrient availability. Additionally, integrating biofuel production with existing energy infrastructure requires significant capital investment and long-term policy support [51].

To address these issues, researchers are focusing on techno-economic optimization and process intensification, such as using genetically engineered strains with higher substrate conversion efficiency, developing low-cost feedstocks such as agricultural or food waste, and applying biorefinery concepts that produce multiple value-added co-products that produce variety of bioproducts including bioplastics or biochemicals to improve profitability [52, 53]. Advances in process modeling, automation, and life-cycle assessment also help identify economic bottlenecks and optimize energy use.

In summary, while scalability and techno-economic feasibility remain key barriers to the widespread adoption of liquid biofuels, technological innovation, systems integration, and supportive economic policies can significantly improve the viability of large-scale biofuel production.

3 Future Outlook for the Development of Sustainable Liquid Biofuel Production

The future of sustainable liquid biofuel production lies in the integration of advanced biological design, process engineering, and systems-level optimization to achieve both economic viability and environmental responsibility. One of the most promising directions is the integration of metabolic engineering with bioprocess optimization, which enables the seamless coupling of genetic-level improvements with process-level efficiency. Through metabolic engineering, microorganisms can be reprogrammed to enhance substrate utilization, redirect carbon flux toward biofuel synthesis, and tolerate harsh industrial conditions. When combined with optimized bioprocess parameters—such as controlled fermentation conditions, reactor design, and adaptive feeding strategies—these engineered strains can achieve higher biofuel yields, productivity, and stability. This synergy between cellular and process optimization creates a dynamic feedback system where experimental data guide further genetic modification, ultimately leading to robust and cost-effective production systems.

However, achieving commercial feasibility remains contingent upon overcoming challenges related to scalability and techno-economic performance. Laboratory-scale breakthroughs often fail to translate directly to industrial applications due to process inefficiencies, high operational costs, and variable biomass quality. To address these challenges, future research should focus on the development of scalable bioreactor configurations, energy-efficient downstream processing, and the use of low-cost, non-food biomass feedstocks such as lignocellulosic waste or microalgae. The implementation of biorefinery concepts, where multiple high-value co-products (e.g., biochemicals, animal feed, and bioplastics) are generated alongside fuels, will further improve the techno-economic feasibility and strengthen the competitiveness of biofuels within the global energy market.

Looking ahead, the ultimate vision is the establishment of carbon–neutral or carbon-negative biofuel production systems that not only replace fossil fuels but also contribute to global carbon mitigation. By coupling biofuel production with CO_2 capture, utilization, and renewable energy integration, it is possible to develop closed-loop, climate-positive systems. Microalgae-based and photosynthetic microbial platforms, in particular, hold great potential for capturing atmospheric CO_2 while generating energy-dense bio-oils and biochemicals. Furthermore, adopting life cycle assessment (LCA) and carbon accounting tools will be essential to ensure that these systems deliver genuine carbon reduction benefits.

In summary, the convergence of genetic engineering, process intensification, and carbon management strategies will define the next generation of sustainable biofuel technologies. As innovation continues across these interconnected domains, liquid biofuels have the potential to evolve from renewable energy alternatives into integral components of a carbon–neutral global bioeconomy.

4 Conclusion

Production of bioenergy and liquid biofuel from renewable resources such as microorganisms has been identified as an alternative approach to partially replace transportation fuel from fossil fuel-based feedstock. Improvement on production of liquid biofuel from microorganisms can be done in different approaches including metabolic pathway engineering, strain improvement and synthetic biology. By harnessing the knowledge on fundamental of microorganisms molecular characteristic and modification through genetic engineering, the limitation of bioenergy production from biomass feedstock can be overcome especially that associated with traditional biofuel production methods. This technology not only offers environmental benefits but also contributes to energy security, rural development, and economic stability. Eventhough this approach shows a promising approach to produce biofuel from renewable biomass, however, it is crucial to address social acceptance, regulatory, ethical, and safety concerns to ensure that genetically engineered bacteria are deployed responsibly and safely in the pursuit of renewable energy solutions. In addition, challenge in scalability and economic aspect need to be considered for future production feasibility.

References

1. Kaygusuz K (2009) Biomass as a renewable energy source for sustainable fuels. Energy Sourc Part A Recov Utiliz Environ Effect 31(6):535–545
2. Wang Y, Zhang Y, Cui Q, Feng Y, Xuan J (2024) Composition of lignocellulose hydrolysate in different biorefinery strategies: nutrients and inhibitors. Molecules 29(10):2275
3. Kucharska K, Rybarczyk P, Holowacz I, Lukajtis R, Glinka M, Kaminnski M (2018) Pretreatment of lignocellulosic materials as substrates for fermentation processes. Molecules 23(11):2937
4. Ojo AO (2023) An overview of lignocellulose and its biotechnological importance in high-value product production. Fermentation 9(11):990
5. Adegboye MF, Ojuederie OB, Talia PM, Bablola OO (2021) Bioprospecting of microbial strains for biofuel production: metabolic engineering, applications, and challenges. Biotechnol Biofuel 14(5)
6. Fujii T, Unoue H, Ishikawa Z (2013) Enhancing cellulase and hemicellulase production by genetic modification of the carbon catabolite repressor gene, creA, in Acremonium cellulolyticus. AMB Express 3(1):73
7. Lin H, Wang Q, Shen Q, Zhan J, Zhao Y (2012) Genetic engineering of microorganisms for biodiesel production. Bioengineered 5(5):292–304
8. Dhokane D, Shaikh A, Yadav A, Giri N, Bandyopdhyay A, Dasgupta S, Bhdra B (2023) CRISPR-based bioengineering in microalgae for production of industrially important biomolecules. Bioeng Biotechnol 11
9. Runguphan P-T-Q-G-Q (2023) Optimizing ethanol production in saccharomyces cerevisiae at ambient and elevated temperatures through machine learning-guided combinatorial promoter modifications. ACS Synth Biol 12(10):2897–2908
10. Nevoigt E (2008) Progress in metabolic engineering of Saccharomyces cerevisiae. Microbiol Mol Biol Rev 379–412

11. Yang S, Fei Q, Zhang Y, Contreras LM, Utturkar SM, Brown SD, Himmei ME, Zhang M (2016) Zymomonas mobilis as a model system for production. Microb Biotechnol 9:699–717
12. Atsumi S, Hanai T, Liao JC (2008) Non-fermentative pathways for synthesis of branched-chain higher alcohols as biofuels. Nature 451(7174):86–86
13. Jang YS, Lee JY, Lee J, Park JH, Im JA, Eom JA, Moon-Ho E, Lee J, Sang-Hyun L, Song H, Jung-Hee C, Seung DY, Lee SY (2012) Enhanced butanol production obtained by reinforcing the direct butanol-forming route in Clostridium acetobutylicum. ASM J 3(5)
14. Li Y, Han D, Hu G, Sommerfeld M, Hu Q (2010) Inhibition of starch synthesis results in overproduction of lipids in Chlamydomonas reinhardtii. Biotechnol Bioeng 107(2):258–268
15. Poliner E, Pulman JA, Zienkiewicz K, Childs K, Benning C, Farre EM (2018) A toolkit for Nannochloropsis oceanica CCMP1779 enables gene stacking and genetic engineering of the eicosapentaenoic acid pathway for enhanced long-chain polyunsaturated fatty acid production. Plant Biotechnol J 16(1):298–309
16. Blombach B, Riester T, Wieschalka S, Ziert C, Jung-Won Y, Wendisch, VF, Eikmanns BJ (2011) Corynebacterium glutamicum tailored for efficient. Appl Environ Microbiol 3300–3100
17. Ducat DC, Way CJ, Silver PA (2011) Engineering cyanobacteria to generate high-value products. Cell Press 29(2):95–103
18. Nawab S, Zahoor, Shah SB, Rahman MU, Keera HA, Rahman I (2025) Optimizing bamboo biomass for sustainable isobutanol production using genetically engineered Escherichia coli. Biochem Eng J 217:109669
19. Lin F, Das D, Lin XN, Marsh ENG (2013) Aldehyde-forming fatty acyl-CoA reductase from cyanobacteria: expression, purification and characterization of the recombinant enzyme. FEBB J 280(19):4773–4781
20. Gong C, Cao L, Fang D, Zhang J, Awasthi MK, Xue D (2022) Genetic manipulation strategies for ethanol production from bioconversion of lignocellulose waste. Biores Technol 252:127105
21. Yang P, Jiang S, Jiang S, Lu S, Zheng Z, Chen J, Wu W, Jiang S (2022) CRISPR-Cas9 approach constructed engineered saccharomyces cerevisiae with the deletion of GPD2, FPS1, and ADH2 to enhance the production of ethanol. J Fungi 8(7)
22. Khattab SMR, Mousa MOM, Nagata T, Watanabe T, Katahira M (2025). Engineering a hypoxia-tolerant Saccharomyces cerevisiae for rapid ethanol production via co-utilization of glucose and acetic acid and redox-enhanced flocculation. Sustain Energy Fuel 14
23. Silvello MAC, Teixeira GS, Bueno D, Goldbeck R (2025) CRISPR/Cas9-engineered thermo-tolerant Saccharomyces cerevisiae JAYET for enhanced 3G-ethanol production from Chlorella vulgaris biomass. Algal Res 90:104275
24. Sukwong P, Sunwoo IY, Jeong DY, Kim SR, Gwi-Taek J, Sung-Koo K (2019) Enhancement of bioethanol production from Gracilaria verrucosa by Saccharomyces cerevisiae through the overexpression of SNR84 and PGM2. Bioproc Biosyst Eng 42:1421–1433
25. Liu L, Fan W, Lei W, Jiang X, Zhang H (2025) Addressing challenges in microbial manufacturing: systematic microbial biotechnology. Innovation 6(6):100871
26. Cho JS, Kim GB, Eun H, Moon CW, Lee SY (2022) Designing microbial cell factories for the production of chemicals. JACS 2(8):1781–1799
27. Wang S, Sun X, Yuan Q (2018) Strategies for enhancing microbial tolerance to inhibitors for biofuel production: a review. Biores Technol 258:202–309
28. E-Bin G, Penglin Y, Haiyan Q, Junhua W (2021) Enhancing bioethanol production by deleting phosphoenolpyruvate synthase and ADP-glucose pyrophosphorylase, and shunting tricarboxylic acid cycle in synechocystis Sp. PCC 6803
29. Yang P, Jiang S, Lu S, Jiang S, Jiang S, Deng Y, Lu J, Wang H and Zhao Y (2022) Ethanol yield improvement in Saccharomyces cerevisiae GPD2 Delta FPS1 Delta ADH2 Delta DLD3 delta mutant and molecular mechanism exploration based on the metabolic flux and transcriptomics approaches. Microb Cell Fact 21(6):160
30. Wen Z, Ledema-Amaro R, Lu M, Jiang Y, Gao S, Jin M, Yang S (2020) Combined evolutionary engineering and genetic manipulation improve low pH tolerance and butanol production in a synthetic microbial Clostridium community. Biotechnol Bioeng 117(7):2008–2022

31. Huang J, Wang X, Chen X, Li H, Chen Y, Hu Z, Yang S (2023) Adaptive laboratory evolution and metabolic engineering of Zymomonas mobilis for bioethanol production using molasses. ACS Synth Biol 12(4):1297–1307
32. Macrommati M, Papanikolaou S, Aggelis G (2023) Improving ethanol tolerance of Saccharomyces cerevisiae through adaptive laboratory evolution using high ethanol concentrations as a selective pressure. Process Biochem 124:280–289
33. Kending C (2019) Synthetic biology and biofuels. In: Encyclopedia of food and agriculture ethics. pp 2304–2312
34. Qumsani AT (2024) The contribution of microorganisms to sustainable development: towards a green future through synthetic biology and systems biology. J Umm Al-Qura Univer Appl Sci 11(4)
35. Tian L, Qi T, Zhang F, Tran VG, Yuan J, Wang Y, He N, Cao M (2025) Synthetic biology approaches to improve tolerance of inhibitors in lignocellulosic hydrolysates. Biotechnol Adv 78:108477
36. Naduthodi MIS, Claassens NJ, D'Adamom S, Oost JVD, Barbosa MJ (2021) Synthetic biology approaches to enhance microalgal productivity. Trends Biotechnol 39(10):1019–1036
37. Shitanaka T, Wang Y, Do S, Yuson J, Khanal SK, Zienkiewicz, K, Zhi-Yan D (2025). Synthetic biology and metabolic engineering of microalgae for sustainable lipid and terpenoid production: an updated perspective. Plant Biotechnol J
38. Saxena S, Singh Y, Soni, M, Solanki K (2024) Biosafety issues associated with genetically modified organisms. In: Biosafety issue associated with genetically modified organisms. Integrated publications, recent trends in molecular biology and biotechnology. pp 43–54
39. Furtado A, Lupoi JS, Hoang NV, Healey A, Singh S, Simmons BA, Henry RJ (2014) Modifying plants for biofuel and biomaterial production. Plant Biotechnol J 12(9)
40. Hon-Choong C, Weng-Wai C, Alwi SRW, Mohammed AH (2014) Issues of social acceptance on biofuel development. J Clean Prod 71:30–39
41. Lucht JM (2015) Public acceptance of plant biotechnology and GM crops. Viruses 30(7)
42. Rozas P, Kessi-Perez EI, Martinez C (2022) Genetically modified organisms: adapting regulatory frameworks for evolving genome editing technologies. Biol Res 55(31)
43. Dijk MV, Goedegebure R, Jan-Peter N (2024) Public acceptance of biomass for bioenergy: the need for feedstock differentiation and communicating a waste utilization frame. Renew Sustain Energy Rev 202:114670
44. Mirsalami SM, Mirsalami M (2025) Advances in genetically engineered microorganisms: transforming food production through precision fermentation and synthetic biology. Fut Food 11:100601
45. Richard TL (2010) Challenges in scaling up biofuels infrastructure. Science 329(5993):793–796
46. Mignogna D, Szabo M, Ceci P, Avino P (2024) Biomass energy and biofuels: perspective, potentials, and challenges in the energy transition. Sustainability 16(16)
47. Naik HN, Kumar DJP, Sankannavar R, Mishra RK (2025) Key technologies for biofuel production from lignocellulosic biomass. In: Biofuel production from lignocellulosic materials. pp 219–251
48. Cerruti E, Gruttola FD, Lauro G, Valentini TD (2020) Assessment of feedstocks and technologies for advanced biofuel production. E3S Web Conf 197:05002
49. Joshi R, Sharma R, Kuila A (2020) Bioreactor design for efficient biofuels production from lignocellulosic biomass. In: Genetic and metabolic engineering for improved biofuel production from lignocellulosic biomass. pp 193–206
50. Kunjapur AM, Eldridge RB (2010) Photobioreactor design for commercial biofuel production from microalgae. Indus Eng Chem Res 49(8)

51. Hu P, Chakraborty S, Kumar A, Stephanopoulos G (2016) Integrated bioprocess for conversion of gaseous substrates to liquids. Appl Biol Sci 113(14):3773–3778
52. Vevizhi G, Balakumar K, Shetti NP, Ahmad EM, Pant KK, Aminabhavi TM (2022) Integrated biorefinery processes for conversion of lignocellulosic biomass to value added materials: paving a path towards circular economy. Biores Technol 343:126151
53. Balan V (2014) current challenges in commercially producing biofuels from lignocellulosic biomass. Int Scholarly Res Notices

Microalgal Secondary Metabolites as Emerging Candidates for the Prevention and Management of Metabolic Diseases: A Review

Nor Shafiqah Nor Shahril, Mohamad Hafizi Abu Bakar, Mohd Asyraf Kassim, and Thiruventhan Karunakaran

Abstract Microalgae are increasingly recognised as sustainable bioresources for producing high-value secondary metabolites with considerable potential in food, nutraceutical and pharmaceutical applications, particularly for the prevention and management of metabolic diseases including obesity, diabetes, inflammation and cancer. In this chapter, the biological properties of major microalgal metabolites are summarised, focusing on antioxidant, antimicrobial, anti-diabetic, anti-inflammatory, anti-obesity and anti-cancer activities that are mechanistically linked to oxidative stress reduction, modulation of carbohydrate-digesting enzymes and regulation of key inflammatory and metabolic signalling pathways. Representative taxa such as *Spirulina platensis*, *Chlorella vulgaris*, *Dunaliella salina* and *Phaeodactylum tricornutum* illustrate the breadth of in vitro and in vivo evidence supporting these effects. Nonetheless, clinical validation, standardised bioassays and scalable bioprocesses remain limited. This review underscores the need for integrated metabolic profiling, mechanistic studies and translational research to harness microalgal secondary metabolites as next-generation therapeutic and nutraceutical agents against metabolic diseases.

Keywords Microalgae · Secondary metabolites · Biological activities · Metabolic diseases

N. S. Nor Shahril · M. H. Abu Bakar (✉) · M. A. Kassim
Bioprocess Technology Division, School of Industrial Technology, Universiti Sains Malaysia, 11800 Gelugor, Penang, Malaysia
e-mail: mhafizi88@usm.my

N. S. Nor Shahril
e-mail: shafiqah17@gmail.com

M. A. Kassim
e-mail: asyrafkassim@usm.my

T. Karunakaran
Centre for Drug Research, Universiti Sains Malaysia, 11800 Gelugor, Penang, Malaysia
e-mail: thiruventhan@usm.my

M. A. Kassim et al. (eds.), *Sustainable Biorefinery*, Green Energy and Technology,
https://doi.org/10.1007/978-981-95-8181-8_13

1 Introduction

Microalgae are unicellular microorganisms mainly living in various types of water bodies such as marine, freshwater, and soil surfaces. Microalgae are microscopic photosynthetic microorganisms that are between 1 and 50 μm, which exist in various colours, characteristics, and shapes [1]. There are about 200,000–800,000 microalgae species in many different genera that have been identified worldwide, however only 50,000 of them have been studied [2]. Microalgae can be cultivated under two main trophic conditions: autotrophic and heterotrophic. In autotrophic cultivation, microalgae rely on photosynthesis, using light typically solar radiation as their primary energy source and inorganic carbon (e.g. CO_2) as the carbon source. By contrast, heterotrophic cultivation involves the growth of microalgae on organic carbon substrates in the absence of light, whereby organic compounds serve as both energy and carbon sources. Under both modes, microalgal cells convert these inputs into a wide array of valuable biochemicals that can be further processed into fine chemicals, biofuels, food ingredients, and bioactive compounds. Furthermore, microalgae can be grown in two conditions either free-living or having symbiotic interactions with other organisms [3].

In some systems, a combination of these nutritional strategies is employed, giving rise to mixotrophic cultivation, in which microalgae simultaneously utilize light and organic carbon sources to enhance growth and metabolite production [4].

In recent years, there has been growing interest in exploring the biological properties of active compounds and ingredients derived from natural sources such as algae and higher plants, collectively known as natural products. Natural bioactive ingredients are often preferred over synthetic counterparts because they generally exhibit lower toxicity profiles and are associated with fewer adverse side effects [5]. Within this context, microalgae have emerged as particularly attractive platforms for the sustainable production of high-value metabolites, owing to several key advantages: they can be cultivated on non-arable land, require relatively modest resource inputs, and are amenable to controlled large-scale production. Compared to terrestrial plants, microalgae can be grown and harvested more rapidly, enabling faster and more flexible production of target compounds. Moreover, their capacity to synthesize a wide range of bioactive molecules under adverse or stress-induced conditions including altered light, salinity, or nutrient availability can be strategically exploited to enrich or induce the formation of novel metabolites with potential therapeutic applications [6].

An expanding body of evidence has demonstrated that microalgae are versatile biofactories capable of producing a wide range of bioactive primary and secondary metabolites, including lipids, proteins, carotenoids, vitamins, and other high-value compounds. It has been reported that more than 15,000 novel molecules have been identified and isolated from diverse microalgal taxa worldwide, underscoring the remarkable chemical diversity of this group [7]. These microalgae-derived metabolites exhibit a broad spectrum of biological activities most notably antioxidant, antimicrobial, anti-diabetic, anti-inflammatory, and anti-obesity effects which are

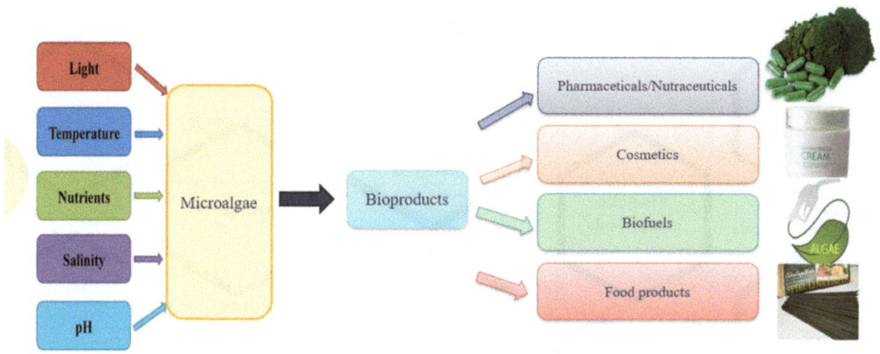

Fig. 13.1 The capability of microalgae in producing metabolites of interest by the alteration of various cultivation parameters towards development of bioproducts

increasingly recognized as relevant to the prevention and management of metabolic and chronic diseases [8]. Owing to these multifunctional properties, microalgal compounds have attracted substantial interest across multiple sectors, including pharmaceuticals, functional foods and nutraceuticals, cosmetics, and agriculture, as illustrated in Fig. 13.1.

2 Microalgal Secondary Metabolites and Their Biological Activities

Microalgae are prolific producers of structurally diverse secondary metabolites, including carotenoids, phenolic compounds, phycobiliproteins, polyunsaturated fatty acids (PUFAs), terpenoids, sterols, peptides, and polysaccharides [1]. These metabolites exhibit a wide spectrum of bioactivities with significant applications in nutraceuticals, pharmaceuticals, functional foods, cosmetics, and bioprocess engineering. Their biological functions are largely attributed to strong redox-modulating capacity, membrane-stabilizing activity, enzyme inhibition, immunomodulation, and metabolic regulation [6]. Figure 13.2 illustrates the major classes of microalgae-derived secondary metabolites and representative molecular structures associated with their diverse biological functions. Microalgae synthesize a wide range of phenolic compounds, including phenolic acids (e.g., benzoic acid) and flavonoids (e.g., quercetin), which are well known for their antioxidant, anti-inflammatory, antimicrobial, and cytoprotective properties. The carotenoid group is represented by astaxanthin and lutein, potent pigments with strong antioxidant, anti-obesity, antidiabetic, and anticancer activities.

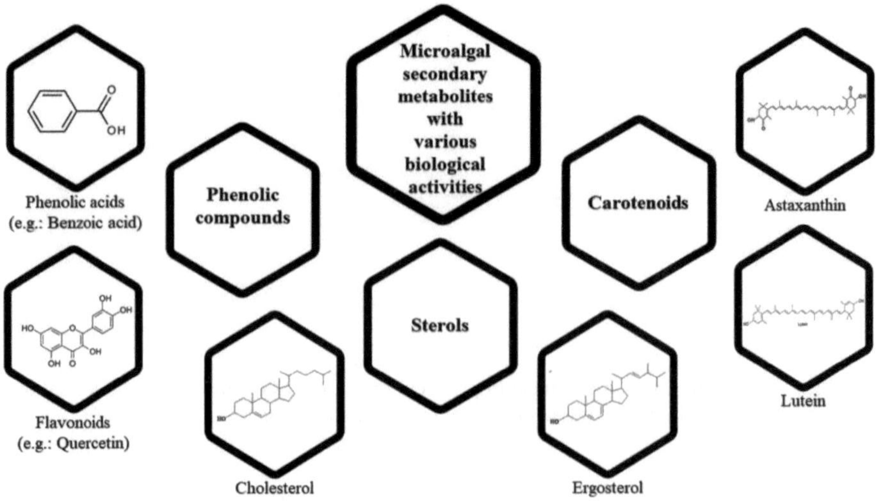

Fig. 13.2 The key secondary metabolites derived from microalgae having various biological activities

2.1 *Antioxidant Activity*

Antioxidants are molecules capable of suppressing or neutralising oxidation processes initiated by other substances, thereby limiting cellular damage. Oxidation involves a series of chemical reactions that generate free radicals, highly reactive species that can injure biomolecules such as lipids, proteins, and DNA, ultimately leading to cellular dysfunction and death [9, 10]. Free radicals are broadly classified into oxygen-derived and nitrogen-derived species. Oxygen-derived free radicals, collectively known as reactive oxygen species (ROS), include superoxide anion and hydroxyl radical, whereas nitrogen-derived free radicals, or reactive nitrogen species (RNS), include nitrogen dioxide and nitric oxide [11, 12]. When the production of ROS and RNS exceeds the capacity of endogenous and exogenous antioxidant defences, a state of oxidative or nitrosative stress ensues, resulting in cumulative cellular and tissue damage. Oxidative stress driven by ROS has been identified as a major contributing factor in the pathogenesis of numerous chronic diseases, including metabolic disorders such as chronic inflammation and cardiovascular disease [13].

The commercially available antioxidants consist of two types: synthetic antioxidants and natural antioxidants. The most common examples of synthetic antioxidants are butylated hydroxyanisole (BHA) and butylated hydroxytoluene (BHT) [14]. Natural antioxidants are preferable since they have a lower risk towards human health such as low toxicity, high stability, and less carcinogenic effects [15]. Natural antioxidants can be obtained from the extracts of fruits, flowers, and microalgae. The most common classes of secondary metabolites possessing antioxidant properties are carotenoids and phenolic compounds [14]. However, there are few additional classes of microalgal secondary metabolites having antioxidant activity including sterols and vitamins [16]. Table 13.1 summarises studies that have investigated the antioxidant properties of microalgae and their associated secondary metabolites.

Numerous studies have demonstrated that microalgae are rich sources of antioxidant secondary metabolites, with phenolic compounds, carotenoids, and chlorophylls being the principal contributors to their bioactivity. Methanolic extracts of phenolic-rich microalgae such as *Galdieria sulphuraria* and *Chlorella marina* consistently exhibit strong radical-scavenging and reducing capacities, as evidenced by exceptionally high DPPH inhibition and ferric reducing antioxidant power, which closely correlate with elevated total phenolic contents [15, 17]. Similarly, mixed ethanol–water extracts of *Chlorella vulgaris* and *Tetraselmis suecica* highlight the combined contribution of phenolics and carotenoids, with xanthophyll-rich species such as *T. suecica* showing antioxidant activities surpassing that of α-tocopherol [14]. Pigment-dominated extracts from *Phormidium autumnale*, *Nannochloropsis* sp., and *Chaetoceros* sp. further confirm the potent antioxidant roles of carotenoids and chlorophylls, where non-polar solvents such as dichloromethane and chloroform were particularly effective in recovering highly active fractions [18, 19].

In cyanobacteria and filamentous microalgae, including *Spirulina platensis*, *Scenedesmus subspicatus*, and *Tetradesmus acuminatus*, antioxidant performance was strongly influenced by extraction solvent, with methanolic and aqueous extracts generally yielding the highest activities due to efficient extraction of phenolics, flavonoids, and tannins [20–24]. Moreover, cultivation under nitrogen stress markedly enhanced antioxidant capacity in *Oscillatoria* sp. and *Nostoc muscorum*, indicating that environmental stress can stimulate the accumulation of antioxidant secondary metabolites such as phenolics, terpenoids, and alkaloids [25]. Collectively, these findings underscore that the antioxidant potential of microalgae is governed by species-specific metabolite profiles, solvent-dependent extraction efficiency, and cultivation strategies, reinforcing their promise as sustainable sources of natural antioxidants for food, nutraceutical, and pharmaceutical applications.

2.2 *Antimicrobial Activity*

Antimicrobial activity refers to the ability of bioactive compounds to inhibit or eliminate pathogenic microorganisms, including bacteria, viruses, and fungi. This area of research has gained substantial scientific interest due to its broad applicability in the development of therapeutic agents, natural preservatives, and other high value bioproducts. Numerous microalgae-derived metabolites including terpenes, fatty acids, peptides, phenolic compounds, indoles, and steroidal glycosides have been reported to possess significant antimicrobial properties [26]. These compounds exert their effects either by killing microbial cells directly or by suppressing their proliferation [27]. As illustrated in Fig. 13.3, antimicrobial agents can act through several major mechanisms, including inhibition of cell wall synthesis, disruption of protein synthesis, impairment of nucleic acid synthesis, or interference with key metabolic pathways [28].

Table 13.1 The antioxidant studies related to secondary metabolites derived from microalgae

Microalgae species	Type of extract (secondary metabolites)	Experimental design	Biological significance	References
Galdieria sulphuraria	Methanol (phenolic compounds)	DPPH	The highest total phenolic contents (312 mg/g gallic acid equivalent) and highest radical scavenging activity by DPPH assay (% 95 ± 0.7)	[15]
Chlorella marina	Methanol (phenolic compounds)	Total antioxidant activity, FRAP, DPPH, Hydrogen peroxide radical scavenging activity, and Nitric oxide radical scavenging assay	The highest total phenolic content (0.78 ± 0.032 mg/g gallic acid equivalent) and highest antioxidant activity; total antioxidant activity (1.03 ± 0.02 mg/g ascorbic acid equivalent), DPPH assay (23.08%), hydrogen peroxide radical scavenging assay (61.33%), nitric oxide radical scavenging assay (25.76%) and ferric reducing antioxidant power (0.73 ± 0.026 mg/g ascorbic acid equivalent)	[17]
Chlorella vulgaris	Ethanol: water (carotenoids and phenolic compounds)	TEAC and FRAP	Carotenoids (total carotenoid content 1.93 ± 0.19) and phenolic compounds (total phenolic contents 1.47 ± 0.16); the antioxidant activity of the extracts; TEAC (19.97 ± 1.48) and FRAP (64.65 ± 5.80)	[14]
Tetraselmis suecica	Ethanol: water (xanthophylls lutein, violaxanthin, neoxanthin, antheraxanthin, and loroxanthin esters)	DPPH	A higher DPPH radical scavenging activity (% 97.72 ± 5.09) compared to the standard; α-tocopherol (% 26.00 ± 1.30)	[14]

(continued)

Table 13.1 (continued)

Microalgae species	Type of extract (secondary metabolites)	Experimental design	Biological significance	References
Phormidiumautumnale	Ethyl acetate, methanol, acetone (carotenoids, and chlorophyll)	Peroxyl radical scavenging assay	Carotenoid extract is 28 (28.1 ± 0.2) times and chlorophyll extract is 85 (84.9 ± 0.7) times more potent than α-tocopherol to scavenge peroxyl radicals	[18]
Nannochloropsis sp.	Hexane, dichloromethane, chloroform, methanol	DPPH, FRAP	Dichloromethane extract has higher antioxidant effects compared to other extracts for both assays; 30.60 ± 4.60 μmol TE/ g DE in DPPH and 470.72 ± 0.60 μmol TE/ g DE in FRAP	[19]
Chaetoceros sp.	Hexane, dichloromethane, chloroform, methanol	DPPH, FRAP	Chloroform extract has higher antioxidant effects compared to other extracts for both assays; 106.68 ± 0.91 μmol TE/ g DE in DPPH and 609.80 ± 0.55 μmol TE/ g DE in FRAP	[19]
Spirulina platensis	Ethanol	DPPH	Fresh samples contain a higher antioxidant effect which resulted 163.61 ppm of IC50 value that are higher than the dried sample (33.075 ppm of IC50) and quercetin (21.616 ppm of IC50)	[20]
Spirulina platensis	Aqueous, methanol, aqueous/ methanol (phenolic compounds)	DPPH, ABTS	Methanol extract resulted in the highest antioxidant activity by having 89.61% of antioxidant activity against DPPH and 99.55% of antioxidant activity against ABTS	[21]

(continued)

Table 13.1 (continued)

Microalgae species	Type of extract (secondary metabolites)	Experimental design	Biological significance	References
Spirulina platensis	Methanol	DPPH, TEAC, FRAP	The total content of phenolic compounds, carotenoids and antioxidant activity (except for antioxidant activity against DPPH) increases as the light illuminance increases during the cultivation of the *Spirulina platensis*	[22]
Scenedesmus subspicatus	Acetone, butanol, DMSO, ethanol, methanol, aqueous	DPPH	The aqueous extract has the highest antioxidant activity among all extracts by showing the percent inhibition (PI) against DPPH 36.7% but lower than the control used (catechin) which is 49.6% of PI	[23]
Tetradesmus acuminatus	Hexane, chloroform, methanol, ethanol, aqueous (flavonoids, phenolic compounds, tannins)	DPPH and hydrogen peroxide scavenging assay	The highest antioxidant activities are shown in methanol and aqueous extracts; the inhibition of DPPH free radicals is 68.17 ± 0.178 (methanol) and 55.96 ± 0.48 (aqueous) at 100 μg/ml while the percentage of inhibition of hydrogen peroxide radicals are 63.70 ± 0.42 (methanol) and 43.86 ± 0.21 (aqueous)	[24]

(continued)

Table 13.1 (continued)

Microalgae species	Type of extract (secondary metabolites)	Experimental design	Biological significance	References
Oscillatoria sp.	Aqueous (phenolic compounds, terpenoids and alkaloids)	DPPH and ABTS	The aqueous extracts of the *Oscillatoria* sp. cultivated under nitrogen stress condition recorded the highest antioxidant capacity; DPPH of 68.00 ± 3.60 and ABTS of 73.60 ± 2.80 which is higher than the same culture cultivated under normal conditions (DPPH of 59.80 ± 0.95 and ABTS of 69.80 ± 1.45)	[25]
Nostoc muscorum	Aqueous (phenolic compounds, terpenoids and alkaloids)	DPPH and ABTS	The aqueous extracts of the *Nostoc muscorum* cultivated under nitrogen stress condition recorded the highest antioxidant capacity; DPPH of 74.00 ± 2.78 and ABTS of 80.30 ± 1.65 at the nitrate concentration of 9 g/L which is higher than the same culture cultivated under normal conditions (DPPH of 69.80 ± 1.22 and ABTS of 72.60)	[25]

The antimicrobial compounds are important for metabolic diseases to reduce or eliminate the risk of infections internally or externally (such as injuries), especially for diabetic patients. Patients with diabetes mellitus often experience foot ulceration which then lead to the microbial infections such as infection by a Gram-positive bacterium, *Staphylococcus aureus* [29]. One of the treatments that can be used is the application of topical antibacterial. Since bioactive compounds derived from microalgae possess antimicrobial activity [27], hence it can be one of the potential bioactive ingredients in order to develop alternative treatments for metabolic diseases. Table 13.2 summarises studies that have investigated the antimicrobial properties of microalgae and their associated secondary metabolites.

Most antimicrobial activity assessments involving microalgal extracts are typically carried out against three major categories of pathogenic microorganisms: Gram-positive bacteria, Gram-negative bacteria, and fungi. For instance, hexane and chloroform extracts of *Coccomyxa onubensis* demonstrated notable antimicrobial potency, producing inhibition zones greater than 14 mm against *Escherichia coli*, *Salmonella*

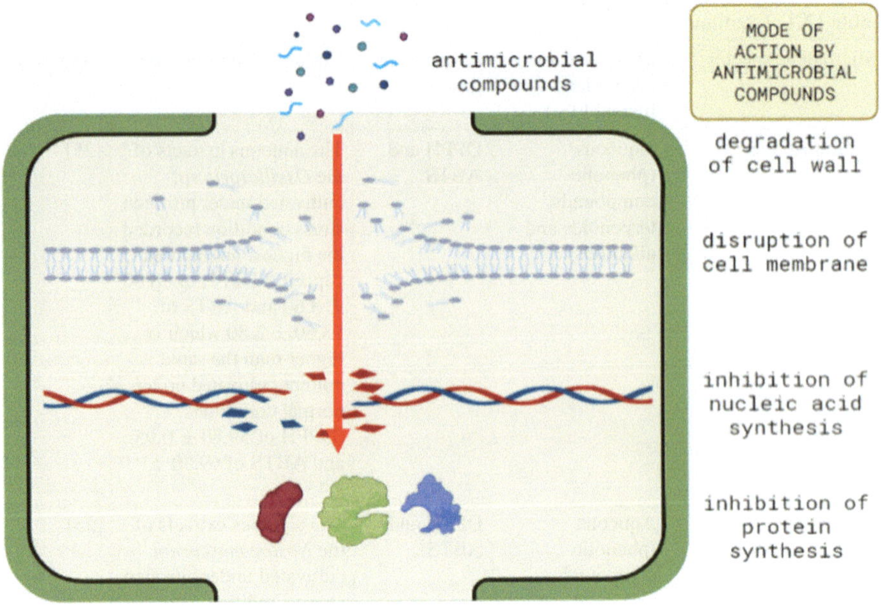

Fig. 13.3 The mechanistic actions of antimicrobial compounds on microbial cells

enterica, *Proteus mirabilis*, *Candida albicans*, and *Staphylococcus aureus* using the agar diffusion method [30]. A previous study reported that extracts from *Scenedesmus* sp. exhibited broad-spectrum antimicrobial activity against *Staphylococcus aureus*, *Alternaria* sp., *Enterococcus faecalis*, *Escherichia coli*, and *Pseudomonas aeruginosa* [31]. In their study, the strongest inhibition was observed against *Alternaria* sp., with a 10 mm zone of inhibition, while the ethanol extract showed the highest antimicrobial effect across all tested microorganisms in the MIC assay. These findings collectively highlight the significant antimicrobial potential of microalgal metabolites and their relevance for future biotechnological and therapeutic applications.

A growing body of evidence demonstrates that microalgae-derived secondary metabolites exhibit broad-spectrum antimicrobial activity against both Gram-positive and Gram-negative bacteria as well as pathogenic fungi, with extraction solvent playing a decisive role in bioactivity. Non-polar and semi-polar extracts from species such as *Coccomyxa onubensis*, *Chlorococcum humicola*, and *Rhizosolenia alata* consistently showed strong antimicrobial effects, with chloroform, hexane, acetone, and benzene extracts producing large zones of inhibition against clinically relevant bacteria including *Escherichia coli*, *Staphylococcus aureus*, *Salmonella* spp., and *Pseudomonas aeruginosa*, as well as fungi such as *Candida albicans* and *Aspergillus* spp. [30, 32, 33]. In contrast, polar extracts particularly ethanol,

Table 13.2 The antimicrobial studies related to secondary metabolites derived from microalgae

Microalgae species	Type of extract (secondary metabolites)	Target microorganism	Experimental method/assay	Biological significance	References
Coccomyxa onubensis	Chloroform, hexane, dichloromethane, diethyl ether	*Escherichia coli, Salmonella enterica, Proteus mirabilis, Candida albicans,* and *Staphylococcus aureus*	Diffusion method	Hexane and chloroform extracts have the highest antimicrobial activity (more than 14 mm of the zone of inhibition)	[30]
Scenedesmus sp.	Ethanol and methanol	*Staphylococcus aureus, Alternaria* sp., *Enterococcus faecalis, Escherichia coli*, and *Pseudomonas aeruginosa*	Diffusion method and Minimal Inhibitory Concentration (MIC) assay	The highest antimicrobial activity on *Alternaria* sp. (10 mm of the zone of inhibition) and ethanol extract has the highest antimicrobial activity on all tested microorganisms through MIC assay	[31]
Scenedesmus bajacalifornicus	Chloroform, acetone, ethanol, methanol, and aqueous	*Escherichia coli, Salmonella typhi, Bacillus subtilis,Aspergillus niger, Clostridium perfringens* and *Candida albicans*	Diffusion method	The aqueous extract showed antimicrobial activity against all tested microorganisms ranging from 10 to 20 mm zone of inhibition	[34]
Chlorococcum humicola	Acetone, benzene chloroform, diethyl ether, ethyl acetate, ethanol, hexane and methanol	*Bacillus subtilis, Staphylococcus aureus, Escherichia coli, Pseudomonas aeruginosa, Salmonella typhimurium, Klebsiella pneumoniae, Vibrio cholerae, Candida albicans, Aspergillus niger* and *Aspergillus flavus*	Diffusion method	Chloroform and benzene extracts showed the highest antibacterial activities against all tested bacteria ranging from 10 to 18 mm of inhibition zone. Benzene extracts have the highest antifungal effects on all tested fungi ranging from 11 to 18 mm of inhibition zone	[32]

(continued)

Table 13.2 (continued)

Microalgae species	Type of extract (secondary metabolites)	Target microorganism	Experimental method/assay	Biological significance	References
Chlorococcum humicola	Purified pigments (beta-carotene, chlorophyll a and chlorophyll b)	*Bacillus subtilis, Staphylococcus aureus, Escherichia coli, Pseudomonas aeruginosa, Salmonella typhimurium, Klebsiella pneumoniae, Vibrio cholerae, Candida albicans, Aspergillus niger* and *Aspergillus flavus*	Diffusion method	Purified beta-carotene has antimicrobial activities on all tested microorganisms except *Pseudomonas aeruginosa* (7 to 15 mm inhibition zone)	[32]
Dunaliella tertiolecta	Methanol	*Staphylococcus aureus, Pseudomonas aeruginosa, Escherichia coli,* and *Klebsiella* spp.	Minimal Inhibitory Concentration (MIC) assay	The extracts showed the highest antimicrobial activity against *Staphylococcus aureus* (MIC50 = 5.6×10^9 cells/ml) and *Pseudomonas aeruginosa* (MIC50 = 4.2×10^9 cells/ml)	[35]
Pseudokirchneriella subcapitata	Methanol	*Staphylococcus aureus, Pseudomonas aeruginosa, Escherichia coli,* and *Klebsiella* spp.	Minimal Inhibitory Concentration (MIC) assay	The extracts showed the highest antimicrobial activity against *Escherichia coli,* and *Klebsiella* spp. with MIC50 value of 5.9×10^9 cells/ml	[35]
Porphyridium sordidum	Methanol	*Bacillus subtilis* and *Escherichia coli*	Diffusion method	The extract has 30% degree of inhibition against *Bacillus subtilis* and 60% inhibition against *Escherichia coli*	[37]

(continued)

Table 13.2 (continued)

Microalgae species	Type of extract (secondary metabolites)	Target microorganism	Experimental method/assay	Biological significance	References
Rhizosolenia alata	Acetone, chloroform, chloroform:methanol, methanol:aqueous, aqueous	*Bacillus subtilis*, *Pseudomonas aeruginosa*, *Staphylococcus aureus*, *Escherichia coli*, *Proteous vulgaris*, *Vibrio cholerae*	Diffusion method	Chloroform and acetone extracts showed the antimicrobial effects on all tested microorganisms ranging from 5 mm until 18 mm zone of inhibition	[33]
Chlorella vulgaris	Chloroform, acetone, ethanol	*Escherichia coli*, *Bacillus* sp., *Klebsiella* sp., *Pseudomonas* sp.	Diffusion method	Ethanol extract showed the maximum antimicrobial activity on all tested microorganisms ranging from 12 mm until 17 mm zone of inhibition	[36]

methanol, and aqueous fractions from *Scenedesmus* spp., *Scenedesmus bajacalifornicus*, *Chlorella vulgaris*, *Dunaliella tertiolecta*, and *Pseudokirchneriella subcapitata* also demonstrated notable antimicrobial potency, as confirmed by diffusion and MIC assays, indicating the involvement of phenolics, pigments, and other polar bioactive compounds [31, 34–36].

Importantly, studies employing purified pigments from *Chlorococcum humicola* revealed that β-carotene and chlorophyll derivatives alone possess intrinsic antimicrobial activity, underscoring the functional role of photosynthetic pigments beyond antioxidant capacity [32]. These findings highlight that microalgae represent a versatile and sustainable source of antimicrobial agents, with species selection, metabolite class, and solvent extraction strategy collectively determining antimicrobial efficacy and spectrum, supporting their potential application in food preservation, pharmaceuticals, and antimicrobial formulations.

2.3 Anti-diabetic Activity

Diabetes mellitus is a chronic metabolic disorder characterised by persistent hyperglycaemia arising from defects in insulin secretion, insulin action, or both. The anti-diabetic potential of microalgal extracts is most commonly evaluated using in vitro enzymatic inhibition assays targeting α-glucosidase and α-amylase, two key enzymes involved in postprandial carbohydrate digestion [9]. α-Glucosidase, located in the brush border of the small intestine, catalyses the hydrolysis of oligosaccharides and disaccharides into absorbable glucose; its overactivity contributes to rapid postprandial hyperglycaemia. Inhibiting α-glucosidase delays carbohydrate breakdown and glucose absorption, thereby attenuating blood glucose spikes. α-Amylase, secreted mainly by the pancreas and salivary glands, hydrolyses α-1,4-glycosidic linkages in starch and glycogen to generate smaller oligosaccharides, which are subsequently converted to glucose. Inhibition of α-amylase slows the initial digestion of complex carbohydrates, complementing α-glucosidase inhibition in moderating glycaemic response [9]. Table 13.3 summarises studies that have investigated the anti-diabetic properties of microalgae and their associated secondary metabolites.

Accumulating evidence indicates that microalgae-derived secondary metabolites and whole biomass possess promising anti-diabetic properties through multiple biochemical and physiological mechanisms. In vitro studies demonstrate that polar extracts from *Scenedesmus bajacalifornicus* exhibit strong anti-diabetic potential, with aqueous fractions showing the highest α-amylase inhibitory activity and enhanced glucose uptake in yeast cells, highlighting the relevance of water-soluble bioactive compounds in regulating carbohydrate metabolism [34]. Carotenoid-rich extracts from *Chlorella* species further contribute to anti-diabetic effects by markedly inhibiting the formation of advanced glycation end-products (AGEs), a key pathogenic process in diabetic complications, with ethyl acetate extracts achieving high inhibition rates in protein–glucose model systems [38]. Similarly, astaxanthin-enriched extracts from *Chlorella zofingiensis* and fucoxanthin isolated

Table 13.3 The anti-diabetic studies related to secondary metabolites derived from microalgae

Microalgae species	Type of extract	Experimental design	Biological significance	References
Scenedesmus bajacalifornicus	Chloroform, ethyl acetate, methanol, ethanol, and aqueous	Alpha-amylase inhibition assay, glucose uptake method	The aqueous extract is the highest antidiabetic properties; highest alpha-amylase inhibitory activity (60.21%) and highest glucose uptake activity in yeast cells (46.12%)	[34]
Chlorella sp.	Ethyl acetate, hexane, aqueous (carotenoids)	AGE formation inhibition method	Ethyl acetate extract has the highest antidiabetic activity with the highest inhibitory rates against the formation of total advanced glycation end-products (AGEs) in the bovine serum albumin-glucose system (ranging from 81.76% to 88.02%)	[41]
Chlorella zofingiensis	Ethyl acetate (astaxanthin)	AGE formation inhibition method	The extracts inhibit the formation of AGE key representative named endogenous N3 -carboxymethyl lysine (CML) in RPE (human retinal pigment epithelial) cells by 17.07%	[42]
Phaeodactylum tricornutum	Methanol:water:hexane (fucoxanthin)	Alpha-glucosidase and alpha-amylase inhibition method	Fucoxanthin extracted from this species has a higher inhibitory effect on alpha-glucosidase (25.93 ± 1.45%) and alpha-amylase (28.38 ± 0.67%) activity at concentration 0.1 mmol/L compared to the standard, acarbose with alpha-glucosidase (19.15 ± 1.29%) and alpha-amylase (25.01 ± 1.38%) activity at the same concentration	[38]

(continued)

Table 13.3 (continued)

Microalgae species	Type of extract	Experimental design	Biological significance	References
Nannochloropsis oculata	Microalgae biomass	Diabetic rats	Diabetic rats fed with microalgae has a lower level of glucose serum concentration, cholesterol, LDL and triglycerides and higher level of insulin serum concentration and HDL-C	[39]
Isochrysis galbana	Microalgae biomass	Diabetic rats	Diabetic rats fed with the microalgae have a lower level of cholesterol, glucose and triacylglycerol	[40]
Nannochloropsis oculata	Microalgae biomass	Diabetic rats	Healthy and diabetic rats fed with microalgae have a higher level of low-density lipoprotein and lower level of high-density lipoprotein	[40]
Spirulina fusiformis	Aqueous (phycocyanin)	Diabetic rats	Diabetic rats fed with microalgal phycocyanin have a lower level of blood glucose after two hours	[43]

from *Phaeodactylum tricornutum* effectively suppressed AGE formation and inhibited α-glucosidase and α-amylase activities, in some cases outperforming the clinical standard acarbose, thereby underscoring the therapeutic relevance of microalgal carotenoids in glycaemic control [39, 40].

In vivo studies further support these findings, as dietary supplementation with microalgal biomass from *Nannochloropsis oculata* and *Isochrysis galbana* in diabetic rat models resulted in reduced serum glucose, cholesterol, low-density lipoprotein, and triglyceride levels, alongside improvements in insulin and high-density lipoprotein concentrations [41, 42]. Moreover, administration of phycocyanin extracted from *Spirulina fusiformis* significantly lowered postprandial blood glucose levels in diabetic rats, highlighting the contribution of phycobiliproteins to glucose homeostasis [43]. Collectively, these studies demonstrate that microalgae exert antidiabetic effects via enzyme inhibition, suppression of glycation processes, enhancement of glucose uptake, and systemic metabolic modulation, reinforcing their potential as functional ingredients or complementary therapeutics for diabetes management.

2.4 Anti-inflammatory Activity

Inflammation is a fundamental host defence response to injury or harmful stimuli, which may be physical, biological, or chemical in nature. In simple terms, it arises when cells or tissues are damaged or invaded by pathogens, triggering a coordinated cascade of vascular and cellular events [44]. This response involves the activation and recruitment of multiple immune cell populations, including macrophages, lymphocytes, and neutrophils, which release a variety of cytokines, chemokines, and other mediators that orchestrate the initiation, amplification, and resolution of the inflammatory process [45]. In the context of metabolic and chronic diseases, persistent low-grade inflammation is now recognised as a key pathogenic driver, and this has stimulated growing interest in microalgae-derived secondary metabolites such as carotenoids, phenolic compounds, and phycobiliproteins as natural modulators of inflammatory pathways [6]. These microalgal compounds have been shown to attenuate pro-inflammatory signalling by suppressing mediators including NO, TNF-α and IL-6, and by downregulating enzymes and transcription factors such as iNOS, COX-2 and NF-κB in immune cells, highlighting their potential as anti-inflammatory candidates for nutraceutical and therapeutic applications. Table 13.4 summarises studies that have investigated the inflammatory properties of microalgae and their associated secondary metabolites.

Multiple studies have demonstrated that microalgae-derived secondary metabolites exhibit significant anti-inflammatory activity, primarily through the modulation of key pro-inflammatory signalling pathways and mediators in macrophage-based cell models. Carotenoid-rich extracts, particularly violaxanthin isolated from *Chlorella ellipsoidea*, effectively suppressed lipopolysaccharide (LPS)-induced inflammation by inhibiting nuclear translocation of the NF-κB p65 subunit in RAW

Table 13.4 The anti-inflammatory studies related to secondary metabolites derived from microalgae

Microalgae species	Type of extract (secondary metabolites)	Experimental model	Biological significance	References
Chlorella ellipsoidea	Ethyl acetate and ethanol (violaxanthin)	Cell culture (RAW 264.7 macrophages cells)	Violaxanthin extracted from this species inhibited LPS-mediated nuclear factor-κB (NF-κB) p65 subunit translocation into the nucleus which involve in pro-inflammatory signalling pathways in RAW 264.7 cells	[50]
Tetraselmis suecica	Methanol, hexane, ethyl acetate, and aqueous	Cell culture (RAW 264.7 macrophages cells)	Ethyl acetate extract showed the highest anti-inflammatory activity by the strong inhibition of various inflammatory mediators such as NO, TNF-α, and IL-6	[46]
Chloromonas reticulata	Ethanol	Cell culture (RAW 264.7 macrophages)	The extract reduced the expression of major regulatory factors of inflammation such as iNOS and COX-2 and inhibited the production of nitric oxide in RAW 264.7 cells	[47]
Micractinium sp.	Ethanol	Cell culture (RAW 264.7 macrophages)	The extracts reduce the expression of main proinflammatory mediators (iNOS and COX-2 genes), the level of nitric oxide (NO) production and the proinflammatory cytokines production (TNF-α and IL-6) in RAW 264.7 macrophages	[48]

(continued)

Table 13.4 (continued)

Microalgae species	Type of extract (secondary metabolites)	Experimental model	Biological significance	References
Nannochloropsis Oculata	Methanol, n-hexane, chloroform, ethyl acetate, aqueous	Cell culture (RAW 264.7 macrophages)	The hexane and chloroform extracts showed the highest anti-inflammatory effect on RAW 264.7 macrophages cells by lowering 40% of the nitric oxide production in cells	[45]
Tetraselmis sp.	Acetone (lutein and zeaxanthin)	Cell culture (RAW 264.7 macrophages)	The acetone extract reduces the nitric oxide overproduction and the production of pro-inflammatory mediators such as iNOS, TNF-alpha, and interleukins	[49]

264.7 macrophages, thereby attenuating downstream pro-inflammatory gene expression [46]. Similarly, solvent extracts from *Tetraselmis suecica* showed marked anti-inflammatory effects, with ethyl acetate fractions most effectively reducing nitric oxide (NO) production and pro-inflammatory cytokines such as TNF-α and IL-6, highlighting the role of semi-polar bioactive compounds in regulating inflammatory responses [47]. Ethanol extracts of *Chloromonas reticulata* and *Micractinium* sp. further demonstrated potent anti-inflammatory potential by downregulating inducible nitric oxide synthase (iNOS) and cyclooxygenase-2 (COX-2) expression, alongside suppressing NO and cytokine production in activated macrophages [48, 49].

In addition, non-polar extracts from *Nannochloropsis oculata*, particularly hexane and chloroform fractions, significantly reduced NO production by approximately 40%, suggesting the involvement of lipid-soluble anti-inflammatory constituents [45]. Complementary findings from *Tetraselmis* sp. revealed that acetone extracts enriched in lutein and zeaxanthin effectively mitigated inflammatory mediator overproduction, including iNOS, TNF-α, and interleukins, reinforcing the importance of xanthophyll carotenoids in anti-inflammatory activity [50]. Collectively, these studies indicate that microalgal carotenoids and other secondary metabolites can modulate inflammatory signalling at multiple regulatory levels, supporting their potential application as natural anti-inflammatory agents in nutraceutical and therapeutic formulations.

2.5 Anti-obesity Activity

Obesity is a complex metabolic disorder characterised by excessive accumulation of body fat, commonly defined clinically as a body mass index (BMI) greater than 30 kg/m^2. Its development is multifactorial, arising from a combination of energy-dense diets, sedentary lifestyle, and genetic predisposition, often in conjunction with other environmental and behavioural factors [51]. In this context, microalgae have emerged as a promising source of anti-obesity therapeutics and nutraceuticals, owing to their repertoire of bioactive compounds including carotenoids, polyunsaturated fatty acids, phenolics, and polysaccharides that can modulate lipid metabolism, adipogenesis, and inflammatory pathways [6]. Table 13.5 summarises studies that have investigated the anti-obesity properties of microalgae and their associated secondary metabolites.

A range of microalgae-derived secondary metabolites and whole biomass have demonstrated pronounced anti-obesity effects through modulation of lipid metabolism, adipogenesis, and systemic metabolic parameters in both in vitro and in vivo models. Carotenoid-rich extracts, particularly β-carotene from *Dunaliella salina*, significantly reduced body and liver weights, visceral fat accumulation, and pancreatic lipase activity in obese rats, while concurrently improving lipid profiles, insulin sensitivity, glucose homeostasis, leptin levels, and liver function enzymes, indicating a multifaceted anti-obesity mechanism [52]. Fucoxanthin, a prominent marine xanthophyll isolated from *Phaeodactylum tricornutum* and *Scenedesmus dimorphus*, consistently suppressed lipid accumulation and adipogenic differentiation in 3T3-L1 adipocytes and reduced body weight gain, adipose tissue mass, and circulating lipid levels in obese rat models, highlighting its role in regulating adipocyte metabolism and lipid utilisation [53, 54]. Aqueous extracts of *Euglena gracilis* Z further demonstrated the capacity to inhibit lipid accumulation and adipogenesis in human adipose-derived stem cells, suggesting translational relevance for human obesity management [55].

In cyanobacterial species, ethanol extracts and biomass of *Spirulina maxima* markedly reduced adipogenesis in adipocytes and mesenchymal stem cells and lowered body weight gain, adipocyte size, and glucose levels in high-fat diet-induced obese rats, effects attributed largely to chlorophyll a and C-phycocyanin [56, 57]. Notably, studies using whole microalgal biomass from *Phaeodactylum tricornutum* confirmed that fucoxanthin retained bioactivity in dietary form, effectively decreasing body weight and adipose tissue mass in obese animals [58]. These findings demonstrate that microalgal carotenoids, pigments, and biomass exert anti-obesity effects via inhibition of adipogenesis, enhancement of lipid metabolism, and improvement of metabolic homeostasis, supporting their potential application as functional food ingredients or nutraceuticals targeting obesity and related metabolic disorders.

Table 13.5 The anti-obesity studies related to secondary metabolites derived from microalgae

Microalgae species	Type of extracts	Experimental model	Biological significance	References
Dunaliella salina	Ethanol (beta-carotene)	Obese rats	Reduce body and liver weights, visceral fat, inhibit pancreatic lipase activity, decrease lipid profile, and increase faecal fat and ameliorate liver function enzymes activity, insulin, blood glucose, and leptin levels	[52]
Phaeodactylum tricornutum	Purified pigment (fucoxanthin)	Cell culture (3T3-L1 adipocytes) and obese rats	Reduce lipid accumulation and adipogenesis	[53]
Euglena gracilis Z	Aqueous	Cell culture (human adipose-derived stem cells (hASCs))	Reduce lipid content and adipogenesis	[55]
Spirulina maxima	Ethanol (Chlorophyll a and C-phycocyanin)	Cell culture (3T3-L1 pre-adipocytes, C3H10T1/2 mesenchymal stem cells) and high-fat diet or obese rats	Reduce adipogenesis in adipocytes and mesenchymal stem cells and reduce final body weight, body weight gain, glucose levels, and adipogenesis in rats	[56]
Phaeodactylum tricornutum	Ethanol (fucoxanthin)	Cell culture (3T3-L1 pre-adipocytes) and obese rats	The extract showed no effects on adipocytes but reduce body weight gain, body fat content in rats	[53]
Spirulina maxima	NA; microalgae biomass	Obese rats	Reduce body weight gain in rats and inhibit the increase of adipocytes' size	[57]

(continued)

Table 13.5 (continued)

Microalgae species	Type of extracts	Experimental model	Biological significance	References
Phaeodactylum tricornutum	NA; microalgae biomass (Fucoxanthin)	Obese rats	The biomass of *Phaeodactylum tricornutum* was fed to the obese rats and the study found that the fucoxanthin present in the biomass has reduced the rats' body weight and adipose tissue body weight	[58]
Scenedesmus dimorphus	Ethanol (fucoxanthin)	Obese rats	The extracted fucoxanthin from *Scenedesmus dimorphus* has decreased the total LDL, cholesterol and triglycerides levels in obese rats. Plus, the extract also increased the weight loss and HDL level in the obese rats	[54]

The overall concept of using microalgal biomass and its derivatives for obesity management, spanning from microalgal cultivation to evaluation in animal and cell-based models, is summarised in Fig. 13.4. Collectively, these in vivo studies highlight fucoxanthin-rich microalgae as promising candidates for the development of functional ingredients targeting obesity and obesity-related dyslipidaemia.

2.6 Anti-cancer Activity

Cancer is a second leading death cause in the world which recorded around 9.6 million cases or 1 in 6 deaths in 2018 [59]. There are five most common types of cancer such as breast cancer, skin melanoma cancer, colon cancer, blood cancer and lung cancer [60]. It has been reported that most of the cancer treatments including chemotherapy, are having side effects which can threaten life as they may not differentiate between normal and cancer cells during the therapy [61, 62]. Hence, exploring the potential of natural products such as bioactive compounds derived from microalgae has

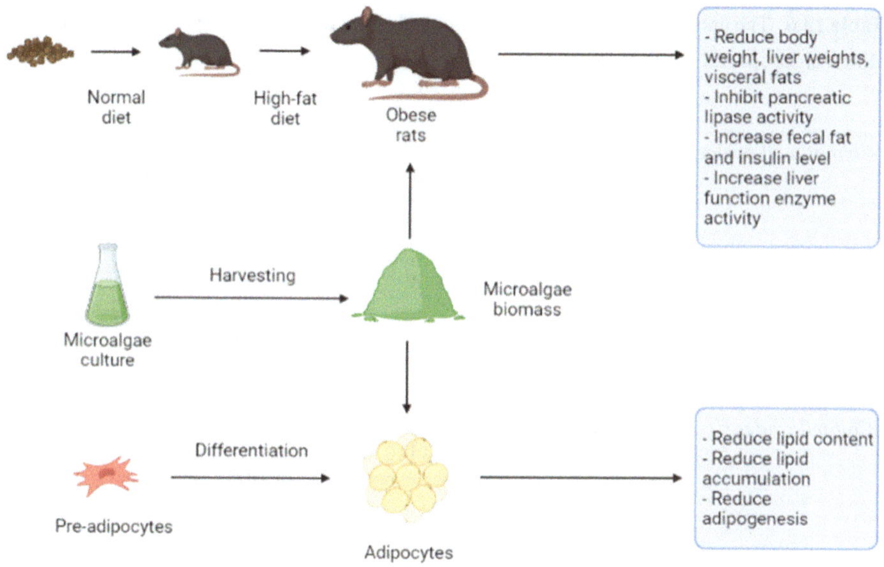

Fig. 13.4 Schematic overview of anti-obesity effects of microalgal biomass in in vivo and in vitro models (Created with BioRender.com)

been a great focus on developing a new alternative cancer treatment. The capability of microalgal metabolites to protect cells from damage suggests their potential applications in the cancer treatment and prevention [63]. These metabolites act as inhibitory agents in inducing apoptosis effects on cancer cells and suppressing the proliferation of the cancer cells [64]. For instances, pigments such as carotenoids and phycocyanin derived from microalgae have potential anti-cancer properties as they can offer genetic damage protection for normal cells, cytotoxic, anti-proliferative and apoptosis effects on cancer cells [65]. Table 13.6 summarises studies that have investigated the anti-cancer properties of microalgae and their associated secondary metabolites.

Extensive in vitro evidence indicates that microalgae-derived secondary metabolites possess notable anti-cancer activities across a broad spectrum of human cancer cell lines, largely mediated through growth inhibition, cytotoxicity, and apoptosis induction. Carotenoid-rich ethanol extracts from *Chlorella ellipsoidea* and *Chlorella vulgaris*, containing violaxanthin, antheraxanthin, zeaxanthin, and lutein, effectively suppressed proliferation of HCT116 human colon cancer cells with comparable IC_{50} values and triggered apoptotic responses, highlighting the anticancer potential of xanthophyll carotenoids [66]. Similarly, purified violaxanthin isolated from *Dunaliella tertiolecta* exerted strong antiproliferative effects on MCF-7 breast cancer cells by inducing early apoptotic biochemical and morphological changes [67]. Methanolic extracts of *Granulocystopsis* sp. demonstrated broad-spectrum cytotoxicity against breast, colorectal, melanoma, and prostate cancer cell lines by disrupting

Table 13.6 The anti-cancer studies related to secondary metabolites derived from microalgae

Microalgae species	Type of extract (secondary metabolites)	Experimental model	Biological significance	References
Chlorella ellipsoidea	Ethanol (violaxanthin, antheraxanthin, zeaxanthin)	Cell culture (HCT116 human colon cancer cells)	The extracts inhibit the growth of HCT116 human colon cancer cells with IC50 values of 40.73 ± 3.71 μg/ml and trigger apoptosis effects in the cell culture	[67]
Chlorella vulgaris	Ethanol (lutein)	Cell culture (HCT116 human colon cancer cells)	The extracts inhibit the growth of HCT116 human colon cancer cells with IC50 values of 40.31 ± 4.43 μg/ml and trigger apoptosis effects in the cell culture	[67]
Dunaliella tertiolecta	Dichloromethane (Violaxanthin)	Cell culture (MCF-7 breast cancer cells)	The purified violaxanthin from dichloromethane extract exerts anticancer effects on MCF-7 cells by showing the antiproliferative activity, inducing biochemical and morphological changes in the early stages of apoptosis of the cells	[66]

(continued)

Table 13.6 (continued)

Microalgae species	Type of extract (secondary metabolites)	Experimental model	Biological significance	References
Granulocystopsis sp.	Methanol	Cell culture (MCF-7 breast cancer cells, HT-29 colorectal cancer cells, SK-MEL-28 skin melanoma cells, DU-145 prostate cancer cells and CCL-81 Vero normal cells)	The methanolic extracts exhibited anti-cancer effects on all cancer cells by causing the membrane integrity loss, inhibiting cell adhesion and proliferation and inducing cell apoptosis	[61]
Micractinium sp.	Ethanol	Cell culture (HCT116 human colon cancer cells)	The extract showed the anti-cancer activity by having a higher cytoxicity on HCT116 cells	[48]
Chloromonas reticulata	Ethanol	Cell culture (HCT116 human colon cancer cells)	The extract showed the anti-cancer activity by inhibiting the proliferation of HCT116 cancer cells	[47]
Chlorella ovalis	Methanol, n-hexane, chloroform, ethyl acetate, aqueous	Cell culture (HL-60 human promyelocytic leukaemia cells, A549 human lung carcinoma cells and B16F10 mouse melanoma cells)	The ethyl acetate extracts presented the highest inhibitory effects on HL-60 human promyelocytic leukaemia cells compared to other cells and extracts by showing 40% of cell viability for concentration of extracts at 50 μg/ml	[45]

(continued)

Table 13.6 (continued)

Microalgae species	Type of extract (secondary metabolites)	Experimental model	Biological significance	References
Amphidinium carterae	Methanol, n-hexane, chloroform, ethyl acetate, aqueous	Cell culture (HL-60 human promyelocytic leukaemia cells, A549 human lung carcinoma cells and B16F10 mouse melanoma cells)	The chloroform extracts presented the highest inhibitory effects on HL-60 human promyelocytic leukaemia cells compared to other cells and extracts by showing 40% of cell viability for concentration of extracts at 50 μg/ml	[45]
Tetradesmus acuminatus	Hexane, chloroform, methanol, ethanol, aqueous (flavonoids, phenolic compounds, tannins)	Cell culture (MCF-7 breast cancer cells)	The methanol, ethanol and aqueous extracts exhibited the significant anti-cancer effects on MCF-7 breast cancer cells by decreasing the percentage of cell viability (from 100 to 65%) when a higher concentration of extracts is used from 3.1 μg/ml to 200 μg/ml	[24]
Cyanophora paradoxa	Dichloromethane, ethanol, aqueous (zeaxanthin and β-cryptoxanthin)	Cell culture (MCF-7 breast cancer cells, A549 human lung adenocarcinoma cells, and A2058 human melanoma cells)	The ethanol and aqueous extracts inhibited the growth of all cell lines and induced apoptosis in all cancer cells at the concentration of extracts 100 μg/ml	[65]

membrane integrity, inhibiting cell adhesion and proliferation, and inducing apoptosis, while sparing normal Vero cells [61]. Ethanol extracts from *Micractinium* sp. and *Chloromonas reticulata* further confirmed selective cytotoxic and antiproliferative effects against HCT116 colon cancer cells [47, 48]. Comparative solvent studies in *Chlorella ovalis* and *Amphidinium carterae* revealed that ethyl acetate and chloroform fractions, respectively, produced the strongest inhibitory effects against HL-60 leukaemia cells, underscoring the importance of solvent polarity in extracting potent anticancer metabolites [45].

In addition, extracts from *Tetradesmus acuminatus* rich in phenolics, flavonoids, and tannins significantly reduced the viability of MCF-7 breast cancer cells in a dose-dependent manner [24], while ethanol and aqueous extracts of *Cyanophora paradoxa* containing zeaxanthin and β-cryptoxanthin effectively inhibited growth and induced apoptosis in breast, lung, and melanoma cancer cell lines [65]. Collectively, these findings demonstrate that microalgal carotenoids, phenolic compounds, and complex metabolite mixtures can target multiple cancer-related pathways, reinforcing the potential of microalgae as sustainable sources of novel anticancer agents for nutraceutical and pharmaceutical development [24, 45, 47, 48, 61, 65–67]. It can be deduced that bioactive compounds derived from microalgae particularly carotenoids and several polar compounds possess anti-cancer effects on various types of cancer cell lines.

3 Conclusion

Microalgae represent one of the most promising biological platforms for the discovery and development of high-value secondary metabolites with significant therapeutic potential. As demonstrated throughout this review, microalgae synthesize a diverse repertoire of bioactive compounds ranging from carotenoids and phenolics to PUFAs, terpenoids, peptides, and polysaccharides that exhibit potent antioxidant, antimicrobial, anti-diabetic, anti-inflammatory, anti-obesity, and anticancer activities. These biological properties position microalgal metabolites as compelling candidates for addressing complex metabolic diseases, in which oxidative stress, chronic inflammation, dysregulated lipid metabolism, and impaired glucose homeostasis are key pathological drivers. Despite extensive in-vitro evidence supporting their therapeutic relevance, the translation of microalgal secondary metabolites into clinically validated interventions remains limited. Critical challenges include optimizing cultivation conditions for consistent metabolite yields, developing efficient and scalable extraction technologies, performing robust structural elucidation, and conducting well-designed in-vivo and pharmacokinetic studies. A systematic pipeline from cultivation and stress-induced metabolite enhancement to purification, mechanistic characterization, and preclinical validation is therefore essential to bridge the gap between laboratory discovery and real-world application.

References

1. De Morais MG, Da Silva Vaz B, De Morais EG, Costa JAV (2015) Biologically active metabolites synthesized by microalgae. Biomed Res Int 1–15. https://doi.org/10.1155/2015/835761
2. Fu W, Nelson DR, Yi Z, Xu M, Khraiwesh B, Jijakli K, Chaiboonchoe A, Alzahmi A, Al-Khairy D, Brynjolfsson S, Salehi-Ashtiani K (2017) Bioactive compounds from microalgae: current development and prospects. In: Studies in natural products chemistry, pp 199–225. https://doi.org/10.1016/B978-0-444-63929-5.00006-1
3. Richmond A (2004) Handbook of microalgal culture, Blackwell Science Ltd
4. de Coêlho DF, Tundisi LL, Cerqueira KS, da Rodrigues JRS, Mazzola PG, Tambourgi EB, de Souza RR (2019) Microalgae: Cultivation aspects and bioactive compounds. Braz Arch Biol Technol 62:1–13. https://doi.org/10.1590/1678-4324-2019180343
5. Nisar B, Sultan A, Rubab SL (2018) Comparison of medicinally important natural products versus synthetic drugs-a short commentary. Nat Prod Chem Res 06. https://doi.org/10.4172/2329-6836.1000308
6. Nor Shahril NS, Abu Bakar MH, Kassim MA, Karunakaran T, Shariff KA (2025) The In vitro antioxidant, anti-obesity, and anti-inflammatory properties of microalgae extracts of *Tetraselmis suecica* cultivated under sodium acetate and sodium nitrate supplementations. J Appl Biotechnol Rep 12:1554–1569
7. Gil-Chávez GJ, Villa JA, Ayala-Zavala JF, Heredia JB, Sepulveda D, Yahia EM, González-Aguilar GA (2013) Technologies for extraction and production of bioactive compounds to be used as nutraceuticals and food ingredients: an overview. Compr Rev Food Sci Food Saf 12:5–23. https://doi.org/10.1111/1541-4337.12005
8. Sathasivam R, Radhakrishnan R, Hashem A, Abd_Allah EF (2019) Microalgae metabolites : a rich source for food and medicine. Saudi J Biol Sci 26:709–722. https://doi.org/10.1016/j.sjbs.2017.11.003
9. Abu Bakar MH, Lee PY, Azmi MN, Lotfiamir NS, Faris Mohamad MS, Nor Shahril NS, Shariff KA, Ya'akob H, Awang K, Litaudon M (2020) In vitro anti-hyperglycemic, antioxidant activities and intestinal glucose uptake evaluation of *Endiandra kingiana* extracts. Biocatal Agric Biotechnol 25 (2020). https://doi.org/10.1016/j.bcab.2020.101594
10. Tay YN, Bakar MHA, Azmi MN, Saad NA, Awang K, Litaudon M, Kassim MA (2020) Inhibition of carbohydrate hydrolysing enzymes, antioxidant activity and polyphenolic content of *Beilschmiedia* species extracts. IOP Conf Ser Mater Sci Eng. https://doi.org/10.1088/1757-899X/716/1/012007
11. Hafizi Abu Bakar M, Kian Kai C, Wan Hassan WN, Sarmidi MR, Yaakob H, Zaman Huri H (2015) Mitochondrial dysfunction as a central event for mechanisms underlying insulin resistance: the roles of long chain fatty acids. Diabetes Metab Res Rev 31. https://doi.org/10.1002/dmrr.2601
12. Abu Bakar MH, Sarmidi MR, Cheng K-K, Ali Khan A, Suan CL, Zaman Huri H, Yaakob H (2015) Metabolomics—the complementary field in systems biology: a review on obesity and type 2 diabetes. Mol Biosyst 11. https://doi.org/10.1039/c5mb00158g
13. Kurutas EB (2016) The importance of antioxidants which play the role in cellular response against oxidative/nitrosative stress: current state. Nutr J 15:1–22. https://doi.org/10.1186/s12937-016-0186-5
14. Goiris K, Muylaert K, Fraeye I, De Brabanter J (2012) Antioxidant potential of microalgae in relation to their phenolic and carotenoid content. https://doi.org/10.1007/s10811-012-9804-6
15. Gürlek C, Yarkent Ç, Köse A, Tuğcu B, Gebeloğlu IK, Öncel SS, Elibol M (2019) Screening of antioxidant and cytotoxic activities of several microalgal extracts with pharmaceutical potential. Health Technol (Berl) 10:111–117. https://doi.org/10.1007/s12553-019-00388-3
16. Sansone C, Brunet C (2019) Promises and challenges of microalgal antioxidant production. Antioxidants 8(7):199. https://doi.org/10.3390/antiox8070199

17. Hemalatha A, Girija K, Parthiban C, Saranya C, Anantharaman P (2013) Antioxidant properties and total phenolic content of a marine diatom, Navicula clavata and green microalgae, Chlorella marina and Dunaliella salina, Pelagia research. Library 4:151–157
18. Rodrigues DB, Menezes CR, Mercadante AZ, Jacob-Lopes E, Zepka LQ (2015) Bioactive pigments from microalgae *Phormidium autumnale*. Food Res Int 77:273–279. https://doi.org/10.1016/j.foodres.2015.04.027
19. Goh S-H, Yusoff FM, Loh SP (2010) A comparison of the antioxidant properties and total phenolic content in a Diatom, *Chaetoceros* sp. and a green microalga, *Nannochloropsis* sp. J Agric Sci 2. https://doi.org/10.5539/jas.v2n3p123
20. Agustini TW, Suzery M, Sutrisnanto D, Ma'ruf WF, Hadiyanto (2015) Comparative study of bioactive substances extracted from fresh and dried *Spirulina* sp. Procedia Environ Sci 23:282–289. https://doi.org/10.1016/j.proenv.2015.01.042
21. Shalaby EA, Shanab SMM (2013) Comparison of DPPH and ABTS assays for determining antioxidant potential of water and methanol extracts of *Spirulina platensis*. Indian J Mar Sci 42:556–564
22. Kepekçi RA, Saygideger SD (2012) Enhancement of phenolic compound production in *Spirulina platensis* by two-step batch mode cultivation. J Appl Phycol 24:897–905. https://doi.org/10.1007/s10811-011-9710-3
23. de Dantas DMM, de Oliveira CYB, Costa RMPB, das Carneiro-da-Cunha MG, Gálvez AO, de Bezerra RS (2019) Evaluation of antioxidant and antibacterial capacity of green microalgae *Scenedesmus subspicatus*. Food Sci Technol Int 25:318–326. https://doi.org/10.1177/1082013218825024
24. Mujeeb MA, Vedamurthy A, Shettar AK, Puranik SI, Ghagane S, Thimmappa SC (2019) In vitro anti-oxidant and anti-cancer activity of *Tetradesmus acuminatus* microalgae extract on MCF-7 human breast cancer cell line. Int J Cancer Res 16:1–9. https://doi.org/10.3923/ijcr.2020.1.9
25. Shanab SMM, Mostafa SSM, Shalaby EA, Mahmoud GI (2012) Aqueous extracts of microalgae exhibit antioxidant and anticancer activities, Asian Pac. J Trop Biomed 2:608–615. https://doi.org/10.1016/S2221-1691(12)60106-3
26. Herrero M, Mendiola JA, Plaza M, Ibanez E (2013) Screening for bioactive compounds from algae. In: Advanced biofuels and bioproducts, pp 833–871. https://doi.org/10.1007/978-1-4614-3348-4
27. Falaise C, François C, Travers MA, Morga B, Haure J, Tremblay R, Turcotte F, Pasetto P, Gastineau R, Hardivillier Y, Leignel V, Mouget JL (2016) Antimicrobial compounds from eukaryotic microalgae against human pathogens and diseases in aquaculture. Mar Drugs 14:1–27. https://doi.org/10.3390/md14090159
28. Kohanski MA, Dwyer DJ, Collins JJ (2010) How antibiotics kill bacteria: from targets to networks. Nat Rev Microbiol 8:423–435. https://doi.org/10.1038/nrmicro2333
29. Rao N, Lipsky BA (2007) Optimising antimicrobial therapy in diabetic foot infections. Drugs 67:195–214. https://doi.org/10.2165/00003495-200767020-00003
30. Navarro F, Forján E, Vázquez M, Toimil A, Montero Z, Ruiz-Domínguez MC, Garbayo I, Castaño M, Vílchez C, Vega JM (2017) Antimicrobial activity of the acidophilic eukaryotic microalga *Coccomyxa onubensis*. Phycological Res 65:38–43. https://doi.org/10.1111/pre.12158
31. Ördög V, Stirk WA, Lenobel R, Bancířová M, Strnad M, Van Staden J, Szigeti J, Németh L (2004) Screening microalgae for some potentially useful agricultural and pharmaceutical secondary metabolites. J Appl Phycol 16:309–314. https://doi.org/10.1023/B:JAPH.0000047789.34883.aa
32. Bhagavathy S, Sumathi P, Jancy Sherene Bell I (2011) Green algae *Chlorococcum humicola*-a new source of bioactive compounds with antimicrobial activity. Asian Pac J Trop Biomed 1:S1. https://doi.org/10.1016/S2221-1691(11)60111-1.
33. Venkatesan R, Karthikayen R, Periyanayagi R, Sasikala V, Balasubramanian T (2007) Antibacterial activity of the marine diatom, *Rhizosolenia alata* (Brightwell, 1858) against human pathogens. Res J Microbiol 2:98–100

34. Patil L, Kaliwal BB (2019) Microalga *Scenedesmus bajacalifornicus* BBKLP-07, a new source of bioactive compounds with in vitro pharmacological applications. Bioprocess Biosyst Eng 42:979–994. https://doi.org/10.1007/s00449-019-02099-5
35. Pane G, Cacciola G, Giacco E, Mariottini GL, Coppo E (2015) Assessment of the antimicrobial activity of algae extracts on bacteria responsible of external otitis. Mar Drugs 13:6440–6452. https://doi.org/10.3390/md13106440
36. Syed S, Arasu A, Ponnuswamy I (2015) The uses of *Chlorella Vulgaris* as antimicrobial agent and as a diet: The presence of bio-active compounds which caters the vitamins, minerals in general. Int J Bio-Sci Bio-Technol 7:185–190. https://doi.org/10.14257/ijbsbt.2015.7.1.19
37. Mudimu O, Rybalka N, Bauersachs T, Born J, Friedl T, Schulz R (2014) Biotechnological screening of microalgal and cyanobacterial strains for biogas production and antibacterial and antifungal effects. Metabolites 4:373–393. https://doi.org/10.3390/metabo4020373
38. Kawee-Ai A, Kim AT, Kim SM (2019) Inhibitory activities of microalgal fucoxanthin against α-amylase, α-glucosidase, and glucose oxidase in 3T3-L1 cells linked to type 2 diabetes. J Oceanol Limnol 37:928–937. https://doi.org/10.1007/s00343-019-8098-9
39. Nasirian F, Sarir H, Moradi-kor N (2019) Antihyperglycemic and antihyperlipidemic activities of *Nannochloropsis oculata* microalgae in Streptozotocin-induced diabetic rats. Biomol Concepts 10:37–43
40. Nuño K, Villarruel-López A, Puebla-Pérez AM, Romero-Velarde E, Puebla-Mora AG, Ascencio F (2013) Effects of the marine microalgae *Isochrysis galbana* and *Nannochloropsis oculata* in diabetic rats. J Funct Foods 5:106–115. https://doi.org/10.1016/j.jff.2012.08.011
41. Sun Z, Peng X, Liu J, Fan KW, Wang M, Chen F (2010) Inhibitory effects of microalgal extracts on the formation of advanced glycation endproducts (AGEs). Food Chem 120:261–267. https://doi.org/10.1016/j.foodchem.2009.10.018
42. Sun Z, Liu J, Zeng X, Huangfu J, Jiang Y, Wang M, Chen F (2011) Protective actions of microalgae against endogenous and exogenous advanced glycation endproducts (AGEs) in human retinal pigment epithelial cells. Food Funct 2:251–258. https://doi.org/10.1039/c1fo10021a
43. Setyaningsih I, Bintang M, Madina N (2015) Potentially Antihyperglycemic from biomass and Phycocyanin of *Spirulina Fusiformis* Voronikhin by in vivo test. Procedia Chem 14:211–215. https://doi.org/10.1016/j.proche.2015.03.030
44. Wang X, Abu Bakar MH, Liqun S, Kassim MA, Shariff KA, Karunakaran T (2025) Targeting metabolic diseases with celastrol: A comprehensive review of anti-inflammatory mechanisms and therapeutic potential. J Ethnopharmacol 344:119560. https://doi.org/10.1016/J.JEP.2025.119560
45. Samarakoon KW, Ko JY, Shah MMR, Lee JH, Kang MC, O-Nam K, Lee JB, Jeon YJ (2013) In vitro studies of anti-inflammatory and anticancer activities of organic solvent extracts from cultured marine microalgae. Algae 28:111–119. https://doi.org/10.4490/algae.2013.28.1.111
46. Jo WS, Choi YJ, Kim HJ, Nam BH, Hong SH, Lee GA, Lee SW, Seo SY, Jeong MH (2010) Anti-inflammatory effect of microalgal extracts from *Tetraselmis suecica*. Food Sci Biotechnol 19:1519–1528. https://doi.org/10.1007/s10068-010-0216-6
47. Suh SS, Hong JM, Kim EJ, Jung SW, Chae H, Kim JE, Kim JH, Kim IC, Kim S (2019) Antarctic freshwater microalga, *Chloromonas reticulata*, suppresses inflammation and carcinogenesis. Int J Med Sci 16:189–197. https://doi.org/10.7150/ijms.30647
48. Suh SS, Hong JM, Kim EJ, Jung SW, Kim SM, Kim JE, Kim IC, Kim S (2018) Anti-inflammation and anti-cancer activity of ethanol extract of antarctic freshwater microalga, *Micractinium* sp. Int J Med Sci 15:929–936. https://doi.org/10.7150/ijms.26410
49. Cheon J, Kim E-A, Kang N, Kim T, Wang L, Heo S-J, Cha S-H (2024) Comparison of the anti-inflammatory effects of carotenoids-rich product from microalgal mass-cultured between natural seawater and magma seawater. Biocatal Agric Biotechnol 62:103420. https://doi.org/10.1016/j.bcab.2024.103420
50. Soontornchaiboon W, Joo SS, Kim SM (2012) Anti-inflammatory effects of violaxanthin isolated from microalga *Chlorella ellipsoidea* in RAW 264.7 macrophages. Biol Pharm Bull 35:1137–1144. https://doi.org/10.1248/bpb.b12-00187

51. Wan-Loy C, Siew-Moi P (2016) Marine algae as a potential source for anti-obesity agents. Mar Drugs 14:1–19. https://doi.org/10.3390/md14120222
52. El-Baz FK, Aly HF (2018) Potential of *Dunaliella salina* microalgae to ameliorate high-fat diet-induced obesity in animals' rodent. Asian J Pharm Clin Res 11: 312–318. https://doi.org/10.22159/ajpcr.2018.v11i3.23398.
53. Koo SY, Hwang J, Yang S, Um J, Hong KW, Kang K, Pan C-H, Hwang KT, Kim SM (2019) Anti-obesity effect of standardized extract of microalga *Phaeodactylum tricornutum* containing fucoxanthin. Mar Drugs 17(5):311.https://doi.org/10.3390/md17050311
54. Choirunisa A, Aprilian R, Wati NS, Effect of fucoxanthin extract from microalgae *Scenedesmus dimorphus* on anti- obesity activity. J Pharm Chem Chem Sci 5:3–6
55. Sugimoto R, Ishibashi-Ohgo N, Atsuji K, Miwa Y, Iwata O, Nakashima A, Suzuki K (2018) Euglena extract suppresses adipocyte-differentiation in human adipose-derived stem cells. PLoS ONE 13:1–14. https://doi.org/10.1371/journal.pone.0192404
56. Seo YJ, Kim KJ, Choi J, Koh EJ, Lee BY (2018) *Spirulina maxima* extract reduces obesity through suppression of adipogenesis and activation of browning in 3T3-L1 cells and high-fat diet-induced obese mice. Nutrients 10:1–15. https://doi.org/10.3390/nu10060712
57. Heo M-G, Choung S-Y (2018) Anti-obesity effects of *Spirulina maxima* in high fat diet induced obese rats via the activation of AMPK pathway and SIRT1. Food Funct 9:4906–4915. https://doi.org/10.1039/C8FO00986D
58. Kim JH, Kim SM, Cha KH, Mok IK, Koo SY, Pan CH, Lee JK (2016) Evaluation of the anti-obesity effect of the microalga *Phaeodactylum tricornutum*. Appl Biol Chem 59:283–290. https://doi.org/10.1007/s13765-016-0151-1
59. World Health Organisation, Cancer (2021). https://www.who.int/health-topics/cancer#tab=tab_1. Last assessed 28 Dec 2025
60. Martínez Andrade KA, Lauritano C, Romano G, Ianora A (2018) Marine microalgae with anti-cancer properties, Mar Drugs 16. https://doi.org/10.3390/md16050165
61. Tavares-Carreón F, De La Torre-Zavala S, Arocha-Garza HF, Souza V, Galán-Wong LJ, Avilés-Arnaut H (2020) In vitro anticancer activity of methanolic extract of *Granulocystopsis* sp., a microalgae from an oligotrophic oasis in the Chihuahuan desert, Peer J 2020:1–21. https://doi.org/10.7717/peerj.8686
62. Varshney A, Singh V (2013) Effects of algal compounds on cancer cell line. J Exp Biol Agric Sci 1:337–352
63. Bule MH, Ahmed I, Maqbool F, Bilal M, Iqbal HMN (2018) Microalgae as a source of high-value bioactive compounds. Front Biosci (Schol Ed) 10:197–216. https://doi.org/10.2741/s509
64. Fedorov SN, Ermakova SP, Zvyagintseva TN, Stonik VA (2013) Anticancer and cancer preventive properties of marine polysaccharides: some results and prospects. Mar Drugs 11:4876–4901. https://doi.org/10.3390/md11124876
65. Baudelet PH, Gagez AL, Bérard JB, Juin C, Bridiau N, Kaas R, Thiéry V, Cadoret JP, Picot L (2013) Antiproliferative activity of *Cyanophora paradoxa* pigments in melanoma, breast and lung cancer cells. Mar Drugs 11:4390–4406. https://doi.org/10.3390/md11114390
66. Pasquet V, Morisset P, Ihammouine S, Chepied A, Aumailley L, Berard JB, Serive B, Kaas R, Lanneluc I, Thiery V, Lafferriere M, Piot JM, Patrice T, Cadoret JP, Picot L (2011) Antiproliferative activity of violaxanthin isolated from bioguided fractionation of *Dunaliella tertiolecta* extracts. Mar Drugs 9:819–831. https://doi.org/10.3390/md9050819
67. Kwang HC, Song YIK, Lee DU (2008) Antiproliferative effects of carotenoids extracted from *Chlorella ellipsoidea* and *Chlorella vulgaris* on human colon cancer cells. J Agric Food Chem 56:10521–10526. https://doi.org/10.1021/jf802111x

Index

M. A. Kassim et al. (eds.), *Sustainable Biorefinery*, Green Energy and Technology,
https://doi.org/10.1007/978-981-95-8181-8

T

V

W

X

MIX
Papier aus verantwortungsvollen Quellen
Paper from responsible sources
FSC® C105338

FSC
www.fsc.org

If you have any concerns about our products,
you can contact us on
ProductSafety@springernature.com

In case Publisher is established outside the EU,
the EU authorized representative is:
Springer Nature Customer Service Center GmbH
Europaplatz 3, 69115 Heidelberg, Germany

Printed by Libri Plureos GmbH
in Hamburg, Germany